Essays in Constructive Mathematics

Harold M. Edwards

Essays in Constructive Mathematics

 Springer

Harold M. Edwards
Courant Institute of Mathematical Sciences
New York University
251 Mercer Street
New York, NY 10012
USA

MSC 2000: 00B15, 03Fxx

Library of Congress Cataloging-in-Publication Data
Edwards, Harold M.
    Essays in constructive mathematics / Harold M. Edwards.
    p. cm.
    ISBN 0-387-21978-1 (alk. paper)
    1. Constructive mathematics.   I. Title.
QA9.56.E39   2004
511.3—dc22   2004049156

ISBN 0-387-21978-1   Printed on acid-free paper.

Printed in the United States of America.   (EB)

9 8 7 6 5 4 3 2 1   SPIN 10985564

Springer is a part of Springer Science+Business Media
springeronline.com

*For Betty
with love*

Contents

Preface

He [Kronecker] was, in fact, attempting to describe and to initiate a new branch of mathematics, which would contain both number theory and algebraic geometry as special cases.—André Weil [62]

This book is about mathematics, not the history or philosophy of mathematics. Still, history and philosophy were prominent among my motives for writing it, and historical and philosophical issues will be major factors in determining whether it wins acceptance.

Most mathematicians prefer constructive methods. Given two proofs of the same statement, one constructive and the other not, most will prefer the constructive proof. The real philosophical disagreement over the role of constructions in mathematics is between those—the majority—who believe that to exclude from mathematics all statements that cannot be proved constructively would omit far too much, and those of us who believe, on the contrary, that the most interesting parts of mathematics can be dealt with constructively, and that the greater rigor and precision of mathematics done in that way adds immensely to its value.

Mathematics came to a fork in the road around 1880. On one side, Dedekind, Cantor, and Weierstrass advocated accepting transfinite "constructions" like those needed to prove the Bolzano–Weierstrass "theorem." On the other, Kronecker argued that no such departure from the standards of proof adhered to by Dirichlet and Gauss was necessary and that the Aristotelian exclusion of completed infinites could be maintained. As we all know, the first group carried the day, and the Dedekind–Cantor–Weierstrass road was the one taken.

The new orthodoxy was consolidated by Hilbert a century ago, and has reigned ever since, despite occasional challenges, notably from Brouwer and Bishop. During this century, the phrase "foundations of mathematics" has come to mean for most working mathematicians the complex of ideas surrounding the axioms of set theory and the axiom of choice, matters that for

Kronecker had no mathematical meaning at all, much less foundational meaning.

Why, a hundred years after this choice was made, and made so decisively, do I believe that the road Kronecker proposed might win new consideration? The advent of computers has had a profound impact on mathematics and mathematicians that has already altered views about the nature and meaning of mathematics in a way favorable to Kronecker. The new technology causes mathematics to be taught and experienced in a much more computational way and directs attention to *algorithms*. In other words, it fosters constructive attitudes. My own preference for constructive formulations was shaped by my experience with computer programming in the 1950s, and computer programming at that time was trivial by today's standards.

No evidence supports the image that is so often presented of Kronecker as a vicious and personal critic of Cantor and Weierstrass—another instance of history being written by the victors. As far as I have been able to discover, Kronecker vigorously opposed the *views* of Cantor and Weierstrass, as well as those of Dedekind, with whom he was on far better terms, but he was not hostile to the men themselves. Moreover, his opposition to their views—which was of course reciprocated—was rarely expressed in his publications. In the rare instances in which he mentioned such issues, he merely stated his belief that the new ways of dealing with infinity that were coming to be accepted were *unnecessary*. Instead of excoriating nonconstructive methods, as legend would have us believe, he concentrated his efforts on backing up his beliefs with concrete mathematical results proved constructively.

No one doubts that Kronecker was one of the giants of nineteenth-century mathematics, but it is often said that he succeeded in his works because he ignored the strictures that he advocated in his philosophy. This view of the relation of Kronecker's mathematics to his philosophy is often ascribed to Poincaré, but as I have written elsewhere [21], this ascription is based on a misinterpretation of a passage [53] in which Poincaré writes about issues unrelated to the treatment of infinity in mathematics. Indeed, no one who has studied Kronecker's works could believe that he accepted completed infinites or made use of nonconstructive arguments. Like many other mathematicians since, he was impatient with the philosophy of mathematics and wanted only to get on with his mathematics itself, but for him "mathematics" was always constructive.

That attitude inspires these essays. My goal has been not to argue against the prevailing orthodoxy, but to show that substantial mathematics can be done constructively, and that such mathematics is interesting, illuminating, and concordant with the new algorithmic spirit of our times. I have given examples of what I mean by constructive mathematics, without trying to define it. The underlying idea is well expressed in the essay of Poincaré mentioned above, in which he says that the guiding principle for both Kronecker and Weierstrass was to "derive everything from the natural numbers" so that the result would "partake of the certainty of arithmetic." I regard the natural numbers not as a completed infinite set but as a means of describing the activity of *counting*. (See Essay 1.1.) The essence of constructive mathematics for me lies in the insistence upon treating infinity, in Gauss's phrase, as a *façon de parler*, a shorthand way of describing ideas that need to be restated in terms of finite calculations when it comes to writing a formal proof.

It will surely be remarked that almost all of the topics treated in the essays come from algebra and number theory. They not only partake of the certainty of arithmetic, as Poincaré says, they *are* arithmetic—what Kronecker called "general arithmetic." (Again, see Essay 1.1.) But there are three exceptions. In Essay 4.4, Newton's polygon is treated as a method of constructing an infinite series, which means, constructively, as an algorithm for generating arbitrarily many terms of the series. Convergence is not an issue because the theory treats the series themselves, not their limits in any sense. In Essay 5.1, a complex root of a given polynomial—a convergent sequence of rational complex numbers whose limit is a root of the polynomial—is found by an explicit construction. Finally, Essay 5.4, which sketches a proof of the spectral theorem for symmetric matrices of integers, necessarily deals with real numbers, that is, with convergent sequences of rationals.

An essay is "a short literary composition on a single subject, usually presenting the personal views of the author." There is nothing literary about these essays, but they do treat their mathematical subjects from a personal point of view. For example, Essay 5.1 explains why the "fundamental theorem of algebra" is misnamed—in a very real sense it isn't even true—and Essay 1.2 explains why Euclid's statement of Proposition 1 of Book 1 of the *Elements*, "On a given finite straight line to construct an equilateral triangle" is better than "Given a straight line segment, there exists an equilateral triangle of which it is one of the sides," the form in which most of Euclid's present-day successors would state it. These are my opinions. To my dismay, it is incessantly borne in on me how few of my colleagues share them and how completely mathematicians today misunderstand and reject them. These compositions try—they essay—to present them in a way that will permit the reader to see past the preconceptions that stand between what I regard as a commonsense attitude toward the study of mathematics and the attitudes most commonly accepted today. They essay to reopen the Kroneckerian road not taken.

Acknowledgments

I am profoundly grateful to Professor David Cox, who provided encouragement when it was sorely needed, and backed it up with sound advice. I also thank Professors Bruce Chandler, Ricky Pollack, and Gabriel Stolzenberg for friendship and for many years of stimulating conversation about the history and philosophy of mathematics.

Most of all, I thank my wife, Betty Rollin, to whom this book is dedicated, for more than I could ever enumerate.

Synopsis

The essays are divided into five parts:

A Fundamental Theorem
Topics in Algebra
Some Quadratic Problems
The Genus of an Algebraic Curve
Miscellany

The fundamental theorem of Part 1 constructs a splitting field for a given polynomial. As is shown in Part 2, the case in which the given polynomial has coefficients in a ring of the form $\mathbf{Z}[c_1, c_2, \ldots, c_\nu]$—a ring of polynomials in some set of indeterminates $c_1, c_2, \ldots, c_\nu$ with integer coefficients—suffices for the apparently more general case of a polynomial $f(x)$ whose coefficients are "algebraic quantities" in a very general sense. For this reason, only polynomials with coefficients in $\mathbf{Z}[c_1, c_2, \ldots, c_\nu]$ are considered in Part 1.

Another way to state the problem "Construct a splitting field for a given polynomial" is "Extend the notion of computation with polynomials with integer coefficients in such a way that the given polynomial can be written as a product of linear factors." Computation in $\mathbf{Z}[c_1, c_2, \ldots, c_\nu]$ involves just addition, subtraction, and multiplication, but it extends to computations involving division in the field of quotients of the integral domain $\mathbf{Z}[c_1, c_2, \ldots, c_\nu]$ in the same way that computation in the ring of integers extends to computation in the field of rational numbers. As Gauss's lemma shows (Essay 1.4), this extension does not affect the factorization of polynomials. A simple *further* extension of $\mathbf{Z}[c_1, c_2, \ldots, c_\nu]$ is effected by "adjoining" *one* root of a *monic, irreducible* polynomial with coefficients in $\mathbf{Z}[c_1, c_2, \ldots, c_\nu]$ to the field of rational functions. This simple construction, which Galois used with amazing success, although with some lack of rigor, is generally known as a "simple algebraic extension" of the field of quotients of $\mathbf{Z}[c_1, c_2, \ldots, c_\nu]$. For the sake of brevity, I have called a field constructed in this way the "root field" of the monic, irreducible polynomial used in its construction (Essay 1.3).

With this specific description of the way in which computations in $\mathbf{Z}[c_1,$ $c_2, \ldots, c_\nu]$ are to be extended, the construction problem to be solved becomes, "Given a polynomial f with coefficients in $\mathbf{Z}[c_1, c_2, \ldots, c_\nu]$, find an auxiliary polynomial g with coefficients in the same ring such that g is monic and irreducible and such that its root field splits f" in the sense that f can be written as a product of linear factors with coefficients in the root field of g.

The problem, then, is, "Given f construct g." The solution in Part 1 is iterative. Suppose that g is a failed attempt at a solution. Thus, the factorization of f over the root field of g contains at least one irreducible factor of degree greater than 1. The iteration needs to construct a *better* attempt at a solution. Specifically, it needs to construct a new auxiliary polynomial, call it g_1, with the property that the factorization of f over the root field of g_1 contains more linear factors than does the factorization of f over the root field of g. If g_1 fails to split f, the same procedure can be applied again to find a g_2 that gives f more linear factors than g_1 did. Since the number of linear factors of f increases with each new g, and since the number of such factors is bounded above by the degree of f, such an iteration must eventually reach a solution of the problem—an attempted g that does not fail.

To make this sketch into an actual iterative construction of a splitting field for f requires two main steps. First, given f and an attempt at g, one needs to be able to *factor f when it is regarded as a polynomial with coefficients not in $\mathbf{Z}[c_1, c_2, \ldots, c_\nu]$ but in its extension, the root field of g*. The difficult step in the construction of a splitting field for f is the algorithmic solution of this factorization problem. The algorithm is set forth in Essay 1.5, with examples, and the proof that it achieves its objective is in Essay 1.6. The relation of the algorithm to Kronecker's solution of the same factorization problem is among the subjects discussed in Essay 1.7. Second, one needs to describe explicitly how to pass from a g that fails to split f to a new g_1 that comes closer to splitting f. The underlying idea of the construction is simple: Because g does not split f, there is an irreducible factor, call it ϕ, of f over the root field of g whose degree is greater than 1. *Adjoin to the root field of g a root of ϕ.* This double adjunction, first of a root of g and then of a root of ϕ, gives a field over which f has more linear factors—a field in which f has more roots— because it contains a root of ϕ, and the root field of g did not. The problem is to write this double adjunction as a simple one—specifically as the field obtained by adjoining a root of a new g_1 with coefficients in $\mathbf{Z}[c_1, c_2, \ldots, c_\nu]$. The construction of such a g_1 is given in Essay 1.8.

Finally, although there are infinitely many polynomials g that split f, there is only one *splitting field* of f in the sense that if g is a *minimal* splitting polynomial of f—one that is itself split by any polynomial that splits f—the root field of g is isomorphic to the root field of any other minimal splitting polynomial of f.

The end result is a theorem that in my opinion deserves the name "Fundamental Theorem of Algebra" much more than the theorem that is and probably always will be known by that name: *Given a polynomial f (in one* variable) *with coefficients in* $\mathbf{Z}[c_1, c_2, \ldots, c_\nu]$ *there is an explicit way to extend rational computations in* $\mathbf{Z}[c_1, c_2, \ldots, c_\nu]$ *so that f factors into linear factors; moreover, any two minimal ways of doing this are isomorphic.* For the relation of this theorem to the "Fundamental Theorem of Algebra" see Essay 5.1.

The theorem of Part 1 that has just been described is implicitly contained—with no hint of a proof—in Lemma III of Galois's treatise [27] on the algebraic solution of equations (see [22]). In this sense, it is the foundation of Galois theory. The connection is explained in Essays 2.1 and 2.3. Essay 2.2 is devoted to justifying Kronecker's assertion that *every field of algebraic quantities is isomorphic to the root field of a polynomial with coefficients in some* $\mathbf{Z}[c_1, c_2, \ldots, c_\nu]$ as that concept was defined in Part 1. This fact is the basis of Kronecker's later view—despite the fact that he had previously given the title *Foundations of an Arithmetical Theory of Algebraic Quantities* to his major publication—that "algebraic quantities" were unnecessary in mathematics and that algebraic questions should be studied using "general arithmetic" instead (see Essay 1.1).

The algorithmic description of fields of algebraic quantities in terms of "adjunction relations" in Essay 2.3 gives a construction of the splitting field of a polynomial that is very close to Chebotarev's in his excellent but little-known book on Galois theory [8].

The construction of the splitting field of a *general* monic polynomial of degree n in Essay 2.4 proves another basic theorem of Galois—another to which Galois gave no hint of a proof—that the Galois group of an nth-degree polynomial $f(x)$ whose coefficients are 'letters' is the full symmetric group. The splitting field is explicitly given by adjunction relations $f_i(\alpha_i) = 0$, where

$$(1) \qquad f_i(x) = \frac{f(x)}{(x - \alpha_1)(x - \alpha_2) \cdots (x - \alpha_{i-1})}$$

is the irreducible polynomial satisfied by a root α_i of $f(x)$ whose coefficients are polynomials in the roots $\alpha_1, \alpha_2, \ldots, \alpha_{i-1}$ already adjoined. (The right side of (1), as it stands, is of course not a polynomial; it becomes one once $i - 1$ roots $\alpha_1, \alpha_2, \ldots, \alpha_{i-1}$ of $f(x)$ have been adjoined and the divisions (1) calls for have been performed.) The degree $n!$ of the extension is of course the product of the degrees of these adjunction relations. The nub of the matter is the proof that each $f_i(x)$ is *irreducible* over the field generated by $\alpha_1, \alpha_2, \ldots, \alpha_{i-1}$. These ideas stem from Kronecker, as does the fundamental theorem of divisor theory in Essay 2.5.

Part 3 deals with different matters altogether. Its primary inspiration is Gauss's proof of quadratic reciprocity in Section 5 of the *Disquisitiones Arithmeticae*, but the proof is recast by translating it from a study of *quadratic forms* and their *composition* to a study of *modules* and their *multiplication*. The "modules" involved are entities of the following type. With "number" meaning a number 0, 1, 2, ..., let a "hypernumber" for a given A mean an expression $y + x\sqrt{A}$ in which x and y are numbers. Such hypernumbers can be added and multiplied, and, when the sizes of the coefficients allow it, subtracted as well. It is natural to assume that the given A is *not a square*, because otherwise, $y + x\sqrt{A}$ would be a number, and nothing new would be found by computing with hypernumbers. It will also be assumed that A is *positive*, but this assumption is made for the sake of simplicity and has no particular justification except that the case of positive A suffices for the proof of quadratic reciprocity.

Let m_1, m_2, ..., m_μ be a list of hypernumbers. Two hypernumbers a and b will be said to be *congruent mod* $[m_1, m_2, \ldots, m_k]$, denoted by $a \equiv b$ mod $[m_1, m_2, \ldots, m_\mu]$, if they can be made equal by adding sums of multiples (with multipliers that are hypernumbers) of m_1, m_2, ..., m_μ to each. This is a simple generalization of Gauss's definition of $a \equiv b$ mod m, which means that a and b can be made equal by adding multiples of m to each. (This form of the definition eliminates the need for negative numbers in the discussion of congruences.) A **module** is a list of hypernumbers $[m_1, m_2, \ldots, m_\mu]$ written between square brackets to indicate that they are to be used to define a congruence relation in this way. Two modules are **equal** if they define the same congruence relation. Essay 3.2 gives all these definitions, along with an algorithm for determining whether two given modules are equal.

When the product of two lists is defined to be the list that contains all products with one term from the first list and one from the second, the module determined by the product depends only on the modules determined by the factors, so the operation gives a way to multiply *modules*. Since the multiplication defined in this way is clearly associative and commutative, the modules for a given A form a *commutative semigroup* in which the module $[1]$ is an identity.

One more level of abstraction is needed for the most interesting construction. Let a module be called **principal** if it can be represented by a list $[y + x\sqrt{A}]$ with just one hypernumber and if, in addition, that hypernumber satisfies $y^2 > Ax^2$. The principal modules are a *subsemigroup* of the semigroup of modules—which means simply that a product of principal modules is principal—so there is an associated quotient structure: Two modules are **equivalent** if they can be made equal by multiplying each by a principal module. (That is, $M_1 \sim M_2$ means there are principal modules P_1 and P_2 for which $M_1 P_1 = M_2 P_2$.) The equivalence classes of modules for a given A defined in this way form a *finite semigroup*, which I call the **class semigroup**. It can be determined for each given A by algorithms described in Essay 3.3. These algorithms are essentially the same as Gauss's methods in Section 5 of the *Disquisitiones* for determining the equivalence classes of binary quadratic forms for a given "determinant" A, but in my opinion they are simpler in conception. Some rudimentary facts about class semigroups are proved in Essay 3.4 that are then used in Essay 3.5 to prove the law of quadratic reciprocity.

The final two essays of Part 3 relate the multiplication of modules to Gauss's composition of binary quadratic forms by showing how, given two binary quadratic forms, the theory of multiplication of modules can be used to *determine whether there is a third binary quadratic form that composes them* in Gauss's sense, and, if so, *to find all such compositions.* Once this connection is clearly made, it seems to me that the module-and-multiplication formulation will have more appeal than the form-and-composition formulation. But whether or not the module-and-multiplication version is preferred over Gauss's, it has the advantage of relating directly to Gauss's, thereby making Gauss's masterpiece more accessible to modern readers whose familiarity with Dirichlet's simplification of it may, for reasons explained in Essay 3.6, be an impediment.

Part 4, on the genus of an algebraic curve, is inspired by Abel's great memoir of 1826 (first published, thanks to the negligence of the Paris Academy, in 1841). The inspiration for the algorithmic method of Part 4 goes back even further, to Newton's method of constructing infinite series expansions of algebraic functions, commonly known today as "Newton's polygon."

The construction used in Part 4 (see Essay 4.1) to describe the genus of a curve is based on ideas of Abel that predate the theory of Riemann surfaces by many years, and it makes no reference to complex numbers, much less to Riemann surfaces. Let an algebraic curve $\chi(x,y) = 0$ be given (where χ is an irreducible polynomial with integer coefficients that contains both x and y and, for simplicity, is monic in y), and let a large number N of points on the curve also be given. Choose a rational function θ on the curve with many zeros (it will have equally many poles, of course) including zeros at all of the N given points. An **algebraic variation** of the N points is a variation of the N points that can be achieved as a variation of the N special zeros of the rational function θ when the coefficients of θ are varied in such a way that the remaining zeros and all of the poles remain unchanged. (It is assumed that the N given zeros are points where x is finite, and the poles are specifically taken to be the poles of x^ν for some large ν, but in fact any set of poles will do, as long as there are enough of them, and the zeros can also be at points where x is infinite.) The N zeros then vary with $N - g$ degrees of freedom, where the number g, the **genus** of the curve, depends only on the curve, not on the other choices.

To make this rough *idea* of the genus of $\chi(x,y) = 0$ into a *definition* requires, of course, that much more be said. Although the natural first step might seem to be the introduction of complex numbers in order to deal rigorously with the zeros and poles of θ on $\chi(x,y) = 0$, the approach taken in Part 4 is quite the opposite: It dispenses with the notion of zeros and poles altogether. Just as the degree of a polynomial determines the number of its roots, general conditions on rational functions θ can be related heuristically to the notions of zeros and poles of θ on the curve $\chi(x,y) = 0$ and their locations. On this basis, one can give a satisfactory description of rational functions θ on $\chi(x,y) = 0$ with prescribed poles of prescribed multiplicity, all in terms of "general arithmetic"—the arithmetic of polynomials with integer coefficients.

Then Abel's description of the genus can be given solid, constructive meaning in terms of the number of free coefficients in θ when conditions are placed on its zeros and poles, a description that uses nothing but general arithmetic.

Essay 4.2 relates Abel's construction to Euler's addition formula for elliptic curves and to the geometric description of addition on an elliptic curve that is so familiar in the present time of great interest in elliptic curves. Essay 4.3 gives the details of the definition of the genus of $\chi(x, y) = 0$ in these terms.

Essay 4.4 is an exposition of the Newton algorithm. The task of the algorithm is to construct a solution y of $\chi(x, y) = 0$ as an infinite series of (possibly) fractional powers of x. In constructive mathematics an infinite series must of course be presented as an *algorithm* that generates the successive terms of the series, and this is what the Newton algorithm does. More precisely, the input to the algorithm is a truncated series solution y of $\chi(x, y) = 0$ and the algorithm generates one further term of the series. In the early stages, a truncated solution may be *ambiguous*, meaning that it may be extended in more than one way, and the algorithm must determine all possible extensions; eventually, as is proved in the essay, a set of *unambiguous* truncated solutions is reached, each of which is prolonged by the algorithm with the addition of one more term in just one way and therefore represents an infinite series solution. (Note that convergence is not an issue, because the series itself—not any kind of limit of the series—is the objective.)

Essay 4.5 gives an algebraic method of evaluating the genus as it was defined in Essay 4.3. Essay 4.6 gives a simpler description of the genus as the dimension of the vector space of **holomorphic differentials** on the curve. These differentials are the ones that have no poles, and they have the property that the algebraic variations of a set of N points on the curve are described by the differential equations

$$\sum_{i=1}^{N} h_j(x_i, y_i)dx_i = 0 \quad \text{for } j = 1, 2, \ldots, g,$$

where the differentials $h_j(x, y)dx$ are a basis of the space of holomorphic differentials. Essay 4.7 uses the holomorphic differentials on a curve to state and prove the Riemann–Roch theorem as a formula for the number of arbitrary constants in a rational function with given poles.

The last essay of Part 4 proves that the genus is a birational invariant, even though the method of Essay 4.5 for computing it depends on the choice of a parameter x on the curve. This important result can be stated as follows: In the terminology of Essay 2.2, the field of rational functions on an algebraic curve is an algebraic field of transcendence degree one. The **genus** as it is defined above depends only on this field itself, not on the particular presentation of it as a root field that is used in the definition. This conclusion is established, in essence, by showing that if x and z are different parameters on the curve then $h\,dx$ is a holomorphic differential (relative to the parameter x)

if and only if $h \cdot \frac{dx}{dz} \cdot dz$ is a holomorphic differential (relative to the parameter z), so the spaces of holomorphic differentials in the two cases are isomorphic.

The miscellany of Part 5 begins with a proof of what is called the Fundamental Theorem of Algebra—in fact two proofs of it. But my main point in that essay is that this theorem is not truly a theorem of algebra at all because it relates in an essential way to the nonalgebraic notion of complex numbers. The next essay gives a proof of the Sylow theorems in the theory of finite groups, and the following two summarize the constructive and algorithmic approach of my 1995 book on linear algebra. My hope is that *Linear Algebra* and the present book will reinforce one another—this book making the case for the clarifying power of algorithmic methods and *Linear Algebra* giving yet another example of that power. The final essay is a further correction to the Kronecker legend.

1

A Fundamental Theorem

Essay 1.1 General Arithmetic

La dernière chose qu'on trouve en faisant un ouvrage c'est de savoir celle qu'il fallait mettre la première. (The last thing one discovers in composing a work is what should be put first.)—Pascal, *Penseés*

Kronecker quoted this saying of Pascal in the first of a series of lectures he gave on the concept of number.[*] It may have been a somewhat rueful reflection on his own experience with his 1881 treatise *Grundzüge einer arithmetischen Theorie der algebraischen Grössen* (Elements of an Arithmetical Theory of Algebraic Quantities) [39], which contains at least two indications that he altered his point of view profoundly while writing it, deciding ultimately that the subject of his title was not the one he should be dealing with at all. In his introduction to the treatise and in point IV of its last section he speaks of reducing the entire theory of "algebraic quantities" to the theory of rational functions of variables. In later works he did just that. For example, in *Ein Fundamentalsatz der allgemeinen Arithmetik* [42] and *Über den Zahlbegriff* [43], both published in 1887, he emphasized the importance of rational computation with polynomials with integer coefficients and again stated that theories of algebraic numbers and algebraic quantities could be reduced to such rational computations. Thus, whether or not it was the last thing he discovered in writing his 1881 treatise, Kronecker came to believe that he should have begun with rational algebra. That is where these essays begin.

Elsewhere Kronecker said, "In mathematics, I recognize true scientific value only in concrete mathematical truths, or, to put it more pointedly, only in mathematical formulas" (see [21]). I would rather say "computations" than "formulas," but my view is essentially the same. Computation, in turn, is an outgrowth of counting, and in this sense mathematics is founded on *numbers,*

---

[*] These lectures, given in the last year of Kronecker's life, 1891, are preserved in the form of a handwritten transcript in the archives of the mathematics library of the University of Strasbourg, and were recently published [45].

Fig. 1.1. Kronecker.

not as abstract "objects" of any kind, but as the system of symbols by which we record the results of counts. We learn to use and understand the symbols 0, 1, 2, 3, . . . at an early age, and most of us understand fairly soon that the important thing is not the actual names or symbols that are used, or even the decimal system on which they are based, but the mere fact that there are agreed-upon symbols and names and an agreed-upon system for counting. In this essay, the word "number" will mean a number 0, 1, 2, . . . in this most basic sense.

The more sophisticated computations that we learn later in life grow out of counting. First, the operations of addition and multiplication are grounded in counting. (Counting first to a then to b is the same as counting to $a + b$. Counting a times to b is the same as counting to ab.) Subtraction—the inverse of addition—of course makes sense only when the number being subtracted is not larger than the one from which it is to be subtracted. However, experience teaches us to widen our horizon in such a way that this limitation on subtraction can be put in the background and for the most part ignored. This can be done very conveniently in the way Kronecker does it in the essay *Über den Zahlbegriff* (On the Concept of Number), mentioned above.

He first introduces "*Buchstabenrechnung*," calculation with letters, in the following way: "The same laws [that govern the addition and multiplication of numbers] needed to be regarded as valid for calculation with letters as soon as letters began to be used to represent numbers whose determination might or should be postponed. With the introduction of the principle of computing with indeterminates *as such,* which originated with Gauss, the special theory of whole numbers broadened into the general arithmetical theory of polynomials with whole number coefficients."[*]

In particular, we can compute with—add and multiply—polynomials in a single indeterminate t whose coefficients are numbers. When we use Gauss's notation $f(t) \equiv g(t) \bmod (t+1)$ to mean that $f(t)$ can be transformed into $g(t)$ using the identity $t+1 \equiv 0 \bmod (t+1)$, we then have a system of computation in which t plays the role of -1. This system of computation is what we mean by *the ring of integers.*

In modern notation, these observations can be abbreviated $\mathbf{Z} = \mathbf{N}[t] \bmod (t+1)$, meaning that an element of $\mathbf{Z}$—an integer—is represented by an expression of the form $a_0 t^n + a_1 t^{n-1} + \cdots + a_n$, where $\mathbf{N}$ denotes the set of numbers $\{0, 1, 2, \ldots\}$, where t is an indeterminate, and where $n, a_0, a_1, \ldots, a_n$ are numbers. Two such representations *by definition* represent the same element of $\mathbf{Z}$ if one can be transformed into the other using $t+1 \equiv 0 \bmod (t+1)$ in conjunction with the usual laws that govern addition and multiplication of numbers. Since $t^2 \equiv t^2 + (t+1) \equiv t^2 + t + 1 \equiv t(t+1) + 1 \equiv 1 \bmod (t+1)$, we can always replace t^2 with 1. Therefore, we can replace t^3 with t, t^4 with t^2 and then with 1, and so forth, to represent any integer by an expression of the form $at + b$, where a and b are numbers. Two such expressions represent the same integer, that is, $at+b \equiv ct+d \bmod (t+1)$, if and only if $at+b+a+c \equiv ct+d+a+c \bmod (t+1)$, which is to say $b+c \equiv d+a \bmod (t+1)$. Since this congruence does not involve t, it is equivalent to the equation $b + c = d + a$. In short, the simple device of computation with polynomials in $t \bmod t + 1$ is all that is needed to describe the usual construction of the ring of integers as ordered pairs of numbers (a, b) subject to the equivalence relation "$(a,b) \equiv (c,d)$ means $a + d = b + c$" when equivalence classes are added and multiplied in the obvious ways. As Kronecker observes, the interpretation of the equation $7 - 9 = 3 - 5$ truly involves this *new meaning* of the equal sign.

In essence, the device of using integers instead of numbers makes it possible to do computations without bothering about whether particular subtractions are possible unless a reason arises to examine that issue. Similarly, it makes possible writing all terms of an equation on one side of the equal sign, which greatly simplifies reasoning by eliminating the need to consider separately cases in which the terms would appear on different sides of the equation if numbers were used instead of integers.

---

[*] Dieselben Gesetze mussten für die sogenannte Buchstabenrechnung als maassgebend angenommen werden, sobald man anfing, die Buchstaben zur Bezeichnung von Zahlen zu verwenden, deren Bestimmung vorbehalten bleiben kann oder soll. Aber mit der *principiellen* Einführung der "Unbestimmten" (indeterminatae), welche von *Gauss* herrührt, hat sich die specielle Theorie der ganzen Zahlen zu der allgemeinen arithmetischen Theorie der ganzen ganzzahligen Functionen von Unbestimmten erweitert.

The fourth rational operation, division, is the most interesting one. The division of *numbers* naturally takes the form of *division with remainder*: Two numbers a and b with $b \neq 0$ determine numbers q and r by the conditions $a = qb + r$ and $r < b$. This operation is of course very closely connected to the original meaning of Gauss's congruence concept $a \equiv r \bmod b$.

In the more general setting of *Buchstabenrechnung*—computing with polynomials in several indeterminates whose coefficients are numbers—the notion of *divisibility* has a clear meaning, but the more useful concept of division with remainder does not. An exception is the case of division by a polynomial with integer coefficients that is *monic* in one of its indeterminates, meaning that $b = x^n$ + terms of degree less than n in x. In this case, for any polynomial a with integer coefficients, there are unique polynomials q and r with integer coefficients for which $a = qb + r$ and for which the degree of r in x is less than the degree n of b in x.

Another role played by division in elementary arithmetic is the *cancellation law of multiplication*: If $ab = ac$ and $a \neq 0$, then $b = c$. This law is obviously valid for numbers—even for integers. Its validity in other contexts, when equality is replaced by some kind of congruence, is often a crucial issue. When it is valid, the ring of congruences classes under addition and multiplication is an **integral domain**. For an integral domain one can construct a *field of quotients*, the set of all formal quotients p/q in which $q \neq 0$, when $p/q = p'/q'$ is defined to mean $pq' = p'q$ and when addition and multiplication are defined by $(p/q) + (p'/q') = (pq' + p'q)/qq'$ and $(p/q) \cdot (p'/q') = pp'/qq'$. However, for the most part fields of quotients themselves—for example, the field of rational numbers—will be avoided. The *potential* of one, the ability to compute with formal quotients of elements of an integral domain, is enough.

Toward the end of his life, Kronecker adopted the term "general arithmetic" (*allgemeine Arithmetik*) for the arithmetic of rings of the form $\mathbf{Z}[c_1, c_2, \ldots, c_\nu]$, which is to say rings of polynomials in some set of indeterminates $c_1, c_2, \ldots, c_\nu$ with integer coefficients (or, in conformity with the approach above, rings $\mathbf{N}[t, c_1, c_2, \ldots, c_\nu] \bmod (t+1)$, where $\mathbf{N}$ denotes the set of numbers). Of course there is little to be said about these rings themselves; rather, the substance of general arithmetic lies in the study of certain further constructions that use them.

Kronecker wrote a number of papers about what he called **module systems**, and he formulated his proof of his *Fundamentalsatz* of general arithmetic in terms of module systems. For me, however, his module systems pose difficulties. For one thing, a module system in its simplest form—namely, a ring of the form $\mathbf{Z}[c_1, c_2, \ldots, c_\nu] \bmod [M_1, M_2, \ldots, M_\mu]$, a polynomial ring in which computations are done modulo a finite number of given relations $M_i \equiv 0$—is not normally a field because division is not normally possible. Therefore, one needs to enlarge the realm of objects with which one computes in some way in order to allow the computations to use the most convenient and natural representations of them—for example to allow the use of $\omega = \frac{-1+\sqrt{-3}}{2}$

in computations with $\sqrt{-3}$. For another thing, this description $\mathbf{Z}[c_1, c_2, \ldots, c_\nu]$ mod $[M_1, M_2, \ldots, M_\mu]$ of module systems opens the door to the problem of determining *whether two such module systems are isomorphic,* which is tantamount to the "ideal membership problem" for rings of the form $\mathbf{Z}[c_1, c_2, \ldots, c_\nu]$, a problem that poses serious difficulties* from a constructive point of view.

For these reasons, I have avoided module systems† and have instead used the concrete notion of a simple algebraic extension of the field of quotients of a ring $\mathbf{Z}[c_1, c_2, \ldots, c_\nu]$; such simple algebraic extensions are described and given the name "root fields" in Essay 1.3. As will be shown in Essay 2.2—using an argument Kronecker himself gave [39, §2]—fields of this type provide a setting for all algebraic computations.

---

* Much serious work has been done on the ideal membership problem and the related problem of constructing Gröbner bases, but since I have not been able to understand this work in a way that is consistent with my notions of constructivity, I am glad that it is unnecessary for the topics I develop here.

† *Specific* module systems in which the problems mentioned above do not arise are used in some of the constructions—for example in Essays 1.5 and 2.4.

Essay 1.2 A Fundamental Theorem

> *Proposition V. Problème. Dans quel cas une équation est-elle soluble par de simples radicaux?* (Proposition V. Problem. In what case is an equation solvable by simple radicals?—É. Galois, [27]

The essays that follow contain applications of general arithmetic to a variety of topics, one of which is the theorem Kronecker stated and proved in his 1887 paper *Ein Fundamentalsatz der allgemeinen Arithmetik* (On a Fundamental Theorem of General Arithmetic) [42]. This theorem states, roughly, that *every polynomial has a splitting field*.

Proposition 1 of Book 1 of Euclid's *Elements* [25] is, in the Heath translation, "On a given finite straight line to construct an equilateral triangle." To modern ears, this seems a strange way to state a proposition. A modern writer would be more likely to say, "Given a finite straight line, there is an equilateral triangle of which it is one of the sides." But Euclid has many such propositions. His propositions fall into two categories, often described as "problems" and "theorems." That Proposition 1 is a "problem" is signaled not only by the form of its statement but also by the fact that its proof ends with "as was to be done" rather than "as was to be proved." Gauss on at least one occasion* concludes a proof with QEF ("quod erat faciendum"—that which was to be done) instead of QED ("quod erat demonstrandum"—that which was to be demonstrated), and Galois in his treatise [27] on the algebraic solution of equations presents eight "propositions," five of which are "theorems," two of which are "problems," and one of which is a "lemma," but today the designation "problem" has disappeared from formal mathematical exposition, and the designation "proposition" has become more or less synonymous with "theorem."

The usual definition of "constructive mathematics" is that it requires existence theorems to be proved constructively—that is, constructive mathematics does not accept as a proof of existence an argument that assumes a *disproof* of existence and derives a contradiction. But the very notion of an "existence theorem" reflects a nonconstructive bias. Is it not ridiculous to say, "Every polynomial has a splitting field," and then to stipulate that "The proof will give an actual construction of a splitting field"? Is it not more reasonable to follow the Euclidean model and say "Given a polynomial, construct a splitting field for it," thereby making clear that the proof is a construction?

In these essays I do follow the Euclidean model, except that I have decided to use the word "theorem" for both types of Euclidean "propositions." Mathematicians today regard "theorems" as the fundamental units of mathematics. Kronecker used the term *Fundamentalsatz* even though his theorem was in truth a construction. I doubt that an effort to give another term like "problem" or "construction" the same status as "theorem" would succeed. And a

---

* See §73 of *Disquisitiones Arithmeticae*. The notation QEF at the end of §73 was omitted from the German translation by Maser.

"fundamental problem" or a "fundamental construction" would be much less imposing than a "fundamental theorem." Thus, these essays, although they will not contain "existence theorems," will contain many theorems that are constructions, including the theorem of this essay.

Gauss's doctoral dissertation of 1799 was devoted to a proof of a theorem very close to what is now called the "fundamental theorem of algebra": *A polynomial of degree n has n complex roots* when they are counted with multiplicities. In the dissertation he sharply criticizes earlier attempts to prove the theorem, saying that they used computations with the roots and that such computations virtually assumed the truth of the theorem to be proved. However, as Bashmakova and Rudakov point out in an essay on the history of the theorem [35], Gauss returned to the theorem in 1815 and gave a new proof that took an approach very similar to the one he had criticized in 1799; he justified *on other grounds* certain limited computations with the roots of the given polynomial, and then used such computations to show that the roots could be described as complex numbers. (For this second proof, see Essay 5.1.)

Kronecker, with his *Fundamentalsatz*, came to the realization that *this is the theorem:* what is important is not the complex numbers but, rather, the fact that computations with the roots can be justified. That is, *given a polynomial with integer coefficients, one can describe a system of computation that extends computations with integers in such a way that the polynomial has a number of roots equal to its degree.*

In a pragmatic sense, Galois had realized the same thing more than fifty years earlier. Lemma III of his treatise on the algebraic solution of equations [27], which was written in 1830–1831 even though it was not published until 1846, is in essence a construction of a splitting field for a given polynomial. Unfortunately, Galois does not prove Lemma III. He does give a construction of a splitting field, but the construction uses computations with the roots! Thus, from a foundational point of view, what Galois proved was that *if* there is any valid way to compute with the roots of a polynomial, *then* computations with the roots can always be done using what is now called a Galois resolvent, as in his Lemma III. Until computations with the roots were validated— until Kronecker's *Fundamentalsatz* was proved—Galois theory was without a general foundation, even though splitting fields could be constructed in specific cases.

For example, in the case of the polynomial $x^3 - 2$, the polynomial $y^6 + 108$ is a Galois resolvent, which is to say that computations in $\mathbf{Q}[y]$ mod $(y^6 + 108)$ extend computations in the field of rational numbers $\mathbf{Q}$ in such a way that $x^3 - 2$ factors into linear factors, as is shown by the formula

$$(1)\quad x^3 - 2 \equiv \left(x - \frac{y^4}{18}\right)\left(x + \frac{y^4 - 18y}{36}\right)\left(x + \frac{y^4 + 18y}{36}\right) \bmod (y^6 + 108).$$

Putting aside for the moment the question of how such a formula might be constructed, one can easily check that it is correct; the product of the last two factors is $\frac{1}{36^2}((36x + y^4)^2 - 324y^2) \equiv \frac{1}{36^2}(36^2x^2 + 72xy^4 + y^2(-108) -$

$324y^2) \equiv x^2 + \frac{x}{18}y^4 - \frac{1}{3}y^2$ mod $(y^6 + 108)$, and multiplication of this result by the first factor gives $x^3 + \frac{x^2}{18}y^4 - \frac{x}{3}y^2 - \frac{x^2}{18}y^4 - \frac{x}{18^2}y^2(-108) + \frac{1}{54}(-108) \equiv x^3 - 2$ mod $(y^6 + 108)$. Otherwise stated, the formula proves that adjunction to the field $\mathbf{Q}$ of a root y of $y^6 + 108 = 0$ gives a field over which $x^3 - 2$ splits into linear factors.

Kronecker states the more general version

$$(2) \quad x^3 - c \equiv \left(x - \frac{y^4}{9c}\right)\left(x + \frac{y^4 - 9cy}{18c}\right)\left(x + \frac{y^4 + 9cy}{18c}\right) \text{ mod } (y^6 + 27c^2)$$

of this formula in his *Fundamentalsatz* paper [42, end of §2]. This version, too, is easy to check even if it is not easy to guess. It proves that adjunction to the field of rational functions in c of a root y of $y^6 + 27c^2$ gives a field over which $x^3 - c$ splits into linear factors.

The general case is developed and proved in the next few essays in the form of the following theorem:

Fundamental Theorem. *Given a polynomial f in x with integer coefficients, construct a polynomial g in y with integer coefficients that is irreducible, monic in y, and has the property that a formula of the form*

$$(3) \qquad\qquad f(x) \equiv a_0 \prod \left(x - \frac{\phi_i(y)}{\psi_i}\right) \text{ mod } g(y)$$

holds, in which the $\phi_i(y)$ are polynomials whose coefficients are integers, the ψ_i are nonzero integers, and a_0 is the leading coefficient of f.

More generally,
Given a polynomial f in x, c_1, c_2, ..., c_ν with integer coefficients, construct a polynomial g in y, c_1, c_2, ..., c_ν with integer coefficients that is irreducible, monic in y, and gives rise to a formula (3), in which the $\phi_i(y)$ are now polynomials in y, c_1, c_2, ..., c_ν with integer coefficients, the ψ_i are nonzero polynomials in c_1, c_2, ..., c_ν with integer coefficients, and a_0 is the leading coefficient of f as a polynomial in x.

To say that $g(y)$ is monic in y means that g has the form $y^n + \cdots$ where the omitted terms all have degree less than n in y. As will be seen in Essay 1.3, this assumption simplifies computations mod $g(y)$. The requirement that g be irreducible guarantees that $\mathbf{Z}[y, c_1, c_2, \ldots, c_\nu]$ mod $g(y, c_1, c_2, \ldots, c_\nu)$ (where ν may be zero) will be an integral domain. In modern parlance, the field of quotients of this integral domain is a splitting field of f as a polynomial with coefficients in the field of rational functions in c_1, c_2, ..., c_ν. It is obtained by adjoining a root y of g to this field of rational functions.

A familiar example of a splitting of this sort, put in an unfamiliar way, is

$$ax^2 + bx + c \equiv a\left(x - \frac{-b+y}{2a}\right)\left(x - \frac{-b-y}{2a}\right) \text{ mod } (y^2 - b^2 + 4ac).$$

Here use is made of the fact that y is a square root of $b^2-4ac \bmod (y^2-b^2+4ac)$ to express the roots of the quadratic polynomial $ax^2 + bx + c$ in the form $(-b \pm y)/2a$. The stated congruence is easily verified by rewriting the right side as $\frac{a}{4a^2}(2ax+b-y)(2ax+b+y) = \frac{1}{4a}(4a^2x^2+4abx+b^2-y^2) \equiv \frac{1}{4a}(4a^2x^2 + 4abx + 4ac) \bmod (y^2 - b^2 + 4ac)$.

Another example that is considerably less simple to verify is

$$x^4 - x^2 - 1 \equiv \left(x - \frac{7y^7 + 220y^5 - 1846y^3 + 9003y}{18966} \right)$$
$$\cdot \left(x + \frac{7y^7 + 220y^5 - 1846y^3 + 9003y}{18966} \right)$$
$$\cdot \left(x - \frac{7y^7 + 220y^5 - 1846y^3 - 9963y}{2 \cdot 18966} \right)$$
$$\cdot \left(x + \frac{7y^7 + 220y^5 - 1846y^3 - 9963y}{2 \cdot 18966} \right)$$
$$\bmod (y^8 - 10y^6 + 47y^4 - 110y^2 + 841).$$

(See Example 1 of Essay 2.3.)

Essay 1.3 Root Fields (Simple Algebraic Extensions)

In brief, one adjoins a root y of $g(y)$ by computing with expressions of the form $p(y)/q$, where $p(y)$ is a polynomial in y and q is nonzero; addition and multiplication are performed in the obvious ways, while division is accomplished by "rationalizing the denominator."

Here the given polynomial $g(y)$ is assumed to be an irreducible polynomial with integer coefficients—or, more generally, with coefficients in the ring $\mathbf{Z}[c_1, c_2, \ldots, c_\nu]$ of polynomials with integer coefficients in some given set c_1, c_2, $\ldots$, c_ν of indeterminates—that is monic in y. The advantage of assuming that $g(y)$ is monic in y is that it means that division with remainder can be used to find, for any polynomial in y, another polynomial congruent to it mod $g(y)$ whose degree is less than $n = \deg g$. Since a polynomial of degree less than $n = \deg g$ can be divisible by $g(y)$ only if it is zero, it follows that *every polynomial in y with coefficients in $\mathbf{Z}[c_1, c_2, \ldots, c_\nu]$ is congruent mod $g(y)$ to one and only one such polynomial whose degree in y is less than n*. Thus, the ring $\mathbf{Z}[y, c_1, c_2, \ldots, c_\nu]$ mod $g(y)$ can be described as the elements of $\mathbf{Z}[y, c_1, c_2, \ldots, c_\nu]$ whose degrees are less than $n = \deg g$, added in the usual way and multiplied by ordinary multiplication of polynomials followed by division with remainder by $g(y)$ to find a polynomial congruent to the product mod $g(y)$ whose degree in y is less than n.

The advantage of the assumption that $g(y)$ is irreducible is that it implies—as will be proved in the next essay—that the ring $\mathbf{Z}[y, c_1, c_2, \ldots, c_\nu]$ mod $g(y)$ defined in this way is an *integral domain*. Therefore, as described in Essay 1.1, there is an associated *field of quotients*. It is this field of quotients of the integral domain $\mathbf{Z}[y, c_1, c_2, \ldots, c_\nu]$ mod $g(y)$ that I will call the **root field** of $g(y)$. Another way to describe it is as the **simple algebraic extension** of the field of rational functions* in c_1, c_2, $\ldots$, c_ν obtained by adjoining a root of $g(y)$.

An element of the root field is represented in the first instance by a quotient of elements of $\mathbf{Z}[y, c_1, c_2, \ldots, c_\nu]$ in which the denominator is not zero mod $g(y)$. Since computations are all to be done mod $g(y)$, one can assume without loss of generality that both numerator and denominator have degree less than the degree n of $g(y)$. Moreover, one can assume without loss of generality that *the denominator does not contain y* by virtue of the following lemma:

Lemma (Basic elimination). *Given polynomials $g(y)$ and $q(y)$ in y with coefficients in $\mathbf{Z}[c_1, c_2, \ldots, c_\nu]$ such that $g(y)$ is irreducible and monic in y and $q(y)$ is nonzero mod $g(y)$, construct polynomials $r(y)$ and $s(y)$ such that $r(y)q(y) + s(y)g(y)$ is nonzero and does not contain y.*

---

* The "field of rational functions" in c_1, c_2, $\ldots$, c_ν is simply the field of quotients of the integral domain $\mathbf{Z}[c_1, c_2, \ldots, c_\nu]$. Note that the word 'function' is not being used in its modern sense here, because p/q has no 'value' at roots of q. When $\nu = 0$, the ring is just the ring of integers $\mathbf{Z}$ and the field is just the field of rational numbers $\mathbf{Q}$.

Proof. Let n be the degree of $g(y)$ and k the degree of $q(y)$. Consider the polynomials in y with coefficients in $\mathbf{Z}[c_1, c_2, \ldots, c_\nu]$ of the form $r(y)q(y) + s(y)g(y)$, where $\deg r < n$ and $\deg s < k$. These polynomials have degree less than $n + k$, so they are described by their $n + k$ coefficients, which depend linearly on the n coefficients of r and the k coefficients of s. Thus, the dependence is described by an $(n + k) \times (n + k)$ matrix of elements of $\mathbf{Z}[c_1, c_2, \ldots, c_\nu]$. If the determinant of this matrix were zero, there would be nonzero polynomials $r(y)$ and $s(y)$ with coefficients in $\mathbf{Z}[c_1, c_2, \ldots, c_\nu]$ for which $r(y)q(y) + s(y)g(y) = 0$, which is impossible because it would imply that $r(y)q(y)$ was a product divisible by $g(y)$ in which neither factor was divisible by $g(y)$ ($q(y) \not\equiv 0 \bmod g(y)$ by assumption, and $r(y) \not\equiv 0 \bmod g(y)$ because $r(y) \neq 0$ and $\deg r < \deg g$), contrary to the fact that *for polynomials with coefficients in* $\mathbf{Z}[y, c_1, c_2, \ldots, c_\nu]$, *irreducible polynomials are prime* (see Essay 1.4). Therefore, the determinant is nonzero, which implies that the system of $n + k - 1$ equations in $n + k$ unknowns that express the condition '$r(y)q(y) + s(y)g(y)$ does not contain y' has a nontrivial solution—i.e., a solution other than the solution $r(y) = s(y) = 0$. (In fact, it has the *explicit* nontrivial solution in which the coefficients of r and s are $D_1, -D_2, D_3, -D_4, \ldots, (-1)^{n+k-1}D_{n+k}$ where D_i is the determinant of the $(n+k-1) \times (n+k-1)$ matrix obtained by omitting the ith column of its $(n + k - 1) \times (n + k)$ matrix of coefficients.) As has already been shown, for such a nontrivial solution $r(y)q(y) + s(y)g(y) \neq 0$.

Thus, given an element of the root field of $g(y)$ represented by $\frac{p(y)}{q(y)}$, the same element of the root field is represented by

$$\frac{r(y)p(y)}{r(y)q(y)} \equiv \frac{r(y)p(y)}{r(y)q(y) + s(y)g(y)} = \frac{r(y)p(y)}{t},$$

where the denominator t does not contain y. In short, every element of the root field of $g(y)$ can be represented by a quotient in which the denominator is independent of y. Since the numerator can be reduced mod $g(y)$ and common factors in numerator and denominator can be canceled, it follows that *an element of the root field of $g(y)$ has a representation in the form*

$$\frac{p_1 y^{n-1} + p_2 y^{n-2} + \cdots + p_n}{q},$$

where $p_1, p_2, \ldots, p_n, q$ are elements of $\mathbf{Z}[c_1, c_2, \ldots, c_\nu]$, where $q \neq 0$, and where no divisor of q other than ± 1 divides all of $p_1, p_2, \ldots, p_n$. Each field element has exactly two such representations, one obtained from the other by reversing the signs of numerator and denominator.

This method of representing elements of the root field of $g(y)$, together with the fact that it is the field of quotients of $\mathbf{Z}[y, c_1, c_2, \ldots, c_\nu]$ mod $g(y)$, describes completely how computations in that field are to be done.

Example: One learns in school how to compute with expressions $\frac{a+b\sqrt{3}}{c}$, where a, b, and c are integers and $c \neq 0$. In other words, one learns about the root field of $y^2 - 3$.

Essay 1.4 Factorization of Polynomials with Integer Coefficients

When you can settle a question by explicit construction, be not satisfied with purely existential arguments.—Hermann Weyl [55]

The naive construction of a splitting field for a given polynomial is simple. Given a polynomial f with coefficients in a field K, let it be factored into its irreducible factors over K. If these factors are all linear, K is a splitting field for f. Otherwise, let f_1 be one of the irreducible factors whose degree is greater than 1, and let K_1 be the field $K[y]$ mod $f_1(y)$. (The irreducibility of f_1 over K implies that $K[y]$ mod $f_1(y)$ is a field, because, as in the last essay, if $p(x) \neq 0$ has degree less than the degree of $f_1(x)$, then $r(x)p(x) + s(x)f_1(x) = 0$ has no nontrivial solution in which $\deg r < \deg f_1$ and $\deg s < \deg p$, which implies that the problem "$r(x)p(x) + s(x)f_1(x)$ does not contain x" has a nonzero solution.) Then K_1 contains K, so every root of f in K is also in K_1, which is to say that f has a least as many linear factors over K as it does over K_1. In fact, it must have at least one more linear factor over K_1, because K_1 contains at least one root of f_1, and K contains none. Thus, there are more linear factors of f as a polynomial with coefficients in K_1 than there were linear factors of f as a polynomial with coefficients in K. If K_1 is not a splitting field for f, the process can be repeated to find another field K_2 over which f has more linear factors than it does over K_1. Since f has at most $\deg f$ linear factors over any field, repetition of this process terminates after at most $\deg f$ steps with a field in which f splits into linear factors.

The problem with this "construction" occurs at the first step. How can one *accomplish* the factorization of a polynomial with coefficients in a given field into irreducible factors? The answer must of course depend on the field of coefficients. If it is a finite field, the list of possible factors is finite, so they can all be tried, but for an infinite field, even for the field of rational numbers, the problem is more subtle. Kronecker presents an algorithm for factoring polynomials with *integer* coefficients early in his *Grundzüge einer arithmetischen Theorie der algebraischen Grössen*. (See also [40].) It is, in essence, the one used in the proof of the following theorem.

Theorem 1. *Given a nonzero polynomial (in any number of indeterminates) with integer coefficients, list its divisors.*

Proof. Consider first the case in which the given polynomial contains no indeterminates, which is to say that it is a nonzero *integer,* call it a. A divisor of a is a nonzero integer, and its absolute value is at most $|a|$. Therefore, a list of divisors of a can be found by listing the $2|a|$ nonzero integers whose absolute values are at most $|a|$, and striking those that do not divide a.

Consider next the case in which the given polynomial contains just one indeterminate, say it is $f(x) = a_0 x^n + a_1 x^{n-1} + \cdots + a_n$, where x is the indeterminate and $a_0, a_1, \ldots, a_n, n$ are integers with $a_0 \neq 0$ and $n > 0$.

Choose* $n+1$ distinct integers r_0, r_1, ..., r_n for which $f(r_i) \neq 0$, and for each $i = 0, 1, \ldots, n$ let $\phi_i(x)$ be the[†] polynomial of degree n with *rational* coefficients determined by the conditions $\phi_i(r_j) = 0$ for $j \neq i$ and $\phi_i(r_i) = 1$. If $g(x)$ is a factor of $f(x)$, then its degree is at most n and its value agrees with the value of $g(r_1)\phi_1(x)+g(r_2)\phi_2(x)+\cdots+g(r_n)\phi_n(x)$ at each r_i. Two polynomials in x with rational coefficients whose degrees are less than $n+1$ and whose values agree for $n+1$ distinct rational values of x are identical. Therefore, the list of polynomials of the form $b_1\phi_1(x)+b_2\phi_2(x)+\cdots+b_n\phi_n(x)$ in which b_i is an integer that divides $f(r_i)$ for each i is a finite list of polynomials with rational coefficients that includes all divisors of $f(x)$—finite because the number of possible values of b_i is finite for each i. To construct a list of divisors of $f(x)$, it will suffice to strike from this list the polynomials that are not divisors of $f(x)$. To do this, first strike all polynomials in the list whose coefficients are not integers. Then test[‡] each of the remaining polynomials to determine whether it divides $f(x)$, and strike it if it does not.

This method of deducing the case of one indeterminate from the case of none can also be used to deduce the case of $k+1$ indeterminates from the case of k. Again $f(x) = a_0x^n + a_1x^{n-1} + \cdots + a_n$, but now $a_0, a_1, \ldots, a_n$ are

---

* Theorem: *A nonzero polynomial $f(x)$ of degree n with coefficients in an integral domain has at most n distinct roots in that integral domain.* Proof: Suppose $f(x)$ is a nonzero polynomial of degree n with coefficients in an integral domain that has $m > 0$ distinct roots $\alpha_1, \alpha_2, \ldots, \alpha_m$ in that domain. Then $n > 0$ and $g(y) = f(y + \alpha_1)$ defines a polynomial of degree n that has m distinct roots 0, $\alpha_2 - \alpha_1, \alpha_3 - \alpha_1, \ldots, \alpha_m - \alpha_1$. Because $g(0) = 0$, $g(y) = yh(y)$, where $h(y)$ has degree $n - 1$. The $m - 1$ roots $\alpha_2 - \alpha_1, \alpha_3 - \alpha_1, \ldots, \alpha_m - \alpha_1$ of $g(y)$ other than zero are roots of $h(y)$, because $(\alpha_i - \alpha_1)h(\alpha_i - \alpha_1) = g(\alpha_i - \alpha_1) = 0$ and the first factor on the left is not zero. If a polynomial of degree n had $n + 1$ distinct roots, this construction would prove that $n > 0$ and would yield a polynomial of degree $n - 1$ with n distinct roots, which is impossible because it would imply an infinite sequence of polynomials of decreasing degrees. Corollary 1: *Given a polynomial of degree n with integer coefficients, any set of $2n + 1$ distinct integers contains at least $n + 1$ that are not roots of it.* Corollary 2: *Let $f(x)$ be a polynomial with coefficients in an integral domain that has degree n and n distinct roots $\alpha_1, \alpha_2, \ldots, \alpha_n$ in the integral domain. Then $f(x) = a(x - \alpha_1)(x - \alpha_2) \cdots (x - \alpha_n)$, where a is the leading coefficient of $f(x)$.* Deduction: The difference between $f(x)$ and $a(x - \alpha_1)(x - \alpha_2) \cdots (x - \alpha_n)$ is a polynomial of degree less than n that has n distinct roots, so it must be zero.

† As was shown in the previous note, $\phi_i(x)$ must be a rational number times $\prod_{j \neq i}(x - r_j)$. The rational multiplier is determined by the condition $\phi(r_i) = 1$ to be the reciprocal of the value of $\prod_{j \neq i}(x - r_j)$ when $x = r_i$.

‡ To test whether $g(x) = c_0x^m + c_1x^{m-1} + \cdots + c_m$ divides a nonzero $f(x) = a_0x^n + a_1x^{n-1} + \cdots + a_n$, determine whether $n \geq m$ and c_0 divides a_0. If not, $g(x)$ does not divide $f(x)$. If so, replace $f(x)$ with $f(x) - \frac{a_0}{c_0}x^{n-m}g(x)$, a polynomial that has degree less than $\deg f$ and that is divisible by $g(x)$ if and only if the original $f(x)$ was. If the new $f(x)$ is zero, it is divisible by $g(x)$. Otherwise, apply the same test to the new $f(x)$.

polynomials in k indeterminates, with $a_0 \neq 0$ and $n > 0$. Again choose $n + 1$ integers r_i for which $f(r_i) \neq 0$ and again construct the polynomials $\phi_i(x)$ with rational coefficients that are one at r_i and zero at all other r_j. Construct the finite list of expressions of the form $b_1\phi_1(x) + b_2\phi_2(x) + \cdots + b_n\phi_n(x)$ in which b_i is a divisor of $f(r_i)$. (Here $f(r_i)$ is a polynomial in k indeterminates with integer coefficients. By the inductive hypothesis, its divisors can be listed.) The required list of divisors is obtained by striking from this finite list the entries that are not polynomials with integer coefficients and those that do not divide $f(x)$.

A polynomial with integer coefficients is a **unit** if it is 1 or -1. The **trivial** divisors of a nonzero polynomial are the units and the polynomial itself times a unit. A nonzero polynomial is **reducible** if its list of divisors given by the theorem contains more than the four trivial divisors, or, what is the same, if it can be written as a product of two polynomials, neither of which is a unit. A nonzero polynomial is **irreducible** if it is not a unit and not reducible.

Corollary. *Given a nonzero polynomial with integer coefficients that is not a unit, write it as a product of irreducible polynomials.*

Deduction. Take the given polynomial as the initial input to the following algorithm:

Input: A product of nonzero polynomials (with integer coefficients) none of them units.

Algorithm: If each input factor is irreducible—which is to say that for each of them the list of divisors given by the theorem contains only the trivial divisors—the algorithm terminates. Otherwise, at least one factor is reducible. Replace one reducible factor in the input product with a representation of that factor as a product of two factors, neither of which is a unit.

If the algorithm terminates, it terminates with a representation of the original polynomial as a product of irreducible polynomials, as required. But it must terminate because the list of all divisors of the original polynomial is finite.

A nonzero polynomial with integer coefficients is **prime** if it is not a unit and if it divides a product of polynomials with integer coefficients only when it divides one of the factors.

Theorem 2. *Irreducible polynomials with integer coefficients are prime.*

When there are no indeterminates in the polynomials, this is the proposition that irreducible integers are prime. Euclid proved it using the following lemma:

Lemma (The Euclidean* algorithm). *Given positive integers f and g, find integers ϕ and ψ for which $\phi f + \psi g$ is positive and divides both f and g.*

Proof. Let a sequence of pairs of numbers (f_n, g_n) be defined by taking (f_1, g_1) to be (f, g) and taking (f_{n+1}, g_{n+1}) to be either $(f_n, g_n - f_n)$ or $(f_n - g_n, g_n)$, depending on whether $f_n \le g_n$ or $f_n > g_n > 0$. When $g_n = 0$, the sequence terminates. A step that changes f_n changes it to a positive value, so each f_n is positive. Similarly, each g_n is nonnegative. A step *reduces* $f_n + g_n$, either by f_n or by g_n. By the principle of infinite descent, therefore, the sequence must terminate. Let $(d, 0)$ be the terminal pair. Then d is positive. Since f_{n+1} and g_{n+1} are sums of multiples of f_n and g_n, both entries of the terminal pair $(d, 0)$ are sums of multiples of the initial pair (f, g). Thus, there are integers ϕ and ψ such that $d = \phi f + \psi g$. A common divisor of f_{n+1} and g_{n+1} is a common divisor of f_n and g_n, so, since d divides both entries of $(d, 0)$, it divides both f and g, which proves the lemma.[†]

The proof of Theorem 2 for polynomials with indeterminates will use a lemma analogous to the one above:

Lemma (The Euclidean[‡] algorithm for polynomials). *Given nonzero polynomials f and g in one indeterminate with coefficients in a field, construct polynomials ϕ and ψ with coefficients in the same field for which $\phi f + \psi g$ divides both f and g.*

Proof. Let a sequence of pairs of polynomials (f_n, g_n) be defined by taking (f_1, g_1) to be (f, g) and taking (f_{n+1}, g_{n+1}) to be the pair derived from (f_n, g_n) in the following way: If $0 \le \deg f_n \le \deg g_n$, set $i = \deg g_n - \deg f_n$ and $(f_{n+1}, g_{n+1}) = (f_n, g_n - \nu x^i f_n / \mu)$, where μ is the leading coefficient of f_n and ν is the leading coefficient of g_n. If $\deg f_n > \deg g_n \ge 0$, set $(f_{n+1}, g_{n+1}) = (f_n - \mu x^i g_n / \nu, g_n)$, where $i = \deg f_n - \deg g_n$ and μ and ν are the leading coefficients of f_n and g_n. The sequence terminates if f_n or g_n is zero (that is, has degree $-\infty$), which must occur eventually, because $\deg f_n + \deg g_n$ decreases with each step. Let d be the nonzero member of the terminal pair. Clearly d, like all members of all pairs of the sequence, has the form $\phi f + \psi g$, where ϕ and ψ are polynomials with coefficients in the field.

The predecessor of (f_{n+1}, g_{n+1}) in the sequence just constructed is either $(f_{n+1}, g_{n+1} + \nu x^i f_{n+1} / \mu)$ or $(f_{n+1} + \mu x^i g_{n+1} / \nu, g_{n+1})$. In both cases, every common divisor of f_{n+1} and g_{n+1} is a common divisor of f_n and g_n. Since d is a common divisor of the polynomials in the terminal pair, it is a common divisor of the polynomials in the initial pair (f, g), and the proof of the lemma is complete.

---

[*] In essence, this is the algorithm Euclid used [25, Book 7, Propositions 1 and 2], although the formulation is entirely different.

[†] For another version of this proof, see Essay 3.2.

[‡] It is "Euclidean" by virtue of the analogy with the previous case, not because Euclid considered anything of the kind.

Proof of Theorem 2. Let f, g, and h be in the ring of polynomials $\mathbf{Z}[x_1, x_2, \ldots, x_m]$ in m indeterminates with integer coefficients and let them satisfy (1) f is irreducible, (2) f divides gh, and (3) f does not divide g. It is to be shown that f must then divide h.

When $m = 0$ the proof is the one that Euclid, in essence, gave [25, Book 7, Prop. 24]. In this case, f, g, and h are integers. Since changing the sign of f or g or h does not change (1), (2), (3), or the desired conclusion, f, g, and h can be assumed to be nonnegative. Then $f > 0$ and $g > 0$, because f is not zero by (1) and g is not zero by (3). The Euclidean algorithm provides a positive $d = \phi f + \psi g$ that divides both f and g. Since f is irreducible and d is positive, $d = 1$ or f. By (3), $d \neq f$. Therefore $\phi f + \psi g = 1$. Thus $\phi fh + \psi gh = h$ is divisible by f, as was to be shown, because f divides both terms on the left.

Next consider the case of Theorem 2 in which f contains fewer than m indeterminates:

A Special Case.   *An irreducible element of $\mathbf{Z}[x_1, x_2, \ldots, x_{m-1}]$ is prime as an element of $\mathbf{Z}[x_1, x_2, \ldots, x_m]$.*

Let f be an irreducible polynomial in $x_1, x_2, \ldots, x_{m-1}$ with integer coefficients, and let $g(x)$ and $h(x)$ be polynomials in $x, x_1, x_2, \ldots, x_{m-1}$ with integer coefficients for which (2) and (3) hold. Then all coefficients of $g(x)h(x)$ are divisible by f, but at least one of the coefficients of $g(x)$ is not divisible by f. Let $g(x) = a_0 x^n + a_1 x^{n-1} + \cdots + a_n$, where the a_i are polynomials in $x_1, x_2, \ldots, x_{m-1}$, and let I be the least index for which a_I is not divisible by f. If $h(x)$ were not divisible by f, then, in the same way, when $h(x)$ was written in the form $h(x) = b_0 x^t + b_1 x^{t-1} + \cdots + b_t$ there would be a least index J for which b_J was not divisible by f. Then the coefficient of x^{I+J} in $g(x)h(x)$ would be $a_I b_J$ plus terms divisible by f (this coefficient is $a_I b_J$ plus terms that are products in which one factor is divisible by f). But $a_I b_J$ would not be divisible by f by the inductive hypothesis, so $g(x)h(x)$ would not then be divisible by f, contrary to hypothesis. Therefore $h(x)$ must be divisible by f, as was to be shown.

The general case, in which f may contain x, can now be deduced from the case of $m - 1$ indeterminates as follows: The Euclidean algorithm gives equations $d(x) = \phi(x)f(x) + \psi(x)g(x)$, $f(x) = q_1(x)d(x)$, $g(x) = q_2(x)d(x)$ where d, ϕ, ψ, q_1, q_2 are polynomials in x with coefficients in the field of rational functions in $x_1, x_2, \ldots, x_{m-1}$. Let δ be a common denominator of all five of these polynomials. (For example, δ could be taken to be the product of all denominators of all coefficients of the five.) Then $D(x) = \delta \cdot d(x)$, $\Phi(x) = \delta \cdot \phi(x)$, $\Psi(x) = \delta \cdot \psi(x)$, $Q_1(x) = \delta \cdot q_1(x)$, and $Q_2(x) = \delta \cdot q_2(x)$ all are in $\mathbf{Z}[x, x_1, x_2, \ldots, x_{m-1}]$ and they satisfy $D(x) = \Phi(x)f(x) + \Psi(x)g(x)$, $\delta^2 \cdot f(x) = Q_1(x)D(x)$ and $\delta^2 \cdot g(x) = Q_2(x)D(x)$. By the special case already proved, each irreducible factor ϵ of δ^2 divides either $Q_1(x)$ or $D(x)$. Therefore, $\delta^2 \cdot f(x) = Q_1(x)D(x)$ can be divided by each of the irreducible factors of δ^2 in succession to find $f(x) = (Q_1(x)/\epsilon_1) \cdot (D(x)/\epsilon_2)$ where $\delta^2 = \epsilon_1 \epsilon_2$. By (1), $D(x)/\epsilon_2$ must be ± 1 or $\pm f(x)$. By (3), $D(x)/\epsilon_2 \neq \pm f(x)$. Therefore,

$D(x) = \pm\epsilon_2$, so $\pm\epsilon_2 h(x) = \Phi(x) \cdot f(x) \cdot h(x) + \Psi(x) \cdot g(x) \cdot h(x)$, which shows that $f(x)$ divides $\epsilon_2 h(x)$, say $\epsilon_2 h(x) = Q_3(x) \cdot f(x)$. By the special case, each irreducible factor of ϵ_2 divides $Q_3(X)$, so $f(x)$ divides $h(x)$, as was to be shown.

Thus Theorem 2 follows by induction.

Corollary 1 (Unique factorization of polynomials with integer coefficients). *If $\phi_1\phi_2\cdots\phi_\mu = \psi_1\psi_2\cdots\psi_\nu$, where the factors on both sides are irreducible polynomials with integer coefficients, then $\mu = \nu$, and the factors can be so ordered that $\phi_i = -\psi_i$ for an even number of values of i, and $\phi_i = \psi_i$ for all others.*

Deduction. Let such an equation $\phi_1\phi_2\cdots\phi_\mu = \psi_1\psi_2\cdots\psi_\nu$ be given in which $\mu \geq 1$. Since Theorem 2 implies that ϕ_1 divides ψ_j for some j, the ψ's can be rearranged to make ϕ_1 divide ψ_1, say $\phi_1 = q_1\phi_1$. Then $\phi_2\phi_3\cdots\phi_\mu = q_1\psi_2\psi_3\cdots\psi_\nu$ is a product of factors, at least $\nu - 1$ of which are irreducible. If μ were less than ν, μ iterations of this step would express 1 as a product of ν factors, at least $\nu - \mu$ of which were irreducible, contrary to the fact that the only factors of 1, the units 1 and -1, are not irreducible. Therefore, $\mu \geq \nu$. For the same reason, $\nu \geq \mu$, so μ and ν must be equal in any such equation. In the first equation $\phi_2\phi_3\cdots\phi_\mu = q_1\psi_2\psi_3\cdots\psi_\nu$ found by the process above, q_1 must therefore have no irreducible factors and therefore must be a unit. Thus, μ steps rearrange the ψ's in such a way that $\psi_i = q_i\phi_i$ for each i, where q_i is a unit and $1 = q_1q_2\cdots q_\mu$. The last equation shows that the number of q's that are -1 is even, and the corollary follows.

Corollary 2 (Gauss's* lemma). *If an element of $\mathbf{Z}[x, x_1, x_2, \ldots, x_{m-1}]$ is reducible over the field of rational functions of $x_1, x_2, \ldots, x_{m-1}$ in the sense that it can be written as a product of two polynomials of positive degree in x with coefficients in this field, then it is reducible as an element of $\mathbf{Z}[x, x_1, x_2, \ldots, x_{m-1}]$.*

Deduction. Let $f(x)$ be reducible over the field of rational functions, say $f(x) = g(x)h(x)$, and let d_1 and d_2 be elements of $\mathbf{Z}[x_1, x_2, \ldots, x_{m-1}]$ that clear the denominators of $g(x)$ and $h(x)$ respectively, in other words, elements such that $G(x) = d_1 g(x)$ and $H(x) = d_2 h(x)$ are in $\mathbf{Z}[x, x_1, x_2, \ldots, x_{m-1}]$. Then $d_1 d_2 f(x) = G(x)H(x)$; this equation can be divided successively by the irreducible factors of $d_1 d_2$ to produce an equation $f(x) = \frac{G(x)}{e_1} \cdot \frac{H(x)}{e_2}$ (where

---

* Gauss's original statement was that a product of *monic* polynomials with rational coefficients can have integer coefficients only if the factors do. The same is true for $m > 1$: A product of monic polynomials whose coefficients are rational functions in $x_1, x_2, \ldots, x_{m-1}$ can have coefficients in $\mathbf{Z}[x_1, x_2, \ldots, x_{m-1}]$ only if the factors do. This statement can be proved in the same way as the statement above: When $g(x)$ and $h(x)$ are monic, $f(x)$ is monic, so its factors $\frac{G(x)}{e_1}$ and $\frac{H(x)}{e_2}$ are monic, which implies that $\frac{d_1}{e_1} = 1$ and $\frac{d_2}{e_2} = 1$ and therefore that $g(x) = G(x)$ and $h(x) = H(x)$. For more on Gauss's lemma, see Essay 2.5.

e_1 and e_2 are integers for which $e_1 e_2 = d_1 d_2$, and both factors have integer coefficients), which shows that $f(x)$ is reducible in $\mathbf{Z}[x, x_1, x_2, \ldots, x_{m-1}]$.

The methods used to prove Corollary 1 prove another proposition:

Proposition. *If $\phi_1 \phi_2 \cdots \phi_\mu = \psi_1 \psi_2 \cdots \psi_\nu$ where the factors on both sides are irreducible polynomials in one indeterminate with coefficients in the root field of some (monic, irreducible) polynomial $g(y)$, then $\mu = \nu$ and the factors can be so ordered that each ϕ_i has the form $\sigma_i \psi_i$ for $i = 1, 2, \ldots, \mu$, where σ_i is a nonzero field element and $\prod \sigma_i = 1$.*

This proposition shows that a factorization of $f(x) \bmod g(y)$ produced by the algorithm of the next essay is determined, except for trivial modifications, by $f(x)$ and $g(y)$.

Essay 1.5 A Factorization Algorithm

Die im Art. 1 aufgestellte Definition der Irreductibilität entbehrt so lange einer sicheren Grundlage, als nicht eine Methode angegeben ist, mittels deren bei einer bestimmten, vorgelegten Function entschieden werden kann, ob dieselbe der aufgestellten Definition gemäss irreductibel ist oder nicht. (The definition of irreducibility given in Art. 1 lacks a firm foundation until a method is given that makes it possible to determine whether a given example does or does not satisfy it.)—L. Kronecker [39, §4]

The naive method of constructing a splitting field that was sketched at the beginning of the preceding essay suggests that the factorization into irreducibles of a polynomial with coefficients in a root field (see Essay 1.3) might be a key tool in the proof of the fundamental theorem. In fact, as later essays will show, it suffices to be able to factor *monic, irreducible* polynomials with *coefficients in* $\mathbf{Z}[c_1, c_2, \ldots, c_\nu]$ *itself*, which is the problem treated by the following theorem:

Theorem. *Given monic, irreducible polynomials f and g in one indeterminate with coefficients in the ring $\mathbf{Z}[c_1, c_2, \ldots, c_\nu]$ of polynomials in c_1, c_2, $\ldots$, c_ν with integer coefficients, factor $f(x)$ as a polynomial with coefficients in the root field of $g(y)$.*

In other words, one is to construct a congruence

$$(1) \qquad f(x) \equiv \phi_1(x,y)\phi_2(x,y) \cdots \phi_k(x,y) \bmod g(y)$$

in which the factors ϕ_i are polynomials in two indeterminates with coefficients in the field of rational functions in c_1, c_2, $\ldots$, c_ν that are monic and of positive degree in x, that have degree in y less than* the degree of g, and that cannot be written, mod $g(y)$, as products of polynomials of lower degree in x.

Notation: The ring of polynomials in c_1, c_2, $\ldots$, c_ν with integer coefficients will be denoted by $R = \mathbf{Z}[c_1, c_2, \ldots, c_\nu]$. Its field of quotients, the field of rational functions in c_1, c_2, $\ldots$, c_ν, will be denoted by K. When $\nu = 0$, R is the ring of integers and K is the field of rational numbers.

This essay is devoted to a *description* of the algorithm for finding the factorization of $f(x)$ mod $g(y)$ in the form of a congruence (1). The *validity* of the algorithm will be proved in the next essay.

The factorization algorithm will make use of computations in the ring $R[x, y]$ mod $(f(x), g(y))$ of polynomials in x, y, c_1, c_2, $\ldots$, c_ν with integer coefficients, where two such polynomials represent the same ring element— by definition—if their difference is a multiple of $f(x)$ plus a multiple of $g(y)$

---

* As far as (1) is concerned, ϕ_i can be replaced by any polynomial that is congruent to it mod $g(y)$, and this condition on the degrees of the ϕ's need not be satisfied. Restricting the degree in this way serves to *determine* the factors $\phi_i(x,y)$ once $f(x)$ and $g(y)$ are given, as is shown by the final proposition of Essay 1.4.

(the multipliers being, of course, in $R[x, y]$). Since $f(x)$ is monic of degree m, $x^m - f(x)$ is a polynomial of degree less than m in x that represents the same ring element as x^m, so any ring element that is represented by a polynomial of degree $m + j$ in x for $j \geq 0$ can also be represented by a polynomial whose degree in x is less than $m + j$ (replace the leading term $\phi(y)x^{m+j}$ in x with $\phi(y)x^j(x^m - f(x))$ while leaving the other terms unchanged). Thus, every ring element can be represented by a polynomial whose degree in x is less than m. In fact, every element of $R[x, y]$ is congruent mod $f(x)$ to one and only one polynomial of degree less than m in x, because an element of $R[x, y]$ whose degree in x is less than m can be a multiple of $f(x)$ only if it is zero. In the same way, any element of $R[x, y]$ is congruent mod $g(y)$ to just one element whose degree in y is less than n. Moreover, since the reduction method can be applied to each coefficient $\phi_i(y)$ of a polynomial $\phi_1(y)x^{m-1} + \phi_2(y)x^{m-2} + \cdots + \phi_m(y)$ that has already been reduced mod $f(x)$, every element of $R[x, y]$ mod $(f(x), g(y))$ is represented by one and only one element of $R[x, y]$ whose degree in x is less than m and whose degree in y is less than n.

Each element of this ring $R[x, y]$ mod $(f(x), g(y))$ is a root of a monic polynomial with coefficients in R. Specifically, if $\phi(x, y)$ is an element of $R[x, y]$, a monic polynomial $\mathcal{F}(z)$ of degree mn with coefficients in R for which $\mathcal{F}(\phi(x, y)) \equiv 0 \bmod (f(x), g(y))$ can be constructed in the following way: For each of the mn monomials $x^i y^j$ in which $0 \leq i < m$ and $0 \leq j < n$, the polynomial $\phi(x, y)x^i y^j$ is congruent mod $f(x)$ and $g(y)$, as was just seen, to a sum of multiples of $x^\alpha y^\beta$, where $0 \leq \alpha < m$ and $0 \leq \beta < n$, in which the multipliers are in R. Thus, the congruence

$$\phi(x, y)x^i y^j \equiv \sum_{\alpha=0}^{m-1} \sum_{\beta=0}^{n-1} M_{ij,\alpha\beta} x^\alpha y^\beta \bmod (f(x), g(y))$$

determines an $mn \times mn$ matrix M of elements $M_{ij,\alpha\beta}$ of R once an ordering of the mn monomials $x^i y^j$ is decided upon. Otherwise stated, M is the matrix that represents multiplication by $\phi(x, y)$ relative to the basis $x^i y^j$ of $R[x, y]$ mod $(f(x), g(y))$ over R. The *characteristic polynomial* of this matrix, which is to say the polynomial $\mathcal{F}(z) = \det(zI - M)$, is monic of degree mn in z; by the Cayley–Hamilton theorem, it satisfies $\mathcal{F}(\phi(x, y)) \equiv 0 \bmod (f(x), g(y))$. (A proof will be given in the next essay.)

Let this construction be applied not to a single polynomial $\phi(x, y)$ but to $tx + uy$, regarded as a polynomial in new indeterminates t and u whose coefficients are in $R[x, y]$. The result is a polynomial $\mathcal{F}(z, t, u)$ in z, t, and u with coefficients in R. Specifically, $\mathcal{F}$ is the characteristic polynomial $\det(zI - M)$ of the $mn \times mn$ matrix M determined by $C \cdot (tx + uy) \equiv MC \bmod (f(x), g(y))$, where C is the column matrix of length mn whose entries are the monomials $x^i y^j$ in which $0 \leq i < m$ and $0 \leq j < n$ arranged in some order and $tx + yu$ is a 1×1 matrix. The entries of M are homogeneous polynomials of degree 1 in t and u with coefficients in R, so the entries of $zI - M$ are homogeneous of degree 1 in z, t, and u. Thus, $\mathcal{F}(z,t,u)$ is homogeneous of degree mn in these indeterminates and has coefficients in R; moreover, it is monic in z.

As was seen in the last essay, the irreducible factors of $\mathcal{F}(z,t,u)$ as a polynomial with integer coefficients (in $3 + \nu$ indeterminates) can be found, say $\mathcal{F}(z,t,u) = \prod \mathcal{F}_i(z,t,u)$, where the $\mathcal{F}_i(z,t,u)$ are irreducible. Because $\mathcal{F}$ is homogeneous, its irreducible factors $\mathcal{F}_i$ are homogeneous. Because $\mathcal{F}$ is monic in z, the leading coefficient of each of its irreducible factors $\mathcal{F}_i$ as a polynomial in z is ± 1, so one can stipulate that each $\mathcal{F}_i$ is monic in z, and this condition determines the $\mathcal{F}_i$ completely. The required factorization

$$(1) \qquad f(x) \equiv \phi_1(x,y)\phi_2(x,y)\cdots\phi_k(x,y) \bmod g(y)$$

contains one factor $\phi_i(x,y)$ for each $\mathcal{F}_i(z,t,u)$. It is constructed as follows:

As will be shown, the degree of $\mathcal{F}_i$ (it is homogeneous in z, t, and u) is a multiple of n, say it is $\mu_i n$. (By symmetry, this degree is also a multiple of m, a fact that is not of interest here.) Substitute $tx + uy$ for z and 1 for t in $\mathcal{F}_i$ and write the result in the form

$$(2) \quad \mathcal{F}_i(x + uy, 1, u) = B_{i,0}u^{\mu_i n} + B_{i,2}u^{\mu_i n-1} + B_{i,2}u^{\mu_i n-2} + \cdots + B_{i,\mu_i n}.$$

Each coefficient $B_{i,j}$ is a polynomial in x, y, c_1, c_2, ..., $c_{\nu-1}$, and c_ν with integer coefficients. The first μ_i of these coefficients are all zero mod $g(y)$, which is to say that reduction mod $g(y)$ gives

$$\mathcal{F}_i(x + uy, 1, u) \equiv \psi_i u^{\mu_i(n-1)} + \cdots \bmod g(y),$$

where the omitted terms are of lower degree in u and $\psi_i \equiv B_{i,\mu_i} \bmod g(y)$. *The factor $\phi_i(x,y)$ of $f(x)$ mod $g(y)$ corresponding to this factor $\mathcal{F}_i$ of $\mathcal{F}$ is*

$$(3) \qquad \phi_i(x,y) \equiv \frac{\psi_i(x,y)}{g'(y)^{\mu_i}} \bmod g(y),$$

where $g'(y)$ is the derivative of $g(y)$. (Implicit in this statement, since $\phi_i(x,y)$ is monic in x, is the statement that $\psi_i \equiv g'(y)^{\mu_i}x^{\mu_i} + \cdots \bmod g(y)$ where the omitted terms have lower degree in x.)

Example 1. $f(x) = x^2 - 2$ and $g(y) = y^2 - 3$. The first step is to find $\mathcal{F}(z,t,u)$ for this f and g. When the monomials $x^\alpha y^\beta$ for $0 \le \alpha < 2$ and $0 \le \beta < 2$ are put in the order 1, x, y, xy, the matrix that represents multiplication by $tx + uy$ becomes

$$\begin{bmatrix} 0 & t & u & 0 \\ 2t & 0 & 0 & u \\ 3u & 0 & 0 & t \\ 0 & 3u & 2t & 0 \end{bmatrix}.$$

Therefore $\mathcal{F}$ is the determinant of

$$\begin{bmatrix} z & -t & -u & 0 \\ -2t & z & 0 & -u \\ -3u & 0 & z & -t \\ 0 & -3u & -2t & z \end{bmatrix},$$

which can be found without too much paper-and-pencil calculation to be $z^4 - (4t^2 + 6u^2)z^2 + 4t^4 - 12t^2u^2 + 9u^4$. This polynomial $\mathcal{F}(z, t, u)$ is irreducible because $\mathcal{F}(z, 1, 1) = z^4 - 10z^2 + 1$ obviously has no root mod 5, so it can only have a factorization of the form $(z^2 + az + b)(z^2 + cz + d) = z^4 - 10z^2 + 1$, and this would imply $a = -c$, $d + ac + b = -10$, and $b = d = \pm 1$, so $a^2 = -ac = b + d + 10 = \pm 2 + 10$, which is impossible. Therefore, $x^2 - 2$ is irreducible mod $y^2 - 3$ (the factorization algorithm produces only the one factor corresponding to $\mathcal{F}$ itself). To determine this factor—which must, of course, be $x^2 - 2$ itself—one computes the coefficient ψ of u^2 in $\mathcal{F}(x + uy, 1, u) = (x + uy)^4 - (4 + 6u^2)(x + uy)^2 + 4 - 12u^2 + 9u^4$ because $\deg \mathcal{F}/\deg g = 2$. (As expected, the coefficient of u^4 is $y^4 - 6y^2 + 9 \equiv 0 \bmod (y^2 - 3)$, and the coefficient of u^3 is $4xy^3 - 12xy \equiv 0 \bmod (y^2 - 3)$.) Because $\psi = 6x^2y^2 - 4y^2 - 6x^2 - 12$, formula (3) gives the factor

$$\frac{6x^2y^2 - 4y^2 - 6x^2 - 12}{g'(y)^2} \equiv \frac{18x^2 - 12 - 6x^2 - 12}{(2y)^2} \equiv \frac{12x^2 - 24}{12} = x^2 - 2 \bmod g(y)$$

as expected.

Example 2. $f(x) = x^2 - 2$ and $g(y) = y^2 - 18$. In this case, $\mathcal{F}(z, t, u)$ is the determinant of

$$\begin{bmatrix} z & -t & -u & 0 \\ -2t & z & 0 & -u \\ -18u & 0 & z & -t \\ 0 & -18u & -2t & z \end{bmatrix},$$

which can be found—the calculation is a variation of the one in Example 1—to be $z^4 - (4t^2 + 36u^2)z^2 + 4t^4 - 72t^2u^2 + 324u^2$. Thus a factorization $\mathcal{F}(z, t, u) = \mathcal{F}_1(z, t, u)\mathcal{F}_2(z, t, u)$ can be found by completing the square to put $\mathcal{F}$ in the form $\mathcal{F}(z, t, u) = (z^2 - 2t^2 - 18u^2)^2 - 144t^2u^2 = (z^2 - 2t^2 - 18u^2 - 12tu)(z^2 - 2t^2 - 18u^2 + 12tu)$. The factor of $f(x) \bmod g(y)$ corresponding to $\mathcal{F}_1(z, t, u)$ is, because in this case $\mu_1 = \deg \mathcal{F}_1/\deg g = 1$, the coefficient of u in $(x + uy)^2 - 2 - 18u^2 - 12u$ divided by $g'(y)$, which is

$$\frac{2xy - 12}{2y} \equiv \frac{2xy^2 - 12y}{2y^2} \equiv \frac{36x - 12y}{36} = x - \frac{y}{3} \bmod (y^2 - 18).$$

(As expected, the coefficient of u^2, which is $y^2 - 18$, is zero mod $g(y)$.) In the same way, the factor of $f(x) \bmod g(y)$ corresponding to $\mathcal{F}_2$ is $x + \frac{1}{3}y$. Indeed, $\left(x - \frac{1}{3}y\right)\left(x + \frac{1}{3}y\right) = x^2 - \frac{1}{9}y^2 \equiv x^2 - 2 \bmod (y^2 - 18)$, so $f(x) = x^2 - 2$ splits mod $g(y) = y^2 - 18$ into linear factors. (If $y = \sqrt{18}$, then $\frac{y}{3} = \sqrt{2}$.)

Example 3. $f(x) = x^2 + c_1 x + c_2$, $g(y) = y^2 - c_1^2 + 4c_2$. The factorization depends on factoring the characteristic polynomial of

$$
\begin{bmatrix}
0 & t & u & 0 \\
-c_2 t & -c_1 t & 0 & u \\
du & 0 & 0 & t \\
0 & du & -c_2 t & -c_1 t
\end{bmatrix},
$$

where $d = c_1^2 - 4c_2$. The computation of this characteristic polynomial is not too onerous. (One method is to use the formula $z^4 - A_1 z^3 + A_2 z^2 - A_3 z + A_4$ for the characteristic polynomial, where A_i is the sum of the $i \times i$ principal minors of the matrix.) The result is

$$
\mathcal{F}(z, t, u) = z^4 + 2c_1 t z^3 + (2c_2 t^2 + c_1^2 t^2 - 2du^2)z^2 + c_1 t(2c_2 t^2 - 2du^2)z
$$
$$
+ c_2^2 t^4 + (2c_2 d - c_1^2 d)t^2 u^2 + d^2 u^4,
$$

a homogeneous polynomial in z, t, and u with coefficients in $\mathbf{Z}[c_1, c_2]$ when $c_1^2 - 4c_2$ is substituted for d. The difficult step is the factorization of $\mathcal{F}(z, t, u)$. When $t = 0$ it is $z^4 - 2du^2 z^2 + d^2 u^4 = (z^2 - du^2)^2$, and when $u = 0$ it is $z^4 + 2c_1 t z^3 + (2c_2 + c_1^2)t^2 z^2 + 2c_1 c_2 t^3 z + c_2^2 t^4 = (z^2 + c_1 z t + c_2 t^2)^2$. Therefore, the factorization of $\mathcal{F}$, if there is one, must be of the form $(z^2 + c_1 tz + c_2 t^2 + ptu - du^2)(z^2 + c_1 tz + c_2 t^2 + qtu - du^2)$, where p and q are in $\mathbf{Z}[c_1, c_2]$. The coefficient of $t^3 u$ is 0 on the one hand and $c_2(p + q)$ on the other, so $p = -q$. Then the coefficient of $t^2 u^2$ is $2c_2 d - c_1^2 d$ on the one hand and $-2dc_2 - p^2$ on the other, so $p^2 = -4c_2 d + c_1^2 d = d^2$, which gives the factorization $(z^2 + c_1 tz + c_2 t^2 + dtu - du^2)(z^2 + c_1 tz + c_2 t^2 - dtu - du^2)$ of $\mathcal{F}(z, t, u)$, where d is the element $c_1^2 - 4c_2$ of $\mathbf{Z}[c_1, c_2]$. Let $\mathcal{F}_1(z, t, u)$ be the first of these factors. Then $\mathcal{F}_1(x + uy, 1, u) = (x + uy)^2 + c_1(x + uy) + c_2 + du - du^2$, and the resulting factor of $f(x) \bmod g(y)$ is

$$
\frac{2xy + c_1 y + d}{2y} \equiv \frac{2xy^2 + c_1 y^2 + dy}{2y^2} \equiv x + \frac{c_1}{2} + \frac{y}{2} \bmod g(y).
$$

(Note that the coefficient of u^2 is $y^2 - d \equiv 0 \bmod g(y)$, as expected.) In the same way, $\mathcal{F}_2$ leads to the factor $x + \frac{c_1}{2} - \frac{y}{2}$. Note that this process merely factors $f(x) \bmod g(y)$. It does *not* construct the polynomial $g(y)$ modulo which $f(x)$ is a product of linear factors.

Example 4. $f(x) = x^3 - 2$ and $g(y) = y^3 - 2$. When the monomials $x^\alpha y^\beta$ with $0 \le \alpha < 3$ and $0 \le \beta < 3$ are put in the order 1, y, y^2, x, xy, xy^2, x^2, $x^2 y$, $x^2 y^2$, the matrix of which $\mathcal{F}(z, t, u)$ is the determinant is* $zI_9 - tM_x - uM_y$, where M_y is the 9×9 matrix of 3×3 blocks in which the diagonal blocks are all equal to

$$
G = \begin{bmatrix}
0 & 1 & 0 \\
0 & 0 & 1 \\
2 & 0 & 0
\end{bmatrix}
$$

---

* Here I_k is the $k \times k$ identity matrix.

and the blocks off the diagonal are zero, and where M_x is the 9×9 matrix of 3×3 blocks that has the form of the matrix G; that is, the first block in the first row is zero, the second block in the first row is the identity matrix I_3, and so forth, the first block in the last row being $2I_3$. Hand computation of this 9×9 determinant is straightforward but tedious. An easier method of hand computation uses the formula $\mathcal{F}(z, 1, u) = \det\left((zI_3 - uG)^3 - 2I_3\right)$ that is proved in the next essay. Since $G^3 = 2I_3$ (this is the main property of G), it follows that $\mathcal{F}(z, 1, u)$ is the product of the determinant of $(zI - uG) - G$ and the determinant of

$$(zI - uG)^2 + (zI - uG)G + G^2 = \begin{bmatrix} z^2 & (-2u+1)z & u^2 - u + 1 \\ 2(u^2 - u + 1) & z^2 & (-2u+1)z \\ 2(-2u+1)z & 2(u^2 - u + 1) & z^2 \end{bmatrix}.$$

The first determinant is $(u+1)^3 \det\left(\frac{z}{u+1}I - G\right) = (u+1)^3 \left(\left(\frac{z}{u+1}\right)^3 - 2\right) =$ $z^3 - 2(u+1)^3$, which gives the factor $z^3 - 2(t+u)^3$ of $\mathcal{F}(z, t, u)$; call it $\mathcal{F}_1(z, t, u)$. The second determinant is $z^6 + 2(-2u+1)^3 z^3 + 4(u^2 - u + 1)^3 - 3 \cdot 2(u^2 - u + 1)(-2u+1)z^3 = z^6 + (-4u^3 + 6u^2 + 6u - 4)z^3 + 4(u^2 - u + 1)^3$, which gives the factor $z^6 + (-4u^3 + 6tu^2 + 6t^2u - 4t^3)z^3 + 4(u^2 - tu + t^2)^3$ of $\mathcal{F}(z, t, u)$; call it $\mathcal{F}_2(z, t, u)$.

The first factor $\mathcal{F}_1(z, t, u)$ is irreducible because its degree is 3, and any factor of $\mathcal{F}$ has degree divisible by 3. That $\mathcal{F}_2(z, t, u)$ is irreducible follows from the irreducibility of $\mathcal{F}_2(z, 1, -1) = z^6 + 108$. (Since $z^6 + 108 \equiv (z^2 + 2)(z^2 + z + 2)(z^2 - z + 2)$ mod 5 and the factors on the right are irreducible mod 5, all factors of $z^6 + 108$ must have even degree. Similarly, $z^6 + 108 \equiv (z^3 - 2)(z^3 + 2)$ mod 7 and the factors on the right are irreducible mod 7, so all factors of $z^6 + 108$ must have degree divisible by 3. Thus, $z^6 + 108$ has no proper factor.) Therefore, $x^3 - 2$ is a product of a factor of degree 1 and a factor of degree 2 mod $y^3 - 2$.

The actual factorization $x^3 - 2 \equiv (x - y)(x^2 + xy + y^2)$ mod $(y^3 - 2)$ is in fact obvious once the factor of degree 2 is known to be irreducible. Its derivation using the algorithm is as follows: The coefficient of u^2 in $\mathcal{F}_1(x + uy, 1, u)$ is $3xy^2 - 6$, so the factor is $\frac{3xy^2 - 6}{3y^2} \equiv \frac{3xy^3 - 6y}{3y^3} \equiv x - y$ mod $g(y)$. (As expected, the coefficient of u^3 is $y^3 - 2 \equiv 0$ mod $g(y)$.) The coefficient of u^4 in $(x + uy)^6 + (-4u^3 + 6u^2 + 6u - 4)(x + uy)^3 + 4(u^2 - u + 1)^3$ is $15x^2y^4 + (-4)(3x^2y) + 6(3xy^2) + 6y^3 + 4 \cdot (3 + 3)$, so the corresponding factor of $f(x)$ mod $g(y)$ is $\frac{30x^2y - 12x^2y + 18xy^2 + 12 + 24}{(3y^2)^2} \equiv \frac{18x^2y + 18xy^2 + 18y^3}{18y} \equiv x^2 + xy + y^2$ mod $(y^3 - 2)$, as expected. (Also as expected, the coefficient of u^6, which is $y^6 - 4y^3 + 4$, and the coefficient of tu^5, which is $6xy^5 + 6y^3 - 12xy^2 - 12$, are both zero mod $(y^3 - 2)$.)

Example 5. $f(x) = x^3 - c$, $g(y) = y^3 - c$. Replacement of 2 with c in the matrix G of Example 4 gives as the two factors of $\mathcal{F}(z, 1, u)$ the polynomials $z^3 - c(u+1)^3$, which is the determinant of

$$\begin{bmatrix} z & -(u+1) & 0 \\ 0 & z & -(u+1) \\ -c(u+1) & 0 & z \end{bmatrix},$$

and $z^6 + c(-2u^3 + 3u^2 + 3u - 2)z^3 + c^2(u^2 - u + 1)^3$, which is the determinant of

$$\begin{bmatrix} z^2 & (-2u+1)z & u^2 - u + 1 \\ c(u^2 - u + 1) & z^2 & (-2u+1)z \\ c(-2u+1)z & c(u^2 - u + 1) & z^2 \end{bmatrix}.$$

The factorization of $\mathcal{F}(z,t,u)$ is derived from these factors by making the factors homogeneous in t. Both factors are irreducible because they are irreducible when $c = 2$. They prove that the obvious factorization $x^3 - c \equiv (x - y)(x^2 + xy + y^2) \bmod (y^3 - c)$ is a factorization into factors that are irreducible mod $y^3 - c$.

Essay 1.6 Validation of the Factorization Algorithm

Let $f(x)$ and $g(y)$ be monic, irreducible polynomials with coefficients in the ring $R = \mathbf{Z}[c_1, c_2, \ldots, c_\nu]$ of polynomials in c_1, c_2, $\ldots$, c_ν with integer coefficients. The last essay showed that if C is a column matrix of length mn whose entries are the monomials $x^\alpha y^\beta$ with $0 \le \alpha < m = \deg f$ and $0 \le \beta < n = \deg g$, arranged in some order, then the congruence

$$MC \equiv C \cdot (tx + uy) \bmod (f(x), g(y))$$

determines an $mn \times mn$ matrix M whose entries are homogeneous linear polynomials in t and u with coefficients in R. (Here $tx + uy$ is to be regarded as a 1×1 matrix, so that the right side of the congruence, like the left, is a matrix product.)

Let $\mathcal{F}(z, t, u) = \det(zI - M)$ be the characteristic polynomial of M (which, by the way, is independent of the choice of the order of the entries of C) and let

$$\mathcal{F}(z, t, u) = \mathcal{F}_1(z, t, u)\mathcal{F}_2(z, t, u) \cdots \mathcal{F}_k(z, t, u)$$

be its factorization into irreducible factors that are monic in z. Clearly $\mathcal{F}$ is homogeneous of degree mn in z, t, and u with coefficients in R, so the $\mathcal{F}_i$ are homogeneous in z, t, and u. It is to be shown that *the degree of each $\mathcal{F}_i(z, t, u)$ is a multiple of n*, say $\deg \mathcal{F}_i = \mu_i n$, that

$$(1) \qquad g'(y)^m f(x) \equiv \psi_1(x, y)\psi_2(x, y) \cdots \psi_k(x, y) \bmod g(y)$$

when $\psi_i(x, y)$ is defined to be the coefficient of $u^{\mu_i(n-1)}$ in $\mathcal{F}_i(x + uy, 1, u)$, and that the factors $\psi_i(x, y)$ on the right are irreducible as polynomials in x with coefficients in the root field of $g(y)$; moreover, it is to be shown that $\psi_i(x, y) \equiv g'(y)^{\mu_i} x^{\mu_i} + \cdots \bmod g(y)$, where the omitted terms have lower degree in x, so that division of (1) by $g'(y)^m$ gives the required factorization $f(x) \equiv \phi_1(x, y)\phi_2(x, y) \cdots \phi_k(x, y) \bmod g(y)$ of $f(x) \bmod g(y)$ into factors $\phi_i(x, y) \equiv \frac{\psi_i(x, y)}{g'(y)^{\mu_i}} \bmod g(y)$ with coefficients in the root field of $g(y)$ that are monic as well as irreducible.

That $\deg \mathcal{F}_i$ is divisible by n can be proved as follows: Set $t = 0$ and $u = 1$ in M to find a matrix, call it M_y, that satisfies $M_y \cdot C \equiv C \cdot y \bmod g(y)$. Let the monomials $x^\alpha y^\beta$ in C be ordered by putting $x^\alpha y^\beta$ ahead of $x^{\alpha'} y^{\beta'}$ if $\alpha < \alpha'$ or if $\alpha = \alpha'$ and $\beta < \beta'$. Then M_y is a matrix of $n \times n$ blocks in which the blocks off the diagonal are all zero and the blocks on the diagonal are all u times the matrix G whose first $n - 1$ rows are the last $n - 1$ rows of the $n \times n$ identity matrix I_n and whose last row is $-b_n, -b_{n-1}, \ldots, -b_1$, where the b_j are the coefficients of $g(y) = y^n + b_1 y^{n-1} + b_2 y^{n-2} + \cdots + b_n$. As is easily* shown, $g(y) = \det(yI_n - G)$. Then $\mathcal{F}(z, 0, u) = \det(zI - uM_y) = \det(zI - uG)^m = u^{mn} \det(\frac{z}{u}I_n - G)^m = u^{mn} g(\frac{z}{u})^m = g(z, u)^m$, where $g(z, u)$

---

* This is the special case $f(x) = x$ and $u = 1$ of the formula $\mathcal{F}(z, 1, u) = \det f(zI - uG)$ proved at the end of this essay.

is the homogeneous polynomial in z and u with coefficients in R for which $g(z) = g(z, 1)$. Thus $\prod \mathcal{F}_i(z, 0, u) = g(z, u)^m$, so, because $g(z, u)$ is irreducible over R and because $\mathcal{F}(z, 0, u)$ and $g(z, u)$ are both monic in z, each $\mathcal{F}_i(z, 0, u)$ is $g(z, u)^{\mu_i}$ for some μ_i, where $\sum \mu_i = m$. Thus, $\deg \mathcal{F}_i = \mu_i n$, as was to be shown.

When $\mathcal{F}_i(x + uy, 1, u)$ is expressed as a polynomial in u whose coefficients are polynomials in x with coefficients in the root field of $g(y)$, the coefficient of $u^{\mu_i(n-1)}$ is $g'(y)^{\mu_i} x^{\mu_i} + \cdots$ mod $g(y)$, where the omitted terms have lower degree in x. In particular, its degree as a polynomial in u is at least $\mu_i(n-1)$. This statement follows from the observation that $\mathcal{F}_i(x + uy, 1, u)$ is a sum of terms $\lambda_{\alpha\beta\gamma}(x + uy)^\alpha 1^\beta u^\gamma$, where $\lambda_{\alpha\beta\gamma}$ is in $\mathbf{Z}[c_1, c_2, \ldots, c_\nu]$ and $\alpha + \beta + \gamma = \mu_i n$. Thus, this polynomial contains no terms whose combined degree in x and u is greater than $\mu_i n$, and the terms whose combined degree in x and u is exactly $\mu_i n$ are the terms of $\mathcal{F}_i(x + uy, 0, u) = g(x + uy, u)^{\mu_i}$. This homogeneous polynomial in x and u with coefficients in the root field of $g(y)$ can also be written in the form

$$u^{\mu_i n} g\left(y + \frac{x}{u}, 1\right)^{\mu_i} \equiv u^{\mu_i n}\left(g(y) + g'(y)\frac{x}{u} + \frac{1}{2}g''(y)(\frac{x}{u})^2 + \cdots\right)^{\mu_i}$$

$$\equiv \left(g(y)u^n + g'(y)xu^{n-1} + \frac{1}{2}g''(y)x^2u^{n-2} + \cdots\right)^{\mu_i}$$

$$\equiv (g'(y)xu^{n-1} + \cdots)^{\mu_i}$$

$$\equiv g'(y)^{\mu_i} x^{\mu_i} u^{\mu_i(n-1)} + \cdots \quad \text{mod } g(y),$$

where the omitted terms all have combined degree $\mu_i n$ in x and u and degree less than $\mu_i(n-1)$ in u. Therefore, the coefficient of $u^{\mu_i(n-1)}$ in $\mathcal{F}_i(x+uy, 1, u)$ is as described above.

When $\mathcal{F}(x + uy, 1, u)$ is expressed as a polynomial in u whose coefficients are polynomials in x with coefficients in the root field of $g(y)$, the coefficient of $u^{m(n-1)}$ is $g'(y)^m f(x)$ mod $g(y)$ and the coefficients of all higher powers of u are zero mod $g(y)$. The main step in the proof of these statements is the proof that, as was stated in the preceding essay, $\mathcal{F}(tx + uy, t, u) \equiv 0$ mod $(f(x), g(y))$. For the proof of this congruence, let M be as above and let $\mathcal{M}^*$ be the adjoint of the matrix $\mathcal{M} = (tx + uy)I_{mn} - M$ of which $\mathcal{F}(tx + uy, t, u)$ is the determinant. (That is, the entry in the ith row of the jth column of $\mathcal{M}^*$ is $(-1)^{i+j}$ times the determinant of the $(mn - 1) \times (mn - 1)$ matrix that remains when the ith column and the jth row of $\mathcal{M}$ are deleted.) Then $\mathcal{M}^* \cdot \mathcal{M} = \mathcal{F}(tx + uy, t, u)I_{mn}$. By the definition of M, all entries of the column matrix MC are zero mod $(f(x), g(y))$. Therefore all entries of $\mathcal{F}(tx + uy, t, u)I_{mn}C$ are zero mod $(f(x), g(y))$. Since 1 is an entry of C, it follows that $\mathcal{F}(tx + uy, t, u) \equiv 0$ mod $(f(x), g(y))$. Thus, $\mathcal{F}(x + uy, 1, u) \equiv 0$ mod $(f(x), g(y))$. Since the combined degree in x and u of any term of $\mathcal{F}(x + uy, 1, u)$ is at most mn, the coefficient of u^{mn-j} has degree at most j in x; when $j < m$ it follows that this coefficient has degree less than m, which means that it is already reduced mod $f(x)$ and therefore, because reduction of it both mod $f(x)$ and mod $g(y)$ gives the result zero, it must be zero mod $g(y)$. Thus, the coefficients of u^k for $k > m(n-1)$ are all zero mod $g(y)$, as was to be shown. As in the previous paragraph, the coefficient of $u^{m(n-1)}$ in $\mathcal{F}(x + uy, 1, u)$ is congruent to $g'(y)^m x^m + \cdots$ mod $g(y)$, where the omitted terms have degree less than m in x. Since $f(x)$ is monic of degree m in x, the polynomial $\mathcal{F}(x + uy, 1, u) - g'(y)^m f(x) u^{m(n-1)}$ is zero mod $(f(x), g(y))$ and the coefficient of $u^{m(n-1)}$ in it has no terms of degree m or greater in x, so this coefficient is zero mod $g(y)$, as was to be shown.

Thus, when the two sides of the congruence $\mathcal{F}(x + uy, 1, u) \equiv \prod_i \mathcal{F}_i(x + uy, 1, u)$ mod $g(y)$ are regarded as polynomials in u whose coefficients are polynomials in x with coefficients in the root field of $g(y)$, the left side is a polynomial of degree $m(n-1)$ whose leading coefficient is $g'(y)^m f(x)$. The ith factor on the left has been shown to have degree at least $\mu_i(n-1)$. If it had greater degree for any i, the degree of the product would be greater than $\sum \mu_i(n-1) = m(n-1)$, which is not the case. (Since the ring of polynomials in x with coefficients in the root field of $g(y)$ is an integral domain, the degree of a product is the sum of the degrees of the factors, and the leading term of a product is the product of the leading terms.) Therefore, $\mathcal{F}_i(x + uy, 1, u) \equiv \psi_i(x, y) u^{\mu_i(n-1)} + \cdots$ mod $g(y)$, where the omitted terms have degree less than $\mu_i(n-1)$ in u, from which the desired congruence $g'(y)^m f(x) \equiv \prod_i \psi_i(x, y)$ mod $g(y)$ follows. Since it has already been shown that $\psi_i(x, y) \equiv g'(y)^{\mu_i} x^{\mu_i} + \cdots$ mod $g(y)$, it remains only to show that the $\psi_i(x, y)$ are irreducible over the root field of $g(y)$. This fact will follow from a quite different description of $\mathcal{F}(z, 1, u)$, namely, as $\det f(zI_n - uG)$, where G is the $n \times n$ matrix described above.

As was noted above, the matrix M_y obtained from M by setting $t = 0$ and $u = 1$ in M is a matrix of $n \times n$ blocks in which the nondiagonal blocks are all zero and the diagonal blocks are all the matrix G described above. Similarly, M_x, the matrix obtained by setting $t = 1$, $u = 0$ in M, is a matrix of $n \times n$ blocks; the first $(m-1)n$ rows are the last $(m-1)n$ rows of I_{mn}, and the last n rows contain the matrices $-a_m I_n, -a_{m-1} I_n, \ldots, -a_1 I_n$, where the a_i are the coefficients of $f(x) = x^m + a_1 x^{m-1} + \cdots + a_m$. What is to be shown is that $\mathcal{F}(z, 1, u)$, which is the determinant of $zI_{mn} - M_x - uM_y$ by the definition of $\mathcal{F}$, is $\det f(zI_n - uG)$. Let $\mathcal{L}$ be the matrix of $n \times n$ blocks in which blocks above the diagonal are zero, blocks on the diagonal are I_n, and blocks i steps below the diagonal are $(zI_n - uG)^i$. By direct computation, $(zI_{mn} - M_x - uM_y)\mathcal{L}$ has the last $(m-1)n$ rows of $-I_{mn}$ in its first $(m-1)n$ rows, and has $f(zI_n - uG)$ in the first n columns of its last n rows. Since the determinant of $\mathcal{L}$ is 1, it follows that $\mathcal{F}(z, 1, u) = \pm \det f(zI_n - uG)$. Since $\mathcal{F}(z, 1, u)$ and $\det f(zI_n - uG)$ are both monic in z, the sign is plus.

Since the matrix G satisfies $g(G) = 0$ (by direct computation,* or, because g is the characteristic polynomial of G, an application of the Cayley–Hamilton theorem), a factorization $f(x) \equiv \theta_1(x,y)\theta_2(x,y)\cdots\theta_l(x,y) \bmod g(y)$ in which each $\theta_i(x,y)$ has coefficients in the field K of rational functions in $c_1, c_2, \ldots,$ c_ν and is monic in x implies $\mathcal{F}(z,1,u) = \det f(zI_n - uG) = \det \prod \theta_i(zI_n - uG, G) = \prod \det \theta_i(zI_n - uG, G)$. Such a formula gives a factorization of $\mathcal{F}(z,1,u)$ over K in which the factors have degrees $\nu_i n$ in z, where ν_i is the degree of θ_i in x. If the factorization $f(x) \equiv \prod \phi_i(x,y) \bmod g(y)$ found above could be further factored, it would follow that $\mathcal{F}(z,1,u)$, and therefore $\mathcal{F}(z,t,u)$, could be factored over K into factors of lower degrees than the factors $\mathcal{F}(z,t,u) = \prod \mathcal{F}_i(z,t,u)$. But by Gauss's lemma (see Essay 1.4) a factorization of the monic polynomial $\mathcal{F}(z,t,u)$ over the field of quotients K of R implies a factorization over R itself. Therefore, since $\mathcal{F}(z,t,u) = \prod \mathcal{F}_i(z,t,u)$ is by definition a factorization that cannot be further factored, the proof is complete.

---

* Let ρ_i denote the ith row of I_n. Then $\rho_1 G^j = \rho_{i+1}$ for $j = 0, 1, \ldots, n-1$, but $\rho_1 G^n = \rho_n G = [-a_0 \quad -a_1 \quad \cdots \quad -a_{n-1}] = \rho_1(-a_0 I - a_1 G - \cdots - a_{n-1}G^{n-1})$, which proves that $\rho_1 \cdot g(G) = 0$. Therefore, $\rho_{1+j} \cdot g(G) = \rho_1 \cdot G^j \cdot g(G) = \rho_1 \cdot g(G) \cdot G^j = 0 \cdot G^j = 0$ for $j = 0, 1, \ldots, n-1$, which is to say $g(G) = I_n g(G) = 0$.

Essay 1.7 About the Factorization Algorithm

The method of factoring $f(x) \bmod g(y)$ in the preceding essays has as an immediate corollary:

The Kronecker–Kneser* Theorem. *Let f and g be monic, irreducible polynomials with coefficients in $\mathbf{Z}[c_1, c_2, \ldots, c_\nu]$, and let*

$$f(x) \equiv \phi_1(x,y)\phi_2(x,y) \cdots \phi_k(x,y) \bmod g(y)$$

and

$$g(y) \equiv \psi_1(x,y)\psi_2(x,y) \cdots \psi_l(x,y) \bmod f(x)$$

be the factorizations of each modulo the other. Then $k = l$, and the factors can be so ordered that $\deg_x \phi_i / \deg_y \psi_i = \deg f / \deg g$ for each i.

Proof. To factor $g(y) \bmod f(x)$, one constructs the characteristic polynomial $\hat{\mathcal{F}}(z,t,u)$ of the matrix $tM_y + uM_x$ for which $(tM_y + uM_x)C \equiv C(ty + ux) \bmod (g(y), f(x))$, where C is the column matrix of length mn that contains the monomials $y^\alpha x^\beta$ in which $0 \le \alpha < n = \deg g$ and $0 \le \beta < m = \deg f$. Because this characteristic polynomial is independent of the order chosen for the entries of C, it is clear that $\hat{\mathcal{F}}(z,t,u) = \mathcal{F}(z,u,t)$. The factorization algorithm proves that $k = l$ is the number of irreducible factors of $\mathcal{F}(z,t,u)$, and the integers $\deg f \deg_y \psi_i = \deg g \deg_x \phi_i$ are the degrees of those factors.

In addition to its aesthetic appeal, this theorem is a powerful tool. See Essay 2.1, where it is used in the proof of Galois's fundamental theorem.

Inevitably, some readers will object that the algorithm is impractical. The construction of $\mathcal{F}(z,t,u)$ is already a formidable task, and the factorization of this polynomial in three indeterminates with integer coefficients is even more daunting. But the practicality of the algorithm is irrelevant, because its purpose is to *prove the existence* of the factorization, not to *effect* it. Once the factorization is known to exist, methods for constructing it can be addressed. A similar situation occurs in the case of the fundamental theorem of algebra (see Essay 5.1); Newton's method is in most cases the best way to *construct* the roots of a polynomial, but other methods are needed to prove that there are roots to be constructed.

Kronecker emphasized the importance of the problem of factoring $f(x) \bmod g(y)$ in a footnote to his 1887 paper "Über den Zahlbegriff" [43, p. 262 of vol.

---

* I called this theorem "Dedekind's reciprocity theorem" in *Galois Theory* [18, p. 66], but I have since learned that it already had the name "Kronecker–Kneser theorem" (see ([8] and [50]). Richard Dedekind discovered it in 1855, but the discovery was not published until Scharlau's paper [59] appeared in 1982. Kronecker included the theorem in his university lectures ([32, p. 309]). He *might* have known of Dedekind's work, but since he does not seem to have cited Dedekind, he probably discovered the theorem independently. The first publication of the theorem was by A. Kneser [36].

IIIa of the republication in *Mathematische Werke*]. In 1881, he had already described an algorithm for such factorizations in the following way:

> It can be assumed that $f(x)$ has no repeated factors, because otherwise one could free it of repeated factors by dividing it by its greatest common divisor with its derivative. One sets $z + uy$ in place of x in $f(x)$, where u is an indeterminate; at the same time, one treats f itself as a function of x and the algebraic quantity y, which may figure in its coefficients. Therefore, denote f by $f(x, y)$ and form the product of all the conjugate expressions $f(z+uy, y)$, that is, all of them that arise when y is replaced by its conjugate values. This product is a polynomial in z whose coefficients are rational functions in c_1, c_2, ..., c_ν [the presence of u in the coefficients is ignored] and therefore, as has been shown, can be decomposed into irreducible factors. If these factors are $F_1(z)$, $F_2(z)$, ..., then, as is easy to see, the greatest common divisors of $f(z + uy, y)$ and $F_i(z)$ for $i = 1$, 2, ...give the irreducible factors of $f(z + uy, y)$, from which the irreducible factors of $f(x)$ itself can be found when $x - uy$ is substituted for z. It remains to remark that substitution of $z + uy$ for x ensures that y actually occurs in the coefficients of f."[†]

[†] Dabei kann angenommen werden, dass die Function $f(x)$ keine gleichen Factoren enthält; denn anderenfalls würde man dieselbe von gleichen Factoren dadurch befreien können, dass man sie durch den grössten Theiler, den die Function $f(x)$ mit ihrer Ableitung gemein hat, dividirt. Man setze nun zuvörderst $z + uy$ an Stelle von x in $f(x)$, wo u eine unbestimmte Grösse bedeutet; man betrachte ferner f selbst als Function von x und der zum Rationalitäts-Bereich gehörigen algebraischen Grösse y welche also auch in den Coëfficienten vorkommen kann, bezeichne demnach die Function f durch $f(x, y)$ und bilde das Product aller mit einander conjugirten Ausdrücke

$$f(z + uy, y),$$

d. h. aller derjenigen, welche entstehen, wenn man die mit y conjugierten algebraischen Grössen an Stelle von y setzt. Dieses Product ist eine ganze Function von z, deren Coëfficienten rationale Functionen der Variabeln c_1, c_2, ..., c_ν sind, kann also nach dem Vorhergehenden in irreductible Factoren zerlegt werden. Sind diese Factoren: $\mathcal{F}_1(z)$, $\mathcal{F}_2(z)$, ..., so bilden, wie leicht zu sehen, die grössten gemeinschaftlichen Theiler von

$$f(z + uy, y) \quad \text{und} \quad \mathcal{F}_h(z)$$

für $h = 1$, 2, ... die irreductibeln Factoren von $f(z+uy, y)$, aus denen die Factoren von $f(x)$ selbst unmittelbar hervorgehen, wenn wieder $x - uy$ an Stelle von z gesetzt wird. Es ist noch zu bemerken, dass die Einführung von $z + uy$ an Stelle von x zu dem Zwecke erfolgt ist, das Vorkommen von y in den Coëfficienten zu sichern. (From §4 of [39]. The translation above is somewhat free, and Kronecker's notation F, $\mathfrak{R}$, $\mathfrak{R}'$, $\mathfrak{R}''$, $\mathfrak{R}'''$, ..., has been changed to f, y, c_1, c_2, ..., c_ν to agree with the notation of these essays.)

My discussion of this subject in [18, §§60–61] shows that I found the exact algorithm Kronecker had in mind—not to mention its validity—far from "easy to see." In retrospect, however, I do see that it is essentially the algorithm of Essay 1.5.

Instead of factoring a polynomial in x alone as in Essay 1.5, Kronecker changes $f(x)$ to $f(z + uy)$, where u is an indeterminate, in order to be sure that the polynomial to be factored *does* involve y. He then forms the "product of its conjugates," by which he surely means (see his §2) the *norm* of $f(z + uy)$ as a polynomial with coefficients in the root field of $g(y)$, which is to say that it is plus or minus the constant term of the polynomial of which $f(z + uy)$ is a root. The polynomial of which $f(z + uy)$ is a root is the characteristic polynomial of the matrix M of elements of $R[z, u]$ defined by $C \cdot f(z+uy) \equiv MC \bmod g(y)$, where C is the column matrix with entries 1, y, $y^2, \ldots, y^{n-1}$ (n being the degree of g). Thus, Kronecker's $F_1(z)$, $F_2(z)$, ... are the irreducible factors of the constant term of the characteristic polynomial of this M. But this is $\pm \det M$, and M is the $n \times n$ matrix $f(zI + uG)$, where G is the matrix determined by $g(y)$ as in Essay 1.6. Thus, the $F_i(z)$ are the irreducible factors of $\det f(zI + uG)$. Since, as was shown in Essay 1.6, $\det f(zI + uG) = \mathcal{F}(z, 1, -u)$, he is saying that the desired irreducible factors $\phi_i(x, y)$ are the greatest common divisors of $f(z+uy)$ with $\mathcal{F}_1(z, 1, -u)$, $\mathcal{F}_2(z, 1, -u), \ldots$, or, better, the greatest common divisors of $f(x)$ with $\mathcal{F}_1(x - uy, 1, -u)$, $\mathcal{F}_2(x - uy, 1, -u), \ldots$. When one changes the sign of u and notes that a common divisor of $f(x)$ and $\mathcal{F}_i(x + uy, 1, u)$ must be independent of u and must therefore divide all coefficients of $\mathcal{F}_i(x + uy, 1, u)$, Kronecker's claim becomes the statement that $\phi_i(x, y)$ is the greatest common divisor of $f(x)$ and the coefficients of $\mathcal{F}_i(x + uy, 1, u)$ when it is expanded in powers of u. Now, $\phi_i(x, y)$ is the greatest common divisor of $f(x)$ and the *leading* coefficient $\psi_i(x, y) \bmod g(y)$ in this expansion, so his claim comes down to the statement that $\phi_i(x, y)$ divides all the other coefficients of $\mathcal{F}_i(x + uy, 1, u)$ when they are regarded as polynomials in x with coefficients in the root field of $g(y)$.

Proposition. *As a polynomial in x and u with coefficients in the root field of $g(y)$, $\mathcal{F}_i(x + uy, 1, u)$ is divisible by $\phi_i(x, y)$.*

Proof. Let $\mathcal{K}$ be the field $K[x, y] \bmod (\phi_i(x, y), g(y))$, which is the ring of polynomials in x with coefficients in the root field of $g(y)$ modulo the irreducible polynomial $\phi_i(x, y)$ with coefficients in this root field. (As before, K denotes the field of rational functions in c_1, c_2, ..., c_ν.) Since $f(x)$ is 0 as a polynomial with coefficients in $\mathcal{K}$ (because $f(x)$ is divisible by $\phi_i(x, y) \bmod g(y)$), and since $\mathcal{F}(z, t, u) \equiv 0 \bmod (f(x), g(y))$ as was shown in Essay 1.6, $\mathcal{F}(z, t, u)$ is zero as a polynomial with coefficients in $\mathcal{K}$. Therefore at least one of its factors $\mathcal{F}_j(z, t, u)$ must be zero as a polynomial with coefficients in $\mathcal{K}$. For any such value of j, $\mathcal{F}_j(x + uy, 1, u)$ must be zero as a polynomial in u with coefficients in $\mathcal{K}$. In particular, $\psi_j(x, y)$ must be zero as an element of $\mathcal{K}$, which is to say that $\psi_j(x, y)$ is divisible by $\phi_i(x, y) \bmod g(y)$. But $\psi_j(x, y)$

is a unit times $\phi_j(x, y)$ and the irreducible factors $\phi_j(x, y)$ of $f(x) \bmod g(y)$ are distinct because $f(x)$ is irreducible. Therefore, $\psi_j(x, y)$ is not divisible by $\phi_i(x, y) \bmod g(y)$ unless $j = i$, and the proposition follows.

Essay 1.8 Proof of the Fundamental Theorem

As before, R will denote the ring $\mathbf{Z}[c_1, c_2, \ldots, c_\nu]$ of polynomials in $c_1, c_2, \ldots,$ c_ν with integer coefficients and K will denote its field of quotients, the field of rational functions of $c_1, c_2, \ldots, c_\nu$. When $\nu = 0$, R is the ring of integers and K is the field of rational numbers. The theorem to be proved was stated in Essay 1.2:

Fundamental Theorem. *Given a polynomial $f(x) = a_0 x^n + a_1 x^{n-1} + \cdots + a_n$ of positive degree n with coefficients in R, construct a monic, irreducible polynomial $g(y)$ with coefficients in R with the property that $f(x)$ is a product of linear factors with coefficients in the root field of $g(y)$.*

In other words, when the factors of $f(x) \bmod g(y)$ are taken to be monic in x, the factorization is to have the form $f(x) \equiv a_0(x - \rho_1(y))(x - \rho_2(y)) \cdots (x - \rho_n(y)) \bmod g(y)$, where a_0 is the leading coefficient of $f(x)$ and the $\rho_i(y)$ are elements of the root field of $g(y)$. Such a polynomial $g(y)$ will be said to **split** $f(x)$. As the proposition at the end of Essay 1.4 implies, the roots $\rho_i(y)$ are determined, as elements of the root field of $g(y)$, by $f(x)$.

Loosely speaking, the root field of $g(y)$ extends computations in R in such a way that the given $f(x)$ with coefficients in R splits into linear factors.

The factorization algorithm of the preceding essays, which assumes that $f(x)$ is *monic and irreducible,* can be used to factor an arbitrary f by taking the change of variable $x_1 = a_0 x$ and writing $a_0^{n-1} f(x) = x_1^n + a_1 x_1^{n-1} + a_0 a_2 x_1^{n-2} + \cdots + a_0^{j-1} a_j x_1^{n-j} + \cdots + a_0^{n-1} a_n$. A factorization of $a_0^{n-1} f(x)$ as a polynomial in x_1 becomes a factorization of $f(x)$ as a polynomial in x when it is divided by the nonzero element a_0^{n-1} of K and the substitution $x_1 = a_0 x$ is made. In this way, the theorem is reduced to the case in which $f(x)$ is monic. The iteration theorem below proves this case of the theorem using the factorization algorithm for monic, irreducible polynomials, which obviously implies a factorization algorithm for arbitrary monic polynomials.

This theorem differs from Kronecker's theorem in *Ein Fundamentalsatz der allgemeinen Arithmetik* [42] in that it specifies that the splitting field is to be described as the root field of $g(y)$, whereas Kronecker left the form of the description open and in fact preferred a "prime module system" of an altogether different type. Nor is the proof below similar to Kronecker's, which constructed specific relations satisfied by the roots in a splitting field. Instead, it constructs a splitting polynomial $g(y)$ for $f(x)$ in an iterative way that follows the naive proof sketched at the beginning of Essay 1.4.

Iteration Theorem. *Given a monic polynomial $f(x)$ with coefficients in R, and given a monic, irreducible polynomial $g(y)$ with coefficients in R that does not split $f(x)$, construct a monic, irreducible polynomial $h(z)$ with coefficients in R for which the factorization of $f(x) \bmod h(z)$ contains more linear factors than does the factorization of $f(x) \bmod g(y)$.*

Proof. The factorization of $f(x)$ mod $g(y)$ is accomplished by applying the factorization algorithm to each of the monic, irreducible factors of $f(x)$ and taking the product of the results. By assumption, at least one of the irreducible factors of $f(x)$ mod $g(y)$ obtained in this way has degree greater than 1. With the notation as before, at least one of the polynomials $\mathcal{F}(z, t, u)$ used in the factorization of $f(x)$ mod $g(y)$ (there is an $\mathcal{F}$ for each irreducible factor of $f(x)$) must, by assumption, have at least one factor $\mathcal{F}_i(z, t, u)$ that gives rise to a monic factor $\phi_i(x, y)$ of $f(x)$ mod $g(y)$ of degree greater than 1. Let $\mathcal{F}_1(z, t, u)$ be such a factor, and let $\phi_1(x, y)$ be the corresponding factor of $f(x)$ mod $g(y)$.

Lemma. *The monic polynomial $h_a(z) = \mathcal{F}_1(z, 1, a)$ with coefficients in R is irreducible for at least one positive integer a less than $(\deg \mathcal{F}_1)^2$.*

Proof. The essence of the construction is a *double adjunction*, first of a root y of $g(y)$ to R to form the root field of $g(y)$ and then of a root x of $\phi_1(x, y)$ to find a simple algebraic extension of the root field of $g(y)$. Let $\mathcal{K}$ be the field obtained in this way. That is, elements of $\mathcal{K}$ are represented by polynomials in x with coefficients in the root field of $g(y)$, and two such polynomials represent the same element of $\mathcal{K}$ when they differ by a multiple of $\phi_1(x, y)$. Clearly, each element of $\mathcal{K}$ has a unique representative of degree less than $\mu_1 = \deg_x \phi_1$; since each element of the root field of $g(y)$ has a unique representative as a polynomial of degree less than $n = \deg g$ in y with coefficients in K, $\mathcal{K}$ is a vector space of dimension $\mu_1 n$ over K. It is obviously a ring in the usual way (to multiply elements, multiply their representatives, and reduce first mod $\phi_1(x, y)$ to reduce the degree in x and then reduce mod $g(y)$). It is a field by virtue of the irreducibility of $\phi_1(x, y)$ mod $g(y)$. (This irreducibility implies that if $q(x, y)$ is a nonzero polynomial of degree less than μ_1 in x with coefficients in the root field of $g(y)$, then $r(x, y)q(x, y) + s(x, y)\phi_1(x, y) \equiv 0$ mod $g(y)$ has only the trivial solution, because irreducible polynomials with coefficients in the root field of $g(y)$ are prime. Therefore, by the argument of the elimination lemma of Essay 1.3, there is a solution of $r(x, y)q(x, y) + s(x, y)\phi_1(x, y) \equiv q_1(y)$, where $q_1(y)$ is a nonzero element of the root field of $g(y)$. Therefore, $q(x, y)$ has a reciprocal $r(x, y)/q_1(y)$ in $\mathcal{K}$ and a fortiori is not a zero divisor.)

Let C_1 denote the column matrix of length $\mu_1 n$ containing the monomials $x^\alpha y^\beta$ in which $0 \le \alpha < \mu_1$ and $0 \le \beta < n$. Let $\mathcal{N}(u)$ be the $\mu_1 n \times \mu_1 n$ matrix of polynomials in u with coefficients in K for which

$$v_1(u)(x + uy)^{\mu_1 n - 1} + v_2(u)(x + uy)^{\mu_1 n - 2} + \cdots + v_{\mu_i n}(u)$$
$$\equiv [\, v_1(u) \quad v_2(u) \quad \cdots \quad v_{\mu_1 n}(u) \,] \cdot \mathcal{N}(u) \cdot C_1 \text{ mod } (\phi_1(x, y), g(y)).$$

In other words, let the entries of the σth row of $\mathcal{N}(u)$ be the coefficients that express the polynomial in u with coefficients in $\mathcal{K}$ represented by $(x + uy)^{\mu_1 n - \sigma}$ with respect to the basis $x^\alpha y^\beta$. The entries of the σth row of $\mathcal{N}(u)$ have degree at most $\mu_1 n - \sigma$ in u, so $\det \mathcal{N}(u)$ has degree at most $1 + 2 + \cdots + (\mu_1 n - 1) = \frac{1}{2} \mu_1 n (\mu_1 n - 1)$ in u.

If $\det \mathcal{N}(u)$ were zero, there would be[*] a nontrivial solution $v_1(u)$, $v_2(u)$, $\ldots$, $v_{\mu_1 n}(u)$ (polynomials with coefficients in K) of the $\mu_1 n \times \mu_1 n$ homogeneous system of linear congruences $v_1(u)(x + uy)^{\mu_1 n - 1} + v_2(u)(x + uy)^{\mu_1 n - 2} + \cdots + v_{\mu_1 n}(u) \equiv 0 \bmod (\phi_1(x, y), g(y))$. (This single congruence is equivalent to $\mu_1 n$ congruences, one for each row $x^\alpha y^\beta$ of C_1.) In other words, there would be a nonzero polynomial $F(z, u)$ with coefficients in K—and therefore one with coefficients in R—whose degree in z was less than $\mu_1 n = \deg \mathcal{F}_1$ and for which reduction of $F(x + uy, u) \bmod (\phi_1(x, y), g(y))$ gave zero. Application of the Euclidean algorithm to $\mathcal{F}_1(z, 1, u)$ and $F(z, u)$ as polynomials in z with coefficients in the field of quotients of $K[u]$ would give polynomials $\alpha(z)$ and $\beta(z)$ in z with coefficients in this field for which $d(z) = \alpha(z)\mathcal{F}_1(z, 1, u) + \beta(z)F(z, u)$ was a common factor of $\mathcal{F}_1(z, 1, u)$ and $F(z, u)$, say $\mathcal{F}_1(z, 1, u) = d(z)q_1(z)$ and $F(z, u) = d(z)q_2(z)$. There would be a polynomial $\Delta(u)$ in u with coefficients in R that cleared the denominators in all three equations, say $D(z, u) = A(z, u)\mathcal{F}_1(z, 1, u) + B(z, u)F(z, u)$, $\Delta(u)\mathcal{F}_1(z, 1, u) = D(z, u)Q_1(z, u)$ and $\Delta(u)F(z, u) = D(z, u)Q_2(z, u)$, where $\deg_z D = \deg d$. Since $\mathcal{F}_1(z, 1, u)$ is irreducible, it would be a factor either of $D(z, u)$ or of $Q_1(z, u)$. It cannot be a factor of $D(z, u)$, because this would imply that it was a factor of $\Delta(u)F(z, u)$, contrary to $\deg_z F < \mu_1 n = \deg_z \mathcal{F}_1$. Nor can it be a factor of $Q_1(z, u)$, because then $D(z, u)$ would divide $\Delta(u)$ and therefore be independent of z, contrary to $D(x + uy, u) = A(x + uy, u)\mathcal{F}_1(x + uy, 1, u) + B(x + uy, u)F(x + uy, u) \equiv 0 \bmod (\phi_1(x, y), g(y))$. Therefore, $\det \mathcal{N}(u) \neq 0$.

Given an integer a, consider the homomorphism ι from $K[z]$ to $\mathcal{K}$ that carries z to $x + ay$ and carries elements of K to themselves. Since ι carries $h_a(z) = \mathcal{F}_1(z, 1, a)$ to $\mathcal{F}_1(x + ay, 1, a)$, which represents the zero[†] element of $\mathcal{K}$, ι defines a homomorphism from $K[z] \bmod h_a(z)$ to $\mathcal{K}$. The matrix of coefficients of ι relative to the basis $x^\alpha y^\beta$ of $\mathcal{K}$ and the basis z^γ $(0 \le \gamma < \mu_1 n)$ of $K[z] \bmod h_a(z)$ is $\mathcal{N}(a)$. If its determinant is nonzero, then ι is an isomorphism. In this case, because $\mathcal{K}$ is a field, $K[z] \bmod h_a(z)$ is a field, which implies that $h_a(z)$ is irreducible over K and therefore, by Gauss's lemma (see Essay 1.4), is irreducible over R.

The degree of $\det \mathcal{N}(u)$ is at most $\frac{1}{2} \mu_1 n (\mu_1 n - 1) < (\mu_1 n)^2 = (\deg \mathcal{F}_1)^2$, so $\det \mathcal{N}(u)$ is zero for fewer than $(\deg \mathcal{F}_1)^2$ integers a, and the lemma follows.

Proof of the Iteration Theorem. Suppose that $h(z) = h_a(z) = \mathcal{F}_1(z, 1, a)$, where a is chosen in such a way that h is irreducible, and let $\mathcal{K}$ be the root

---

[*] If the rank of $\mathcal{N}(u)$ is $\mu_1 n - 1$, any nonzero row of the adjoint matrix $\mathcal{N}(u)^*$ of $\mathcal{N}(u)$ is such a solution. Otherwise, choose an $(r + 1) \times (r + 1)$ subsystem of the original $\mu_1 n \times \mu_1 n$ system whose rank r is the same as that of the original system. A nonzero row of the adjoint of the matrix of coefficients of the chosen subsystem, filled out with zeros, is a nontrivial solution of the original system.

[†] In fact, according to the proposition of Essay 1.7, each coefficient of $\mathcal{F}_1(x + uy, 1, u)$ as a polynomial in u is zero mod $(\phi_1(x, y), g(y))$.

field of $h(z)$. It is to be shown that the factorization of $f(x)$ mod $h(z)$ contains more linear factors than does the factorization of $f(x)$ mod $g(y)$. In other words, it is to be shown that the root field of $h(z)$ contains more roots of $f(x)$ than does the root field of $g(y)$.

As was shown in the proof of the lemma, $K[z]$ mod $h(z)$ is isomorphic to $\mathcal{K}$, which is by definition $K[x, y]$ mod $(g(y), \phi_1(x+ay, y))$. Since this last field contains a root y of $g(y)$, it contains a field isomorphic to $K[y]$ mod $g(y)$, so it certainly contains at least as many roots of $f(x)$ as $K[y]$ mod $g(y)$ does. But $f(x) \equiv \phi_1(x, y)\phi_2(x, y) \cdots \phi_k(x, y)$ mod $g(y)$ implies that $f(x + ay) \equiv 0$ mod $(g(y), \phi_1(x+ay, y))$, so $\mathcal{K}$ contains a root of $f(x)$—namely, the element represented by $x + ay$—that is not in the subfield corresponding to $K[y]$ mod $g(y)$ (because it is reduced mod $(g(x), \phi_1(x + ay, y))$ and contains x). Thus, $f(x)$ has at least one more root in the new field than it did in the old, as was to be shown.

Proof of the fundamental theorem. Given $f(x)$, start with $g(y) = y$ and apply the following algorithm. If $g(y)$ splits $f(x)$, the algorithm terminates. Otherwise, use the iteration theorem to construct $h(z)$ such that the factorization of $f(x)$ mod $h(z)$ has more linear factors than does the factorization of $f(x)$ mod $g(y)$. Replace $g(y)$ with $h(z)$ and repeat. Since the number of linear factors of $f(x)$ mod $g(y)$ increases with each step and can never exceed $\deg f$, the algorithm must terminate after at most $\deg f$ steps with a polynomial that splits $f(x)$.

Essay 1.9 Minimal Splitting Polynomials

The theorem of the preceding essay puts the statement "every polynomial has a splitting field" in a very specific and concrete form: Given a polynomial $f(x)$ in one indeterminate with coefficients in $R = \mathbf{Z}[c_1, c_2, \ldots, c_\nu]$, the iterative algorithm that proves the fundamental theorem in the last essay constructs a monic, irreducible polynomial in one indeterminate $g(y)$, with coefficients in the same ring R, with the property that all factors of the factorization of $f(x) \bmod g(y)$ are linear in x.

The splitting field of a polynomial, as opposed to *a* splitting field, is the field generated by the roots. In other words, it is the field implicit in the assertion (see Essay 1.2) that there is a valid way to do rational computations with the roots of a polynomial. A specific and concrete description of *the* splitting field of $f(x)$ is given by an amended version of the theorem of the preceding essay:

Theorem. *Given a polynomial $f(x) = a_0 x^n + a_1 x^{n-1} + \cdots + a_n$ with coefficients in R, construct a monic, irreducible polynomial $g(y)$ with coefficients in R that splits $f(x)$ and is itself split by any polynomial that splits $f(x)$.*

In particular, $g(y)$ *splits itself.* Galois wrote,[*] "...every equation depends on an auxiliary equation with the property that all the roots of this new equation are rational functions of one another," which is to say that the polynomial in the new equation splits itself. He went on to write, "...this remark is a mere curiosity; in fact, an equation which has this property is not in general any easier to solve than any other," but it is hard to understand how Galois could call his observation a "mere curiosity," because, as Essay 2.1 will explain, his brilliant insight into the algebraic solution of equations is based on the existence of such an "auxiliary equation" and on the fact that the solution of such an equation can be analyzed using the automorphisms of its splitting field. A polynomial that splits itself is called a **Galois polynomial**.

Proof. The construction that proved the theorem of the preceding essay in fact proves this stronger theorem, because *the polynomial it constructs is split by any polynomial that splits $f(x)$.* Since any polynomial that splits $f(x)$ also splits the identity polynomial $g(y) = y$, the proof of this statement comes down to proving the following statement: *Let a monic, irreducible polynomial with coefficients in R split both $f(x)$ and $g(y)$; then it also splits any polynomial $h(z)$ constructed by the iteration theorem of the preceding essay.* Since $h(z)$ has the form $\mathcal{F}_1(z, 1, a)$ for some integer a, it will suffice to prove that any polynomial $g_1(v)$—monic, irreducible, and with coefficients in R—that splits the monic, irreducible polynomials $f(x)$ and $g(y)$ also splits the polynomial $\mathcal{F}(z, t, u)$ that is constructed from $f(x)$ and $g(y)$ by the factorization algorithm in Essay 1.5.

---
[*] Quoted from my translation of Galois's memoir in Appendix 1 of [18].

Let $\mathcal{K}$ denote the root field of $g_1(v)$. Since $g_1(v)$ splits $f(x)$, $\mathcal{K}$ contains $m = \deg f$ distinct (because f is irreducible, it is relatively prime to its derivative, so it has no multiple roots) roots of f; call them $a_1, a_2, \ldots, a_m$. Similarly, $\mathcal{K}$ contains $n = \deg g$ distinct roots of g; call them $b_1, b_2, \ldots, b_n$. That $g_1(v)$ splits $\mathcal{F}(z, t, u)$ will be proved by proving that

$$\mathcal{F}(z, t, u) = \prod(z - a_i t - b_j u),$$

where the product is over all mn pairs (i, j) in which $1 \leq i \leq m$ and $1 \leq j \leq n$. The number mn of factors on the right is the degree of the homogeneous polynomial on the left, and they are distinct (to say that two polynomials of the form $z - a_i t - b_j u$ are equal means that their coefficients are equal, which occurs only when both i and j have the same values in them), so the formula will be proved if it is proved that each $z - a_i t - b_j u$ is a factor of $\mathcal{F}(z, t, u)$. In other words, what is to be shown is that $\mathcal{F}(a_i t + b_j u, t, u) = 0$. To put it yet another way, the determinant of the $mn \times mn$ matrix $(a_i t + b_j u)I_{mn} - tM_x - uM_y$ is to be shown to be zero for all pairs (i, j). Here M_x is by definition the $mn \times mn$ matrix of elements of R for which $M_x C \equiv Cx \bmod (f(x), g(y))$, where C is a column matrix of length mn whose entries are $x^\alpha y^\beta$ for $0 \leq \alpha < m$, $0 \leq \beta < n$. Thus, each of the mn entries of the column matrix $M_x C$ differs from the corresponding entry of Cx by a polynomial of the form $\phi(x, y)f(x) + \psi(x, y)g(y)$. Because a_i is a root of $f(x)$ and b_j is a root of $g(y)$, it follows that $M_x C_{ij}$ and $C_{ij} a_i$ are equal as matrices whose entries are in $\mathcal{K}$, where C_{ij} is C with a_i substituted for x and b_j substituted for y. In the same way, $M_y C_{ij} = C_{ij} b_j$. Therefore $((a_i t + b_j u)I_{mn} - tM_x - uM_y)C_{ij} = t(a_i I_{mn} - M_x)C_{ij} + u(b_j I_{mn} - M_y)C_{ij} = 0$. Since $1 = a_i^0 b_j^0$ is one of the entries of C_{ij}, the determinant of $(a_i t + b_j u)I_{mn} - tM_x - uM_y$ must be zero, as was to be shown.

A polynomial that splits $f(x)$ and is split by any other polynomial that splits $f(x)$ is a **minimal splitting polynomial** of $f(x)$. There are infinitely many minimal splitting polynomials of $f(x)$, but only one splitting field. In other words, the root fields of two minimal splitting polynomials are isomorphic, as follows from the fact that each polynomial splits the other. This observation is the key to the notion of the Galois group of $f(x)$, which is introduced in the first essay of Part 2.

Topics in Algebra

Essay 2.1 Galois's Fundamental Theorem

Théorème. *Soit une équation donnée, dont a, b, c, ... sont les m racines.
Il y aura toujours un groupe de permutations des lettres a, b, c, ... qui
jouira de la propriété suivante:*
1° *que toute fonction des racines, invariables par les substitutions de ce
groupe, soit rationellement connue;*
2° *réciproquement, toute fonction des racines, déterminable rationellement, soit invariable par ces substitutions.*—É. Galois [27] (English translation [18, p. 104])

A **Galois polynomial** is a monic, irreducible polynomial that splits itself,
and a **Galois field** is the root field of a Galois polynomial. Here the Galois
polynomial, call it g, is assumed to have its coefficients in a ring of the form
$R = \mathbf{Z}[c_1, c_2, \ldots, c_\nu]$, the polynomials with integer coefficients in some set c_1,
$c_2, \ldots, c_\nu$ of indeterminates, so the associated Galois field extends the field
of quotients K of R, which is the field of rational functions in $c_1, c_2, \ldots, c_\nu$.
As is explained in Essay 1.3, computations in the Galois field are done with
expressions of the form $p(y)/q$, where $p(y)$ is a polynomial in y with coefficients
in R whose degree is less than $\deg g$, and q is a nonzero element of R. Such
an expression can also be regarded as a polynomial in y with coefficients in
K, in which case the notation $K[y] \bmod g(y)$ becomes a natural one for the
root field of $g(y)$.

Because g splits itself, $g(x) \equiv \prod_{i=1}^{n}(x - \rho_i(y)) \bmod g(y)$, where n is the
degree of g and $\rho_1(y), \rho_2(y), \ldots, \rho_n(y)$ are elements of the field. The polynomials $\rho_i(y)$ represent distinct elements of the root field of $g(y)$. When they
are reduced mod $g(y)$, they are determined by $g(y)$. One of them is y. Substitution of one of the roots $\rho_i(y)$ in place of y gives an automorphism of
$K[y] \bmod g(y)$. (Since $g(\rho_i(y)) \equiv 0 \bmod g(y)$, $\phi(y) \equiv \psi(y) \bmod g(y)$ implies
$\phi(\rho_i(y)) \equiv \psi(\rho_i(y)) \bmod g(y)$, so substitution of $\rho_i(y)$ for y gives a homomorphism of $K[y] \bmod g(y)$ to itself. It is represented by an $n \times n$ matrix

Fig. 2.1. Galois.

of elements of K, so to prove that it is one-to-one and onto it will suffice to prove $\phi(\rho_i(y)) \equiv 0 \bmod g(y)$ only when $\phi(y) \equiv 0 \bmod g(y)$. But if $\phi(y) \not\equiv 0 \bmod g(y)$, then $\phi(y)$ has a reciprocal mod $g(y)$, so $\phi(\rho_i(y))$ has a reciprocal mod $g(y)$ and is therefore not zero mod $g(y)$.) Conversely, since an automorphism of the field must carry the root y of g to another root of g, *there are precisely n automorphisms of $K[y]$ mod $g(y)$*, and an automorphism is determined by the root $\rho_i(y)$ of g to which it carries y. The group of these automorphisms is the **Galois group** of the field. More generally, the **Galois group** of a polynomial $f(x)$ is the Galois group of a minimal splitting polynomial of $f(x)$ (see Essay 1.9).

The modern version of the fundamental theorem of Galois theory states that the subgroups of the Galois group of a given Galois polynomial $g(y)$ correspond one-to-one to subfields of the Galois field $K[y]$ mod $g(y)$ that contain K. This is nearly Galois's statement of his Proposition 1. Galois was thinking of the field as being built up by a succession of what he called "adjunctions" (see Essay 2.3), and the main problem was to determine, after certain adjunctions had been made, which elements of the field could then be "determined rationally." He said (see above) that there is a subgroup of the group of automorphisms that leaves fixed the elements that can be determined rationally and leaves *only* these fixed. (He described the automorphisms in terms of the way that they permute the roots of some polynomial of which the Galois field is the splitting field. For him, the elements of the Galois field were rational functions of the roots, so an automorphism of the field was tantamount to the permutation of the roots that it effected.) In more modern terminology, he was saying that the subfield generated by the adjoined elements x_1, x_2, $\ldots$,

x_m contains exactly the elements that are left fixed by the automorphisms that leave all of the x's fixed:

Galois's Theorem. *Let x_1, x_2, ..., x_m be elements of a Galois field $K[y]$ mod $g(y)$. The obvious necessary condition for another element z of the field to be expressible as a polynomial in x_1, x_2, ..., x_m with coefficients in K—namely, that z be unmoved by any automorphism of the field that leaves each of x_1, x_2, ..., x_m unmoved—is also sufficient.*

Proof. Suppose first that x_1, x_2, ..., x_m are all in K, or, what is effectively the same, that $m = 0$. Then all automorphisms leave the x's unmoved, and what is to be shown is that all automorphisms leave z unmoved only if z is in K. Let z be expressed as a polynomial in y of degree less than $n = \deg g$ with coefficients in K, say $z(y) = a_1 y^{n-1} + a_2 y^{n-2} + \cdots + a_n$. Consider the polynomial $\zeta(X) = a_1 X^{n-1} + a_2 X^{n-2} + \cdots + a_n - z(y)$, a polynomial in the indeterminate X with coefficients in the Galois field. (All coefficients but the constant term $a_n - z(y)$ are in the smaller field K.) To say that substitution of any root $\rho_i(y)$ in place of y in $z(y)$ does not change $z(y)$ mod $g(y)$ is to say that $a_1(\rho_i(y))^{n-1} + a_2(\rho_i(y))^{n-2} + \cdots + a_n - z(y) = z(\rho_i(y)) - z(y) \equiv 0$ mod $g(y)$. In other words, each $\rho_i(y)$ is a root of $\zeta(X)$ in the Galois field. Since the degree of $\zeta(X)$ is less than n, the fact that it has n distinct roots in the root field of $g(y)$ (that is, n distinct linear factors $X - \rho_i(y)$) implies* that it is zero. In particular, its constant term $a_n - z(y)$ must be zero, so $z(y) = a_n$ is in K, as was to be shown.

   Suppose next that $m = 1$, and let $x(y)$ denote the one given field element. Let the list of images $x(\rho_i(y))$ of $x(y)$ under the Galois group consist of σ distinct field elements, each occurring $\tau = n/\sigma$ times, and let $\phi(X) = \prod(X - x(\rho_i(y)))$ be the polynomial of degree σ in the indeterminate X with coefficients in the Galois field obtained by taking the product of the *distinct* factors $X - x(\rho_i(y))$. For any $j = 1, 2, \ldots, n$, changing y to $\rho_j(y)$ in $\phi(X)$ merely permutes the factors and does not change $\phi(X)$. Therefore, each coefficient of $\phi(X)$ is unchanged by the automorphisms in the Galois group, so, by the case $m = 0$ that has already been proved, $\phi(X)$ is a polynomial with coefficients in K. Therefore, an element a_0 in R can be found for which $a_0\phi(X)$ has coefficients in R. If $x(y)$ is replaced by $a_0 x(y)$, the new $x(y)$ is a root of the monic polynomial $a_0^{\sigma-1}\phi(X/a_0)$, and an element of the Galois field can be expressed as a polynomial in the old $x(y)$ with coefficients in K if and only if it can be expressed as a polynomial in the new one with coefficients in K. Therefore, one can assume without loss of generality that $\phi(X)$ is monic in X with coefficients in R. It is irreducible, because a monic factor of $\phi(X) = \prod(X - x(\rho_i(y)))$ over K is a product of some subset of its factors $X - x(\rho_i(y))$ over the Galois field, so, if it is not 1, it has at least one $x(\rho_i(y))$ as a root, which implies that it has all $x(\rho_i(y))$ as roots (because an automorphism carries roots to roots, and the automorphisms act

---

* See the first footnote of Essay 1.4.

transitively on the $\rho_i(y)$) and is therefore $\phi(X)$ itself. If $g(Y)$ is the Galois polynomial that defines the field, the statement that $\phi(X)$ has $\sigma = \deg \phi$ roots in the Galois field is the statement that $g(Y)$ splits $\phi(X)$. Therefore, by the Kronecker–Kneser theorem,[*] the factorization of $g(Y)$ mod $\phi(X)$ consists of σ factors, each of degree $\tau = n/\sigma$, say $g(Y) \equiv \prod \Psi_i(Y, X)$ mod $\phi(X)$, where each Ψ_i is a polynomial in two indeterminates with coefficients in K. Since $\phi(x(y)) \equiv 0$ mod $g(y)$, $g(Y) \equiv \prod \Psi_i(Y, x(y))$ mod $g(y)$ (the difference of $g(Y)$ and $\prod \Psi_i(Y, X)$ is a multiple of $\phi(X)$ so it is a polynomial in X and Y in which the substitution of $x(y)$ for X gives a polynomial in y and Y that is zero mod $g(y)$), so each $\Psi_i(Y, x(y))$ has precisely τ distinct roots $\rho_i(y)$ mod $g(y)$, and the roots of g mod $g(y)$ are partitioned in this way into σ sets of τ each. Let $\Psi_1(Y, X)$ be the factor for which $\Psi_1(y, x(y)) \equiv 0$ mod $g(y)$. Suppose now that $z(y)$ is unchanged by all automorphisms $y \mapsto \rho_i(y)$ that leave $x(y)$ unmoved; i.e., $z(\rho_i(y)) \equiv z(y)$ mod $g(y)$ for every i for which $x(\rho_i(y)) \equiv x(y)$ mod $g(y)$. Division of $z(Y)$ by $\Psi_1(Y, X)$—where capital letters are used to emphasize that the division is division of a polynomial in Y with coefficients in K by a monic polynomial in Y with coefficients in $K[X]$ and has nothing to do with operations in the Galois field—gives an equation $z(Y) = q(Y, X)\Psi_1(Y, X) + r(Y, X)$, where the degree of the remainder $r(Y, X)$ in Y is less than τ. Since $\Psi_1(\rho_i(y), x(y)) \equiv 0$ mod $g(y)$ for each of the τ values of i for which $x(\rho_i(y)) \equiv x(y)$ mod $g(y)$ (because the automorphism $y \mapsto \rho_i(y)$ carries $\Psi_1(y, x(y))$ mod $g(y)$ to $\Psi_1(\rho_i(y), x(y))$ and $\Psi_1(y, x(y)) \equiv 0$ mod $g(y)$ by the definition of $\Psi_1(Y, X)$), and since by assumption $z(\rho_i(y)) \equiv z(y)$ mod $g(y)$ for each of them, $r(\rho_i(y), x(y))$ is the same element of the Galois field for each of τ distinct values of i, which is to say that $r(Y, x(y))$, as a polynomial in Y with coefficients in the Galois field, has the same value $z(y)$ for τ distinct values $\rho_i(y)$ of Y. Because the degree of r in Y is less than τ, the argument given in the case $m = 0$ shows that $r(Y, x(y)) - z(y)$ is the zero polynomial (in Y with coefficients in the Galois field), which gives the required expression $z(y) = r(0, x(y))$ of $z(y)$ as a polynomial in $x(y)$ with coefficients in K.

Finally, the general case follows from the case $m = 1$ just proved once one proves the lemma below, because a polynomial in $u_1 x_1 + u_2 x_2 + \cdots + u_m x_m$, where $u_1, u_2, \ldots, u_m$ are integers, can obviously be expressed as a polynomial in $x_1, x_2, \ldots, x_m$.

Lemma (Theorem of the primitive element). *Let $x_1, x_2, \ldots, x_m$ be elements of a Galois field $K[y]$ mod $g(y)$. Construct integers $u_1, u_2, \ldots, u_m$ for which the automorphisms of the Galois field that leave all of $x_1, x_2, \ldots, x_m$ unmoved coincide with the automorphisms that leave $x = u_1 x_1 + u_2 x_2 + \cdots + u_m x_m$ unmoved.*

Proof. Consider first the case $m = 2$. Let ν be the number of automorphisms $y \mapsto \rho_i(y)$ for which $x_1(\rho_i(y)) \neq x_1(y)$. For each of these ν automorphisms,

---

[*] See Essay 1.7.

let z_i be the element $\frac{x_2(y) - x_2(\rho_i(y))}{x_1(\rho_i(y)) - x_1(y)}$ of the Galois field, and let u be an integer that is not equal to any such z_i. (The number of z's is at most $n-1$, so there is sure to be such an integer u among 1, 2, ..., n.) Then $x = ux_1 + x_2$ has the required property, because an automorphism that does not move x cannot move x_1 (by the choice of u, $u \cdot x_1(\rho_i(y)) + x_2(\rho_i(y)) \neq u \cdot x_1(y) + x_2(y)$ when $x_1(\rho_i(y)) \neq x_1(y)$) and therefore cannot move x_2 (because $u \cdot x_1(\rho_i(y)) + x_2(\rho_i(y)) = u \cdot x_1(y) + x_2(y)$ and $x_1(\rho_i(y)) = x_1(y)$ imply $x_2(\rho_i(y)) = x_2(y)$).

If the lemma is true for m, it is true for $m+1$, because one can use the inductive hypothesis first to find integers u_1, u_2, ..., u_m such that the only automorphisms that leave $u_1 x_1 + u_2 x_2 + \cdots + u_m x_m$ unmoved also leave all of x_1, x_2, ..., x_m unmoved, and then use the case $m = 2$ to find integers U_1 and U_2 such that the only automorphisms that leave $U_1(u_1 x_1 + u_2 x_2 + \cdots + u_m x_m) + U_2 x_{m+1}$ unmoved leave both $u_1 x_1 + u_2 x_2 + \cdots + u_m x_m$ and x_{m+1} unmoved and therefore leave all x's unmoved.

Essay 2.2 Algebraic Quantities

> *...the deep meaning of Kronecker's view, according to which the absolutely algebraic fields* [finite fields or algebraic number fields of finite degree] *are the natural ground-fields of algebraic geometry, at any rate as long as purely algebraic methods are being used.*—André Weil [62]

Kronecker asserted, in substance, in §2 of his treatise *Grundzüge einer arithmetischen Theorie der algebraischen Grössen* [39] that *the field of rational functions in any finite set of algebraic quantities is isomorphic to a root field* as that term is defined in Essay 1.3.

Because the term "algebraic quantity" is vague, this is not an assertion that can be *proved*. However, it can be deduced from a few natural assumptions about "algebraic quantities" as follows.

Suppose first that just one "algebraic quantity" q is involved. For any polynomial $F(X)$ with integer coefficients, $F(q)$ should be a meaningful algebraic quantity, because one must surely be able to add, subtract, and multiply algebraic quantities. Moreover, one should be able to determine when two such polynomials in q have the same value, or, what is the same, able to determine which such polynomials in q are zero. If only the zero polynomial $F(X)$ satisfies $F(q) = 0$, then the field of rational functions in q is simply the field of rational functions in a single indeterminate, which is the root field of the polynomial $y - c_1$ with coefficients in $\mathbf{Z}[c_1]$. Otherwise, there is an *irreducible* polynomial $F(X)$ with integer coefficients for which $F(q) = 0$, provided one makes the natural assumption that a product of "algebraic quantities" can be zero only when one of the factors is zero. (Of course, $F(q) = 0$ implies that F is not a unit.) Let $F(q) = a_0 q^n + a_1 q^{n-1} + \cdots + a_n$ be zero, where the coefficients a_i are integers and $F(y)$ is irreducible, and let f be the monic polynomial with integer coefficients defined by $f(y) = a_0^{n-1} F(y/a_0)$. Then, because rational functions of q are rational functions of $a_0 q$ and conversely, and because $a_0 q$ is a root of f, the field of rational functions in q is isomorphic to the root field of $f(y)$, which completes the "proof" of Kronecker's assertion in the case of a single algebraic quantity.

Suppose now that the assertion is true for a set $q_1, q_2, \ldots, q_m$ of m algebraic quantities and let an $(m+1)$st algebraic quantity q_{m+1} be given. Let Ω and Ω_0 denote the fields of rational functions in $q_1, q_2, \ldots, q_{m+1}$ and in $q_1, q_2, \ldots, q_m$, respectively. By the inductive hypothesis, Ω_0 is isomorphic to the root field of some monic, irreducible polynomial $g(y)$ with coefficients in $\mathbf{Z}[c_1, c_2, \ldots, c_\nu]$ for some ν. An element of Ω can be expressed as a quotient of polynomials in q_{m+1} with coefficients in Ω_0. As before, there are two cases, depending on whether there is a nonzero polynomial $F(X)$ with coefficients in Ω_0 for which $F(q_{m+1}) = 0$. If not, Ω is the field of quotients of the ring $\mathbf{Z}[c_1, c_2, \ldots, c_\nu, q_{m+1}]$ mod $g(y)$, which is the root field of $g(y)$ when it is regarded as a polynomial with coefficients in $\mathbf{Z}[c_1, c_2, \ldots, c_{\nu+1}]$ and $c_{\nu+1}$ is identified with q_{m+1}. If so, then q_{m+1} is a root of a polynomial with coefficients in the root field of $g(y)$, say $F(q, y) = 0$, where F is a polynomial in two indeterminates with coefficients in $\mathbf{Z}[c_1, c_2, \ldots, c_\nu]$ and y is a root of $g(y)$ in a splitting field of g; then q is a root of $\mathcal{F}(X) = \prod_{i=1}^{n} F(X, y_i)$, where n is the degree of g and the y_i are the roots of g in the splitting field, so that $\mathcal{F}(X)$ has coefficients that are symmetric polynomials in $y_1, y_2, \ldots, y_n$ and are therefore in $\mathbf{Z}[c_1, c_2, \ldots, c_\nu]$. The given field Ω is then isomorphic to a subfield of the splitting field of $g(X)\mathcal{F}(X)$, namely, the subfield obtained by adjoining first a root y of g and then a root q of $F(X, y)$. By the theorem of the primitive element, this field is the root field of a monic, irreducible polynomial with coefficients in $\mathbf{Z}[c_1, c_2, \ldots, c_\nu]$. In either case, then, Ω is a root field, as was to be shown.

The definition of an "algebraic quantity" is problematic, because algebraic quantities by their very nature do not exist in isolation; they are "things on which rational computations can be performed," and therefore are items in entire systems of computation, which is to say, elements of entire fields. All in all, it seems best to adopt Kronecker's later view (see Essay 1.1) and to abandon the notion of algebraic quantities in favor of a formulation of the subject in terms of "general arithmetic" and to make the above assertion a definition:

An **algebraic field** is a field that can be described as the field of quotients of an integral domain of the form $\mathbf{Z}[c_1, c_2, \ldots, c_\nu]$ mod F, where F is a nonconstant irreducible element of $\mathbf{Z}[c_1, c_2, \ldots, c_\nu]$. (If F is constant and irreducible, this field of quotients is the field of rational functions in $c_1, c_2, \ldots, c_\nu$ with coefficients in the field with p elements for some prime p, a field that does not really seem to merit the name "algebraic field." The field of rational functions in $c_1, c_2, \ldots, c_\nu$ with integer coefficients is an algebraic field; for example, it is the field of quotients of $\mathbf{Z}[c_1, c_2, \ldots, c_\nu, c_{\nu+1}]$ mod $c_{\nu+1}$.)

Note that there is no stipulation that the irreducible polynomial be *monic* in one of its indeterminates. Such a stipulation would involve the indeterminates asymmetrically. As was seen above, one can adjoin a root of a polynomial $F(X) = a_0 X^n + a_1 X^{n-1} + \cdots + a_n$ by adjoining a root of the monic polynomial $f(X) = a_0^{n-1} F(X/a_0)$, from which it follows easily that every algebraic field can be presented as the root field of an irreducible element of $\mathbf{Z}[c_1, c_2, \ldots, c_\nu]$ that is monic and of positive degree in c_i for some i.

A set of elements $u_1, u_2, \ldots, u_k$ of an algebraic field is **algebraically independent** if no nonzero polynomial ϕ in k indeterminates with integer coefficients satisfies $\phi(u_1, u_2, \ldots, u_k) = 0$. If the polynomial F that determines an algebraic field involves c_ν, then the indeterminates $c_1, c_2, \ldots, c_{\nu-1}$ represent algebraically independent elements of the algebraic field. (The construction of this field can be described as the adjunction of a root of a monic irreducible polynomial to the field of rational functions of $c_1, c_2, \ldots, c_{\nu-1}$.) Therefore, such an algebraic field always contains a set of $\nu - 1$ algebraically independent elements. The **transcendence degree** of an algebraic field is the maximum number of elements in an algebraically independent subset. The following proposition implies that an algebraic field defined by some irreducible F in $\mathbf{Z}[c_1, c_2, \ldots, c_\nu]$ has transcendence degree $\nu - 1$:

Proposition. *Let a nonconstant irreducible element F of $\mathbf{Z}[c_1, c_2, \ldots, c_\nu]$ be given. Assume that F involves c_ν and let u be an element of the algebraic field F determines. For a given number $l < \nu$, one can determine whether the elements $c_1, c_2, \ldots, c_l, u$ are algebraically independent, and if they are, one can describe the algebraic field as being determined by an irreducible element F_1 of $\mathbf{Z}[d_1, d_2, \ldots, d_\nu]$, where the first $l+1$ indeterminates $d_1, d_2, \ldots, d_{l+1}$ are identified with $c_1, c_2, \ldots, c_l, u$.*

Proof. Assume without loss of generality that F is monic in c_ν. Every element of the algebraic field can be written in exactly one way in the form $\sum_{i=1}^{N} \eta_i(c_1, c_2, \ldots, c_{\nu-1})c_\nu^{N-i}$, where N is the degree of F in c_ν and the $\eta_i(c_1, c_2, \ldots, c_{\nu-1})$ are rational functions in the first $\nu - 1$ indeterminates. When the $N+1$ elements $1, u, u^2, \ldots, u^N$ of the field are written in this form, one can construct a linear dependence of these elements of the field over the field of rational functions in $c_1, c_2, \ldots, c_{\nu-1}$, and from this one can construct a polynomial in ν indeterminates with integer coefficients in which substitution of $c_1, c_2, \ldots, c_{\nu-1}, u$, regarded as elements of the given root field, results in zero. One of the irreducible factors of this polynomial, call it G, must have the same property. This G is determined, up to sign, by u. Since $c_1, c_2, \ldots, c_l, u$ are algebraically independent if and only if at least one of the indeterminates $c_{l+1}, c_{l+2}, \ldots, c_{\nu-1}$ occurs in G, what is to be shown is that when this is the case, the given algebraic field can be described as stated in the proposition.

Suppose c_h occurs in G for some $h > l$. Assume without loss of generality that $h = \nu - 1$. The field of quotients of $\mathbf{Z}[u, c_1, c_2, \ldots, c_{\nu-1}]$ mod G is an algebraic field that can be identified with a subfield of the given algebraic field. It can also be described as the field obtained by adjoining $c_{\nu-1}$ to the field of rational functions in $u, c_1, c_2, \ldots, c_{\nu-2}$. Since the given algebraic field can then be obtained by adjoining c_ν to this field, the desired conclusion follows from the theorem of the primitive element, which implies that the given algebraic field, which can be obtained by two successive adjunctions to the field of rational functions in $u, c_1, c_2, \ldots, c_{\nu-2}$, can be obtained by a single adjunction.

Thus, given k algebraically independent elements of an algebraic field, one can successively alter the presentation of the field so that the first k indeterminates of the polynomial that describes it represent these k field elements. In particular, because the ν indeterminates of a defining relation do satisfy an algebraic relation, $k < \nu$.

In conclusion, every algebraic field has a well-defined transcendence degree, and an algebraic field of transcendence degree ν is one that can be presented as the root field of an irreducible polynomial g with coefficients* in $\mathbf{Z}[c_1, c_2, \ldots, c_{\nu+1}]$ that is monic in $c_{\nu+1}$. An **algebraic number field** is an algebraic field of transcendence degree zero.

---

* The coefficient ring $\mathbf{Z}[c_1, c_2, \ldots, c_\nu]$ may include indeterminates that do not occur in g. For example, $g(c_1) = c_1^2 + 2$ can be regarded as having coefficients in $\mathbf{Z}[c_1]$ or in $\mathbf{Z}[c_1, c_2]$. In the first case its root field is $\mathbf{Q}(\sqrt{-2})$, which has transcendence degree 0; in the second, its root field is the field of rational functions in c_2 with coefficients in $\mathbf{Q}(\sqrt{-2})$ which has transcendence degree 1.

Essay 2.3 Adjunctions and the Factorization of Polynomials

Cela posé, nous appellerons rationelle *toute quantité qui s'exprimera en fonction rationnelle des coefficients de l'équation et d'un certain nombre de quantités* adjointes *à l'équation et convenues arbitrairement.*—É. Galois, [27] (English translation, [18, p. 101])

Loosely speaking, the assertion of the preceding essay states that any algebraic computation can be regarded as taking place inside the root field of some polynomial. Since the theorem of Part 1 implies that every such field is a subfield of a Galois field—the root field of a *Galois* polynomial—*every algebraic computation can be regarded as taking place inside a Galois field.*

For example, the factorization of $f(x)$ mod $g(y)$, where f and g are monic and irreducible, can be described, now that the theorem of Part 1 and Galois's fundamental theorem have been proved, in the following way: In a Galois field in which fg splits, adjoin to* K a root b of g to obtain a subfield $K(b)$ of the Galois field that contains a root of g. The factorization problem is to determine the irreducible factors of f when elements of $K(b)$ are permitted as coefficients. Over the Galois field, $f(x)$ is a product of linear factors, say $f(x) = \prod(x - a_i)$. Because $f(x)$ is irreducible, it is relatively prime to its derivative, which implies that the roots a_i of $f(x)$ are *distinct*. The automorphisms of the Galois field that leave b unmoved partition these roots into *orbits,* two roots being in the same orbit if and only if there is an automorphism that carries one to the other without moving b. Thus, $f(x) = \prod \phi_j(x)$, where the factors $\phi_j(x)$ on the right are the products of the factors $x - a_i$ over all roots a_i in one orbit. Each factor $\phi_j(x)$ of $f(x)$ found in this way is unchanged by the automorphisms that leave b fixed—such automorphisms merely permute the factors of $\phi_j(x)$—so by Galois's theorem each $\phi_j(x)$ can be expressed in the form $\phi_j(x, b)$, where ϕ_j is a polynomial in two indeterminates with coefficients in K that is monic in x. Each $\phi_j(x, b)$ is irreducible over $K(b)$, because any monic factor other than 1 over $K(b)$ must be divisible by at least one of the monic factors $x - a_i$ of $\phi_j(x, b)$ over K and therefore must, by the definition of $\phi_j(x, b)$, be divisible by all such factors, which implies that it is divisible by $\phi_j(x, b)$ (because its linear factors are distinct). In conclusion, $f(x) = \prod \phi_j(x, b)$ is the unique factorization of $f(x)$ into monic factors irreducible over $K(b)$.

Similarly, the factorization $\mathcal{F}(z, t, u) = \prod(z - a_i t - b_j u)$, where the product is over all mn pairs (a_i, b_j) in which a_i is a root of f and b_j is a root of g in a splitting field for fg (see Essay 1.9), implies that the factorization of $\mathcal{F}$ into factors irreducible over R is obtained by grouping[†] together factors $z - a_i t - b_j u$ that are in the same orbit under the action of the Galois group. The Galois

---

* As before, K is the field of quotients of the ring $R = \mathbf{Z}[c_1, c_2, \ldots, c_\nu]$ in which f and g have their coefficients.

† Here the notion of "grouping" is very close to Galois's original use of the word "group" in a similar context.

group acts transitively on the b_j (because g is irreducible), and the a_i for which $z - a_i t - b_j u$ are in a given orbit for any one b_j are, as was just shown, the roots a_i of one of the irreducible factors $\phi_k(x, b_j)$ of $f(x)$ when b_j is adjoined. From this it follows that the factorization $f(x) \equiv \prod \phi_k(x, y) \bmod g(y)$ implies that the factorization $\mathcal{F}(z, t, u) = \prod \mathcal{F}_k(z, t, u)$ of $\mathcal{F}$ into factors irreducible over R is found by defining the factors $\mathcal{F}_k(z, t, u)$ to be $\prod(z - a_i t - b_j u)$, where, for a given k, the product is over all roots b_j of g and, for each b_j, over all roots a_i of f for which $\phi_k(a_i, b_j) = 0$. Thus, the degree of $\mathcal{F}_k$ is $\mu_k n$, where $\mu_k = \deg_x \phi_k$. Moreover, $\mathcal{F}_k(x + uy, 1, u) = \prod((x - a_i) + u(y - b_j))$ when the product is over all $\mu_k n$ pairs (a_i, b_j) for which $\phi_k(a_i, b_j) = 0$. When $\mathcal{F}_k(x + uy, 1, u)$ is expanded as a polynomial in u whose coefficients are polynomials in x and y with coefficients in R, the terms of degree greater than $\mu_k n - n$ in u are zero $\bmod g(y)$, because, as is clear from the product representation, it is a sum of terms, each of which contains all n factors $y - b_j$ of $g(y)$. (The portion $x - a_i$ of fewer than n factors is used in forming the term, so all distinct portions $u(y - b_j)$ are used at least once.) By the same token, the coefficient $\psi_k(x, y)$ of $u^{\mu_k(n-1)}$ in $\mathcal{F}_k(x + uy, 1, u)$ (in the notation of Essay 1.5) contains only n terms that are nonzero $\bmod g(y)$; explicitly,

$$\psi_k(x, y) \equiv \sum_{j=1}^{n} \phi_k(x, b_j) \left(\frac{g(y)}{y - b_j} \right)^{\mu_k} \bmod g(y).$$

Therefore, $\psi_k(x, y)$ has degree μ_k in x and its leading coefficient is

$$\sum_{j=1}^{n} \left(\frac{g(y)}{y - b_j} \right)^{\mu_k} \equiv \left(\sum_{j=1}^{n} \frac{g(y)}{y - b_j} \right)^{\mu_k} = g'(y)^{\mu_k} \bmod g(y).$$

(The first step follows from $\frac{g^2(y)}{(y - b_j)(y - b_k)} \equiv 0 \bmod g(y)$ when $j \neq k$, and the second follows from the observation that $\sum \frac{g(y)}{y - b_j}$ has degree $n - 1$ and agrees with $g'(y)$ when $y = b_j$.) Similarly, $\mathcal{F}(x + uy, 1, u) \equiv f(x) g'(y)^m u^{m(n-1)} + \cdots \bmod g(y)$, which implies $f(x) g'(y)^m \equiv \prod \psi_k(x, y) \bmod g(y)$. By unique factorization $\bmod g(y)$, it follows that each $\psi_k(x, y)$ is of the form $g'(y)^{\mu_k} \phi_l(x, y) \bmod g(y)$ for some l. That $l = k$ follows from the observation that if (a_i, b_j) is a pair for which $\phi_l(a_i, b_j) = 0$, then substitution of (a_i, b_j) for (x, y) in $\psi_k(x, y) \equiv \sum_{\iota=1}^{n} \phi_k(x, b_\iota) \left(\frac{g(y)}{y - b_\iota} \right)^{\mu_k} \bmod g(y)$ gives 0 when $k = l$, so $\psi_l(x, y)$ cannot divide $\phi_l(x, y) \bmod g(y)$ unless $k = l$. Therefore, $\psi_k(x, y) \equiv g'(y)^{\mu_k n} \phi_k(x, y) \bmod g(y)$, as the algorithm of Essay 1.5 asserts.

Galois theory can be used to describe computations in a subfield $K(\alpha_1, \alpha_2, \ldots, \alpha_n)$ of a Galois field in the following way: First, if μ_1 is the number of distinct images of α_1 under the Galois group, then the product of the linear polynomials $x - S\alpha_1$ over all *distinct* images $S\alpha_1$ of α_1 under the Galois group is a monic polynomial in x with coefficients in K that is *irreducible* over K by the argument above. Call it $f_1(x)$. Then $f_1(\alpha_1) = 0$ provides a relation that can be used to replace any element of the Galois field expressed as a polynomial in α_1 with coefficients in K, say $\psi(\alpha_1)$, by another expression of the same element of the same form in which the degree of the polynomial is less than μ_1; one has only to divide $\psi(\alpha_1)$ by $f_1(\alpha_1)$ regarded as a monic polynomial with coefficients in K to find $\psi(\alpha_1) = q(\alpha_1)f_1(\alpha_1) + r(\alpha_1)$ and then to note that $\psi(\alpha_1) = r(\alpha_1)$ as elements of the Galois field.* In this way, *the relation $f_1(\alpha_1) = 0$ can be used to find the unique representation of any element of $K(\alpha_1)$ as a polynomial in α_1 with coefficients in K of degree less than μ_1.* Similarly, if μ_2 is the number of distinct images of α_2 under the elements of the Galois group that leave α_1 unmoved, the product of the μ_2 distinct linear polynomials of the form $x - S\alpha_2$, where $S\alpha_2$ is the image of α_2 under an element S of the Galois group that leaves α_1 unmoved, is a monic polynomial $f_2(\alpha_1, x)$ of degree μ_2 with coefficients in the field $K(\alpha_1)$ that is irreducible over this field and that satisfies $f_2(\alpha_1, \alpha_2) = 0$. Using the relation $f_2(\alpha_1, \alpha_2) = 0$ one can (division with remainder by a monic polynomial) find, for any polynomial in α_1 and α_2 with coefficients in K, another polynomial representing the same element of $K(\alpha_1, \alpha_2)$ whose degree in α_2 is less than μ_2. By the irreducibility of f_2, the only way that a polynomial in α_1 and α_2 of degree less than μ_2 in α_2 can be zero is for the coefficient of each power of α_2 to be zero. Division by $f_1(\alpha_1)$ to reduce the degree in α_1 does not increase the degree in α_2 and proves that *each element of $K(\alpha_1, \alpha_2)$ is represented by one and only one polynomial in α_1 and α_2 with coefficients in K whose degree in α_1 is less than μ_1 and whose degree in α_2 is less than μ_2.* Continuation of this process leads to the following description of the Galois field:

Proposition. *Given elements α_1, α_2, ..., α_n of a Galois field $K[y]$ mod $g(y)$, and given an element β of this Galois field that is unmoved by the elements of the Galois group that leave all of α_1, α_2, ..., α_n unmoved, express β as a polynomial in α_1, α_2, ..., α_n with coefficients K whose degree in α_i for each i is less than the number of distinct images of α_i under elements of the Galois group that leave all of α_1, α_2, ..., α_{i-1} unmoved. This polynomial is determined by β. It will be called the representation in* **canonical form** *of β as an element of the subfield $K(\alpha_1, \alpha_2, ..., \alpha_n)$ of the Galois field generated by α_1, α_2, ..., α_n over K.*

Computations in $K(\alpha_1, \alpha_2, ..., \alpha_n)$ can be done by dealing with polynomials in canonical form; addition is the usual addition of polynomials, while multiplication is the usual multiplication of polynomials *followed* by reduction to canonical form. The reduction of a polynomial to canonical form uses n relations $f_1(\alpha_1) = 0$, $f_2(\alpha_1, \alpha_2) = 0$, ..., $f_n(\alpha_1, \alpha_2, ..., \alpha_n) = 0$ that will be called **adjunction relations**. The relation $f_i = 0$ can be described either as $\alpha_i^{\mu_i} = \phi_i(\alpha_1, \alpha_2, ..., \alpha_i)$, where $\phi_i(\alpha_1, \alpha_2, ..., \alpha_i)$ is the expression of $\alpha_i^{\mu_i}$ in canonical form, or as the statement that $x = \alpha_i$ is a root of the

---

* Alternatively, one can iteratively replace $\alpha_1^{\mu_1}$ with $\alpha_1^{\mu_1} - \phi(\alpha_1)$ until the degree is less than $\mu_1 = \deg \psi$.

polynomial $\prod(x - S\alpha_i)$, where the product is over all distinct images of α_i under elements S of the Galois group that leave $\alpha_1, \alpha_2, \ldots, \alpha_{i-1}$ fixed and where the coefficients of this polynomial are written in canonical form. The adjunction relations are naturally used in reverse order—that is, $f_n = 0$ is used to reduce the degree in α_n, then $f_{n-1} = 0$ is used to reduce the degree in α_{n-1} without increasing the degree in α_n, then $f_{n-2} = 0$ is used to reduce the degree in α_{n-2} without increasing the degree in either α_n or α_{n-1}, and so forth, to end with a polynomial in canonical form.

The theorem of the primitive element (Essay 2.1) implies, of course, that the field $K(\alpha_1, \alpha_2, \ldots, \alpha_n)$ described in this way can be described by a single adjunction $K(\beta)$ with a single adjunction relation $f(\beta) = 0$. However, a construction of a field by a number of simple adjunctions may be preferable to a single adjunction, because it may describe the field more simply. For example, the classical question of whether a given equation can be solved by radicals simply asks whether a splitting field for a given a polynomial $f(x)$ can be described by adjunction relations of the special form $\alpha_i^{\mu_i} = \phi_i(\alpha_1, \alpha_2, \ldots, \alpha_{i-1})$ in which the right side does not involve α_i.

Example 1. The splitting field of $f(x) = x^4 - x^2 - 1$. This polynomial is irreducible over $\mathbf{Z}$, because a factorization would have the form $(x^2 + Ax + 1)(x^2 + Bx - 1)$, and $A + B = 0$, $-1 + AB + 1 = -1$, and $-A + B = 0$ would all hold. The factorization of $f(x) \bmod f(y)$ can be found by the elementary calculation $x^4 - x^2 - 1 \equiv x^4 - x^2 - y^4 + y^2 \equiv (x^2 - y^2)(x^2 + y^2) - (x^2 - y^2) \equiv (x - y)(x + y)(x^2 + y^2 - 1) \bmod (y^4 - y^2 - 1)$. Thus, adjunction of a root a of $f(x)$ to $\mathbf{Q}$ gives $f(x)$ either 2 or 4 roots, depending on whether $x^2 + a^2 - 1$ splits over $\mathbf{Q}(a)$.

If one *assumes* that $f(x)$ is not a Galois polynomial—which is to say that $x^2 + y^2 - 1$ is irreducible mod $y^4 - y^2 - 1$—this factorization of $f(x) \bmod f(y)$ implies that the adjunction relations

$$a^4 = a^2 + 1,$$
$$b = -a,$$
$$c^2 = 1 - a^2,$$
$$d = -c,$$

describe the splitting field.

These relations show that the permutation (ab) is in the Galois group, because they easily imply the relations $b^4 = b^2 + 1$, $a = -b$, $c^2 = 1 - b^2$, $d = -c$ that result when a and b are interchanged and c and d are unmoved. Similarly, the four-cycle $(acbd)$ is in the Galois group because the adjunction relations also imply all the relations $c^4 = c^2 + 1$, $d = -c$, $b^2 = 1 - c^2$, $a = -b$ obtained when the roots of $f(x)$ in the splitting field are permuted in this way. These two permutations generate a group of 8 motions (a dihedral group). Since the field described by the adjunction relations obviously has degree 8 over $\mathbf{Q}$, these 8 motions account for the entire Galois group.

As for the proof that $f(x)$ is not a Galois polynomial—which comes down to the statement that $1 - a^2$ is not a square in the field $\mathbf{Q}(a)$ obtained by adjoining a root of $a^4 - a^2 - 1$ to the rationals—the means to do it are given by the algorithms of Part 1, but the computations they require are not easily done with pencil and paper. As is shown in Essay 1.6, $x^2 + y^2 - 1$ is irreducible mod $y^4 - y^2 - 1$ if and only if the determinant of $(zI_4 - uG)^2 + G^2 - I_4$ is irreducible in $\mathbf{Z}[z, u]$, where G is the 4×4 matrix whose first 3 rows are the last 3 rows of I_4 and whose last row is 1, 0, 1, 0. This determinant, call it $\mathcal{F}_3(z, 1, u)$, is $z^8 + (-2u^2 - 2)z^6 + (-u^4 + 16u^2 - 1)z^4 + (2u^6 - 12u^4 - 12u^2 + 2)z^2 + (u^8 + 6u^6 + 11u^4 + 6u^2 + 1)$. It is irreducible because $\mathcal{F}_3(z, 1, 2) = z^8 - 10z^6 + 47z^4 - 110z^2 + 841$ is in fact irreducible. (The proof can be accomplished using the primitive methods of Essay 1.4, but of course a computer algebra package will find the answer much more quickly.) Therefore, the algorithms of Part 1 lead to the conclusion that $\mathcal{F}_3(z, 1, 2)$ is a minimal splitting polynomial of $x^4 - x^2 - 1$.

An easier method of proving that $f(x)$ is not a Galois polynomial using an altogether different method is to observe that $f(x)$ has the *real* root

$$\sqrt{\frac{1 + \sqrt{5}}{2}},$$

so that $\mathbf{Q}(a)$ can be embedded in the field of real numbers; relative to this embedding, $1 - a^2$ is a negative number, so it is not a square in $\mathbf{Q}(a)$.

Once the adjunction relations are known, the computation of $\mathcal{F}_3(z, 1, 2)$ can be accomplished as follows: This polynomial is the polynomial of which $a + 2c$ and its conjugates—a set of 8 distinct elements of the splitting field—are the roots. Therefore, it is the product of the eight conjugates of $z - a - 2c$ under the Galois group. The product of this linear polynomial and its conjugate $z - c - 2a$ under $(ac)(bd)$ is $z^2 - 3(a + c)z + 5ac + 2$. The motion $(ab)(cd)$ carries $z - a - 2c$ to $z + a + 2c$ and $z - c - 2a$ to $z + c + 2b$ and their product to $z^2 + 3(a + c)z + 5ac + 2$; thus, the product of all four factors is $z^4 + (10ac + 4 - 9(a+c)^2)z^2 + (5ac+2)^2 = z^4 - (5+8i)z^2 - 21 + 20i$, where use is made of the identity $a^2c^2 = a^2(1-a^2) = -a^4 + a^2 = -1$ to write i for ac. Since (ab) carries i to $-i$, the product of all 8 conjugates is therefore the norm of $z^4 - (5+8i)z^2 - 21 + 20i$, which is the polynomial $z^8 - 10z^6 + 47z^4 - 110z^2 + 841$ found above.

The explicit factorization of $x^4 - x^2 - 1$ in the field $\mathbf{Q}[z]$ mod $(z^8 - 10z^6 + 47z^4 - 110z^2 + 841)$ can be found using a method described by Galois (see [18, §37]). Because $(X - b)(X - c)(X - d) = (X^4 - X^2 - 1)/(X - a) = X^3 + aX^2 + (a^2 - 1)X + (a^3 - a)$, a polynomial relation between a and $t = a + 2c$ is given by $0 = (t - a - 2c)(t - a - 2b)(t - a - 2d) = 8\left(\frac{t-a}{2} - b\right)\left(\frac{t-a}{2} - c\right)\left(\frac{t-a}{2} - d\right) = 8\left(\left(\frac{t-a}{2}\right)^3 + a\left(\frac{t-a}{2}\right)^2 + (a^2 - 1)\frac{t-a}{2} + a^3 - a\right) = (t-a)^3 + 2a(t-a)^2 + 4(a^2 - 1)(t - a) + 8(a^3 - a) = 5a^3 + 3ta^2 - (t^2 + 4)a + t^3 - 4t$. In other words, a is a root of the polynomial $5Y^3 + 3tY^2 - (t^2 + 4)Y + t^3 - 4t$ in Y with coefficients in the splitting field. Since it is also a root of $Y^4 - Y^2 - 1$, it is a root of the greatest common divisor of these two polynomials, which can be found using the Euclidean algorithm. Explicitly, if $A = Y^4 - Y^2 - 1$ and $B = 5Y^3 + 3tY^2 - (t^2 + 4)Y + t^3 - 4t$, then a is a root of A, B, $C = yB - 5A = 3tY^3 - (t^2 - 1)Y^2 + (t^3 - 4t)Y + 5$, $D = 3tB - 5C = (14t^2 - 5)Y^2 - (8t^3 - 8t)Y + (3t^4 - 12t^2 - 25)$, $E = \left(3tYD - (14t^2 - 5)C\right)/5 = (-2t^4 + t^2 + 1)Y^2 + (-t^5 + 5t^3 - 19t)Y - 14t^2 + 5$, and $F = \left((-2t^4 + t^2 + 1)D - (14t^2 - 5)E\right)/3t = (10t^6 - 33t^4 + 97t^2 - 29)Y + (-2t^7 + 9t^5 + 79t^3 - 59t)$. Thus, a is expressed rationally in terms of t as

$$a = \frac{2t^7 - 9t^5 - 79t^3 + 59t}{10t^6 - 33t^4 + 97t^2 - 29},$$

and so are $b = -a$, $c = \frac{t-a}{2}$, and $d = -c$. The explicit splitting is therefore

$$x^4 - x^2 - 1 \equiv \left(x - \frac{2t^7 - 9t^5 - 79t^3 + 59t}{10t^6 - 33t^4 + 97t^2 - 29}\right)\left(x + \frac{2t^7 - 9t^5 - 79t^3 + 59t}{10t^6 - 33t^4 + 97t^2 - 29}\right)$$
$$\times\left(x - \frac{4t^7 - 12t^5 + 88t^3 - 44t}{10t^6 - 33t^4 + 97t^2 - 29}\right)\left(x + \frac{4t^7 - 12t^5 + 88t^3 - 44t}{10t^6 - 33t^4 + 97t^2 - 29}\right)$$
$$\mod(t^8 - 10t^6 + 47t^4 - 110t^2 + 841).$$

To express the four roots as polynomials in t with rational coefficients, one must 'rationalize the denominator' $10t^6 - 33t^4 + 97t^2 - 29$, which is simple in theory but lies beyond the range of hand computation. Doing the algebra on a computer (I used Maple), one can obtain

$$(674t^6 - 6363t^4 - 7501t^2 + 179117)(10t^6 - 33t^4 + 97t^2 - 29)$$
$$= (6740t^4 - 18472t^2 - 301153)(t^8 - 10t^6 + 47t^4 - 110t^2 + 841) + 248075280.$$

Multiplication of numerator and denominator of a by $674t^6 - 6363t^4 - 7501t^2 + 179117$ and simplification then gives

$$a = \frac{7t^7 + 220t^5 - 1846t^3 + 9003t}{18966}.$$

Because $c = \frac{a-t}{2}$, the factorization

$$2^2 \times 18966^4(x^4 - x^2 - 1) \equiv (18966x - (7t^7 + 220t^5 - 1846t^3 + 9003t))$$
$$\times(18966x + (7t^7 + 220t^5 - 1846t^3 + 9003t))$$
$$\times(2 \cdot 18966x - (7t^7 + 220t^5 - 1846t^3 - 9963t))$$
$$\times(2 \cdot 18966x + (7t^7 + 220t^5 - 1846t^3 - 9963t))$$
$$\mod(t^8 - 10t^6 + 47t^4 - 110t^2 + 841)$$

then follows easily.

Example 2. The splitting field of $f(x) = x^4 + 3x^2 + 7x + 4$. In this case, my derivation of the adjunction relations is rather long and ad hoc. The end result is that the Galois group is the alternating group, which makes the first two adjunction relations and the last easy. The third one is the hard one:

$$a^4 + 3a^2 + 7a + 4 = 0,$$
$$b^3 + ab^2 + a^2b + a^3 + 3a + 3b + 7 = 0,$$
$$c = \frac{1}{133}(-24a^3b^2 + 84a^3b - 36a^3 + 42a^2b^2 - 14a^2b + 63a^2$$
$$-22ab^2 + 210ab - 299a - 63b^2 + 154b - 294),$$
$$d = -a - b - c = \frac{1}{133}(24a^3b^2 - 84a^3b + 36a^3 - 42a^2b^2 + 14a^2b - 63a^2$$
$$+22ab^2 - 210ab + 166a + 63b^2 - 287b + 294).$$

Essay 2.4 The Splitting Field of $x^n + c_1 x^{n-1} + c_2 x^{n-2} + \cdots + c_n$

> *Dans le cas des équations algébriques, ce groupe n'est autre chose que l'ensemble des $1 \cdot 2 \cdot 3 \cdots m$ permutations possibles sur les m lettres, puisque dans ce cas, les fonctions symétriques sont seules déterminables rationnellement.*—É. Galois, [27] (English translation, [18, p. 104])

Theorem. *Construct the splitting field of the polynomial $f(x) = x^n + c_1 x^{n-1} + c_2 x^{n-2} + \cdots + c_n$ in which the coefficients c_1, c_2, ..., c_n are indeterminates.*

As the preceding essay explains, a natural way in which to describe the splitting field of a polynomial is to give the relations that tell how to adjoin each new root to the field obtained by adjoining the roots that precede it. Given the splitting field of any polynomial $f(x)$, one can write

$$(1) \quad (x - \alpha_i)(x - \alpha_{i+1}) \cdots (x - \alpha_n) = \frac{f(x)}{(x - \alpha_1)(x - \alpha_2) \cdots (x - \alpha_{i-1})},$$

where α_1, α_2, ..., α_n are the roots of $f(x)$ in the splitting field, arranged in some order. The right side can be regarded as specifying a monic polynomial of degree $n - i + 1$ with coefficients in the field obtained by adjoining $\alpha_1, \alpha_2, \ldots, \alpha_{i-1}$, namely, the polynomial that results from simple division of the monic polynomial in the numerator by the monic polynomial in the denominator, because the numerator does not involve any α's at all, and the denominator involves just $\alpha_1, \alpha_2, \ldots, \alpha_{i-1}$. As the left side shows, α_i is a root of this monic polynomial. Since the adjunction relation is the *irreducible* monic polynomial with coefficients in the field obtained by adjoining $\alpha_1, \alpha_2, \ldots, \alpha_{i-1}$ of which α_i is a root, it must be an irreducible factor of the polynomial indicated by the right-hand side of formula (1). The key to the theorem above is that these polynomials themselves are irreducible, so they are the adjunction relations that describe the splitting field.

Galois, with his customary terseness, said this when he wrote (see above) that "In the case of algebraic equations, this group is none other than the set of all $1 \cdot 2 \cdot 3 \cdots n$ permutations of the n letters, because in this case the symmetric functions are the only ones that can be determined rationally"; the order of the Galois group is the product of the degrees of the adjunction relations, so Galois's statement that the order of the Galois group is $n \cdot (n-1) \cdot (n-2) \cdots 1$ implies that the polynomial in (1) is the adjunction relation at each step. How Galois might have justified his assertion can only be guessed. The proof that follows is inspired by Kronecker.*

The adjunction relations are certain polynomials in $\alpha_1, \alpha_2, \ldots, \alpha_n$ and $c_1, c_2, \ldots, c_n$ that are zero as elements of the splitting field of $f(x) = x^n + c_1 x^{n-1} + c_2 x^{n-2} + \cdots + c_n$. In theory they can of course be found by

---

* See Kronecker, [37, Sec V]. Also [39, §12].

constructing the splitting field as the root field of a Galois polynomial with coefficients in $\mathbf{Z}[c_1, c_2, \ldots, c_n]$ and then doing the computations sketched in Essay 2.3. The objective is to find them *without* constructing the splitting field.

Let R be the ring of polynomials in $2n$ indeterminates $a_1, a_2, \ldots, a_n, c_1, c_2, \ldots, c_n$ with integer coefficients. Imagine that a splitting field of $f(x) = x^n + c_1 x^{n-1} + c_2 x^{n-2} + \cdots + c_n$ has been constructed and that an order has been chosen for the roots $\alpha_1, \alpha_2, \ldots, \alpha_n$ of $f(x)$ in the splitting field. Substituting α_i for a_i for $i = 1, 2, \ldots, n$ while sending each c_i to itself defines a homomorphism from R to the splitting field. The objective is to determine the kernel of this homomorphism, and, more generally to determine conditions under which two elements ϕ and ψ of R have the same image in the splitting field. The following proposition gives sufficient conditions for this to be the case:

Proposition 1. *Let σ_i for $i = 1, 2, \ldots, n$ be the elementary symmetric polynomials in $a_1, a_2, \ldots, a_n$, which is to say the coefficients of x^{n-1}, x^{n-2}, $\ldots$, 1, respectively, in the polynomial $\prod_{i=1}^{n}(x + a_i)$. Then the elements c_i and $(-1)^i \sigma_i$ of R have the same image in the splitting field of $f(x) = x^n + c_1 x^{n-1} + c_2 x^{n-2} + \cdots + c_n$ under any homomorphism from R to the splitting field constructed in the way that was just described. Consequently, elements ϕ and ψ of R have the same image in the splitting field whenever their difference is a sum of multiples of the polynomials $c_i - (-1)^i \sigma_i$.*

Proof. Change x to $-x$ in the definition of the elementary symmetric polynomials and multiply by $(-1)^n$ to obtain $x^n - \sigma_1 x^{n-1} + \sigma_2 x^{n-2} + \cdots + (-1)^n \sigma_n = (x - a_1)(x - a_2) \cdots (x - a_n)$. Under a homomorphism of R into the splitting field that comes from ordering the roots $\alpha_1, \alpha_2, \ldots, \alpha_n$ of $f(x)$ and sending a_i to α_i for each i while sending each c_i to itself, this polynomial is carried to $f(x)$. Since $f(x) = x^n + c_1 x^{n-1} + c_2 x^{n-2} + \cdots + c_n$ has this same image under such a homomorphism, the corresponding coefficients $(-1)^i \sigma_i$ and c_i must also have the same image, as was to be shown.

Let $A_i = c_i - (-1)^i \sigma_i$ for $i = 1, 2, \ldots, n$. Then such homomorphisms send $A_1, A_2, \ldots, A_n$ all to zero. Therefore, if ϕ and ψ are elements of R for which $\phi - \psi = \sum D_i A_i$, where $D_1, D_2, \ldots, D_n$ are elements of R, then ϕ and ψ must have the same image under any such homomorphism, and the proof is complete.

Let the statement that ϕ and ψ satisfy $\phi - \psi = \sum D_i A_i$ for some elements $D_1, D_2, \ldots, D_n$ of R be abbreviated $\phi \equiv \psi \bmod A$. Since this congruence relation is consistent with addition and multiplication, a ring $R \bmod A$ is defined in this way.

Proposition 2. *As a polynomial with coefficients in the ring $R \bmod A$, the polynomial $(x - a_i)(x - a_{i+1}) \cdots (x - a_n)$, call it $f_i(x)$, for $i = 1, 2, \ldots, n$, is congruent to*

(2) $\phi_{i,0}(x) + \phi_{i,1}(x)c_1 + \phi_{i,2}(x)c_2 + \cdots + \phi_{i,n-i}(x)c_{n-i} + c_{n-i+1} \bmod A$

where $\phi_{i,j}(x)$ is defined by

(3) $$\phi_{i,j}(x) = \sum a_1^{e_1} a_2^{e_2} \cdots a_{i-1}^{e_{i-1}} x^{e_i},$$

for $i \geq 1$ and for $j = 0, 1, 2, \ldots, n - i$, the sum being over all monomials in which the exponents $e_1, e_2, \ldots, e_i$ are nonnegative integers whose sum is $n - i - j + 1$. (In particular, $\phi_{i,j}(x)$ is monic of degree $n - i - j + 1$ in x. When $i = 1$, it is simply x^{n-j}.)

Proof. To say that two elements of $R[x]$ (polynomials in x, a_1, a_2, $\ldots$, a_n, c_1, c_2, $\ldots$, c_n with integer coefficients) are congruent mod A means, of course, that their difference can be written in the form $\sum_{j=1}^{n} D_j A_j$ where D_1, D_2, $\ldots$, D_n are in $R[x]$. The case $i = 1$ of (2) follows from the observation that the right side minus the left side is $\sum_{j=1}^{n} x^{n-j} A_j$. When the formula is proved for one value of i, one knows in particular, because

$$(x - a_i)f_{i+1}(x) = f_i(x) = f_i(x) - f_i(a_i),$$

that $(x - a_i)f_{i+1}(x)$ is congruent to

$$(\phi_{i,0}(x) - \phi_{i,0}(a_i)) + (\phi_{i,1}(x) - \phi_{i,1}(a_i))\, c_1 + \cdots + (\phi_{i,n-i}(x) - \phi_{i,n-i}(a_i))\, c_{n-i}$$

mod A, say their difference is $\sum_{j=1}^{n} D_{i,j} A_j$, where the elements $D_{i,1}$, $D_{i,2}$, $\ldots$, $D_{i,n}$ of $R[x]$ are defined in this way. Each of the $n - i + 1$ summands $(\phi_{i,\iota}(x) - \phi_{i,\iota}(a_i))c_\iota$ is divisible by $x - a_i$, as is $(x - a_i)f_{i+1}(x)$, so $\sum_{j=1}^{n} D_{i,j} A_j$ is divisible by $x - a_i$. The quotient is $\sum E_{i,j} A_j$, where $E_{i,j}$ is found by striking from $D_{i,j}$ all terms that are not divisible by x and dividing what is left by x. In particular, $f_{i+1}(x)$ and

(4) $$\frac{\phi_{i,0}(x) - \phi_{i,0}(a_i)}{x - a_i} + \frac{\phi_{i,1}(x) - \phi_{i,1}(a_i)}{x - a_i} \cdot c_1 + \cdots + \frac{\phi_{i,n-i}(x) - \phi_{i,n-i}(a_i)}{x - a_i} \cdot c_{n-i}$$

are congruent mod A (their difference is $\sum E_{i,j} A_j$). But this is the statement to be proved, because, $\phi_{i,n-i}(x)$ being monic of degree 1 in x, the numerator of the last term is $x - a_i$, while the numerators of the other terms are $(x - a_i)\phi_{i+1,j}(x)$, as follows easily from

$$x^e - a^e = (x - a) \left(x^{e-1} + x^{e-2}a + \cdots + a^{e-1} \right).$$

Let an element of R be said to be in **canonical form** if its degree in a_i is at most $n - i$ for each $i = 1, 2, \ldots, n$. The following proposition makes computations in the ring $R \bmod A$ possible:

Proposition 3. *Each element of R is congruent mod A to one and only one element in canonical form.*

Proof. Let $T_i = \phi_{i,0}(a_i) + \phi_{i,1}(a_i)c_1 + \phi_{i,2}(a_i)c_2 + \cdots + \phi_{i,n-i}(a_i)c_{n-i} + c_{n-i+1}$. Then $T_i = a_i^{n-i+1} + \cdots + c_{n-i+1}$, where the omitted terms have degree at most $n-i$ in a_i, and do not contain a_j for $j > i$ or c_j for $j \geq n-i+1$. Moreover, $T_i \equiv 0 \bmod A$ (because $f_i(a_i) = 0$).

Division of a given element ϕ of R by $T_n = a_n + \cdots$ regarded as a monic polynomial in a_n leaves a remainder that is congruent to $\phi \bmod A$, call it ϕ_1, from which a_n has been eliminated. Division of ϕ_1 by $T_{n-1} = a_{n-1}^2 + \cdots$ regarded as a monic polynomial in a_{n-1} leaves a remainder that is congruent to $\phi_1 \equiv \phi \bmod A$, call it ϕ_2, in which the degree of a_{n-1} is at most 1 and a_n has not been reintroduced. Continuing in this way—on the ith step dividing ϕ_{i-1} by T_{n+1-i} regarded as a monic polynomial in a_{n+1-i} and calling the remainder ϕ_i—produces a sequence $\phi = \phi_0$, ϕ_1, ϕ_2, ..., ϕ_n of polynomials congruent to $\phi \bmod A$. Since the degree in a_i is reduced to at most $n-i$ by the $(n+1-i)$th step and is not increased by any subsequent step, ϕ_n is in canonical form. Thus, every element ϕ of R is congruent mod A to an element ϕ_n in canonical form.

Any element ψ of R is congruent mod A to an element from which the c's have been eliminated, because division of ψ by T_1 regarded as a monic polynomial in c_n leaves a remainder that does not contain c_n, then division of this remainder by T_2 regarded as a monic polynomial in c_{n-1} leaves a remainder that contains neither c_{n-1} nor c_n, and so forth. (In other words, at the ith step, one substitutes $c_{n+1-i} - T_i$ in place of c_{n+1-i} to obtain an element that is unchanged mod A in which c_{n+1-i} is no longer present and no c with a larger index is reintroduced.)

When the input to the first algorithm is a polynomial in $a_1, a_2, \ldots, a_n$ alone and the input to the second algorithm is a polynomial in canonical form, these two algorithms are *inverse* to one another and establish a one-to-one correspondence between polynomials in $a_1, a_2, \ldots, a_n$ and polynomials in canonical form in which corresponding polynomials are congruent mod A, because the first algorithm produces a sequence of equations

$$\phi = Q_1 T_n + \phi_1,$$
$$\phi_1 = Q_2 T_{n-1} + \phi_2,$$
$$\cdots$$
$$\phi_{n-1} = Q_n T_1 + \phi_n,$$

in which, when ϕ contains none of $c_1, c_2, \ldots, c_n$, ϕ_i contains none of c_{i+1}, $c_{i+2}, \ldots, c_n$ for $i = 1, 2, \ldots, n-1$, and ϕ_n is in canonical form. Thus, ϕ_i is the remainder when ϕ_{i+1} is divided by T_{n-i} regarded as a monic polynomial in c_{i+1}, as was to be shown.

To say that a polynomial ϕ in $a_1, a_2, \ldots, a_n$ alone is congruent to zero mod A means that it has the form $\phi = \sum D_j A_j$; this means that $\phi = 0$ because substitution of $(-1)^i \sigma_i$ for c_i in this equation leaves the left side unchanged and makes the right side zero.

Thus, two elements of R in canonical form are congruent mod A if and only if the two polynomials in $a_1, a_2, \ldots, a_n$ alone to which they correspond are congruent mod A, which is true if and only if these two polynomials are equal. This shows that two elements of R in canonical form are congruent mod A only if they are equal and completes the proof of the proposition.

Proof of the Theorem. The formula $x^n + c_1 x^{n-1} + c_2 x^{n-2} + \cdots + c_n \equiv (x - a_1)(x - a_2) \cdots (x - a_n)$ mod A gives an explicit factorization of this polynomial with indeterminate coefficients into linear factors over an integral domain that contains $\mathbf{Z}[c_1, c_2, \ldots, c_n]$ as a subring. (Elements of R that do not contain a_1, $a_2, \ldots, a_n$ are in canonical form, so they are congruent mod A only if they are equal.) Thus, the field of quotients of this integral domain is a splitting field of the polynomial. Since it is generated over $\mathbf{Z}[c_1, c_2, \ldots, c_n]$ by the roots of the polynomial in the ring, it is a minimal splitting field and is therefore *the* splitting field.

Corollary 1.   *The Galois group of $x^n + c_1 x^{n-1} + c_2 x^{n-2} + \cdots + c_n$, where the c's are indeterminates, permutes the roots in the splitting field in all $n!$ possible ways.*

Deduction. Each element of the integral domain R mod A has a unique representation in the form $\sum B_{e_1, e_2, \ldots, e_n}(c_1, c_2, \ldots, c_n) a_1^{e_1} a_2^{e_2} \cdots a_n^{e_n}$, where the coefficients $B_{e_1, e_2, \ldots, e_n}(c_1, c_2, \ldots, c_n)$ are in $\mathbf{Z}[c_1, c_2, \ldots, c_n]$ and the monomials $a_1^{e_1} a_2^{e_2} \cdots a_n^{e_n}$ range over all $n!$ such monomials in which $e_i \leq n - i$ for each i, which shows that the degree of the splitting field as an extension of the field of rational functions in $c_1, c_2, \ldots, c_n$ is $n!$ and therefore that $n!$ is the order of the Galois group.

Corollary 2.   *A polynomial in $a_1, a_2, \ldots, a_n$ that is unchanged by all $n!$ permutations of $a_1, a_2, \ldots, a_n$ has one and only one representation as a polynomial with integer coefficients in the elementary symmetric polynomials $\sigma_1, \sigma_2, \ldots, \sigma_n$ in $a_1, a_2, \ldots, a_n$.*

Deduction. Let ϕ be a given polynomial in $a_1, a_2, \ldots, a_n$ that is unchanged by permutations of the a's. When ϕ is regarded as an element of $R = \mathbf{Z}[a_1, a_2, \ldots, a_n, c_1, c_2, \ldots, c_n]$ it is congruent mod A to one and only one element in canonical form; call it ϕ_n. Since ϕ and $A_1, A_2, \ldots, A_n$ are unchanged by permutations of the a's, ϕ_n is unchanged by permutations of the a's. Since ϕ_n does not contain a_n, it cannot contain any a. Therefore ϕ_n is a polynomial $\Phi(c_1, c_2, \ldots, c_n)$ in the c's alone. Since it is congruent mod A to ϕ and to no other element of $\mathbf{Z}[a_1, a_2, \ldots, a_n]$, it follows that ϕ is equal to $\Phi(-\sigma_1, \sigma_2, \ldots, (-1)^n \sigma_n)$ and to no other polynomial in the σ's with integer coefficients.

Example. The adjunction relations that describe the splitting field of $x^5 + c_1 x^4 + c_2 x^3 + c_3 x^2 + c_4 x + c_5$, where $c_1, c_2, \ldots, c_5$ are indeterminates, are

$$\alpha_1^5 + c_1 \alpha_1^4 + c_2 \alpha_1^3 + c_3 \alpha_1^2 + c_4 \alpha_1 + c_5 = 0,$$

$$\left(\alpha_2^4 + \alpha_2^3 \alpha_1 + \alpha_2^2 \alpha_1^2 + \alpha_2 \alpha_1^3 + \alpha_1^4\right) + c_1 \left(\alpha_2^3 + \alpha_2^2 \alpha_1 + \alpha_2 \alpha_1^2 + \alpha_1^3\right)$$
$$+ c_2 \left(\alpha_2^2 + \alpha_2 \alpha_1 + \alpha_1^2\right) + c_3 (\alpha_2 + \alpha_1) + c_4 = 0,$$

$$\left(\alpha_3^3 + \alpha_3^2 \alpha_2 + \alpha_3^2 \alpha_1 + \alpha_3 \alpha_2^2 + \alpha_3 \alpha_2 \alpha_1 + \alpha_3 \alpha_1^2 + \alpha_2^3 + \alpha_2^2 \alpha_1 + \alpha_2 \alpha_1^2 + \alpha_1^3\right)$$
$$+ c_1 \left(\alpha_3^2 + \alpha_3 \alpha_2 + \alpha_3 \alpha_1 + \alpha_2^2 + \alpha_2 \alpha_1 + \alpha_1^2\right) + c_2 (\alpha_3 + \alpha_2 + \alpha_1) + c_3 = 0,$$

$$\left(\alpha_4^2 + \alpha_4 \alpha_3 + \alpha_4 \alpha_2 + \alpha_4 \alpha_1 + \alpha_3^2 + \alpha_3 \alpha_2 + \alpha_3 \alpha_1 + \alpha_2^2 + \alpha_2 \alpha_1 + \alpha_1^2\right)$$
$$+ c_1 (\alpha_4 + \alpha_3 + \alpha_2 + \alpha_1) + c_2 = 0,$$

$$\alpha_5 + \alpha_4 + \alpha_3 + \alpha_2 + \alpha_1 + c_1 = 0,$$

where $\alpha_1, \alpha_2, \ldots, \alpha_5$ are the roots of the equation in a splitting field. (In the above construction, α_i is the element of R mod A represented by a_i.)

Essay 2.5 A Fundamental Theorem of Divisor Theory

In 1894, when Richard Dedekind's theory of ideals was first beginning to gain widespread recognition, Adolph Hurwitz published a paper [33] in which he proposed a new approach to the theory. Unfortunately for him, Dedekind did not accept his proposal as a friendly amendment. In fact, Dedekind replied [13] rather sharply that he too had discovered the approach Hurwitz described—he had even published a paper [12] on it in an out-of-the-way journal—but that he had firmly decided against it on philosophical grounds.

Neither Hurwitz nor Dedekind seems to have been petty or disagreeable as a general rule, but both of them felt they had a vital stake in the matter. Hurwitz replied [34] that he was not persuaded by Dedekind's philosophical objections to his proposal and that the published theorem to which Dedekind had referred was in fact a special case of a theorem Kronecker had published [41] long before Dedekind had, although Hurwitz himself had only lately learned of Kronecker's theorem.

The interesting part of this story is that three eminent mathematicians—Kronecker, Dedekind, and Hurwitz, in that order—discovered and focused on this one theorem and its importance for the theory of ideals (or, as Kronecker termed it, the theory of divisors), yet the theorem is not widely known today and is rarely included in modern treatments of the theory.

Dedekind regarded the theorem as a generalization of Gauss's lemma [28, Art. 42]: *If the product of two monic polynomials with rational coefficients has integer coefficients, then the factors must have integer coefficients.* He restated Gauss's lemma in the following form: *If the product of two polynomials* (not necessarily monic) *with rational coefficients has integer coefficients, then the product of any coefficient of the first and any coefficient of the second is an integer.* This statement obviously implies Gauss's lemma (1 is a coefficient of both polynomials in Gauss's case). The reverse implication can be proved fairly easily, so the two statements are essentially the same. The advantage of Dedekind's version is that it is true with *algebraic numbers* and *algebraic integers* in place of rational numbers and rational integers. Thus, Dedekind's theorem states that *if the product of two polynomials whose coefficients are algebraic numbers has coefficients that are algebraic integers, then the product of any coefficient of the first and any coefficient of the second is an algebraic integer.*

Another way to put it is the following: Let $a_0, a_1, \ldots, a_m$ and $b_0, b_1, \ldots, b_n$ be two sequences of algebraic numbers. There are two ways to "multiply" them, call them the "polynomial" way, in which the product is the sequence $a_0 b_0, a_0 b_1 + a_1 b_0, a_0 b_2 + a_1 b_1 + a_2 b_0, \ldots, a_m b_n$ of coefficients of the product $(a_0 x^m + a_1 x^{m-1} + \cdots + a_m)(b_0 x^n + b_1 x^{n-1} + \cdots + b_n)$ of polynomials of which the given sequences are the coefficients, and the "pairwise" way, in which the product is the sequence of products $a_i b_j$, where $0 \le i \le m$ and $0 \le j \le n$, arranged in some order. Each term of the polynomial product is a sum of terms of the pairwise product, so if all terms of the pairwise product sequence are algebraic integers, the terms of the polynomial product sequence are too. Dedekind's theorem states the converse: *if all terms of the polynomial product of two sequences of algebraic numbers are algebraic integers, then so are all terms of the pairwise product.*

Here an **algebraic number** is a "quantity" x that satisfies an equation $x^N + A_1 x^{N-1} + \cdots + A_N = 0$ in which the coefficients A_i are rational numbers. After this definition is made, one can prove that in fact x is an algebraic number if the coefficients A_i are merely algebraic numbers. An **algebraic integer** is an algebraic number that satisfies such an equation in which the A_i are integers. Again, one can prove that in fact x is an algebraic integer if the A_i are merely algebraic integers.

Dedekind's version of Gauss's lemma is then the statement that each term $a_i b_j$ of the pairwise product satisfies an equation of the form $(a_i b_j)^N + A_1 (a_i b_j)^{N-1} + A_2 (a_i b_j)^{N-2} + \cdots + A_n = 0$ in which the coefficients A_i can be chosen to be algebraic integers whenever all terms of the polynomial product are algebraic integers. Kronecker made the stronger statement [41] that one can give (theoretically, at least) explicit formulas in which the A's are expressed as polynomials with integer coefficients in the terms of the polynomial product and are therefore algebraic integers whenever the terms of the polynomial product are algebraic integers. Stated in this way, the theorem becomes a very concrete theorem of "general arithmetic":

Theorem. *Let a_0, a_1, ..., a_m, b_0, b_1, ..., b_n be indeterminates and let $R = \mathbf{Z}[a_0, a_1, \ldots, a_m, b_0, b_1, \ldots, b_n]$ be the ring of polynomials in these indeterminates with integer coefficients. Let c_0, c_1, ..., c_{m+n} be the elements of R defined by*

$$c_i = \sum_{j+k=i} a_j b_k.$$

For each product $a_j b_k$ of one a and one b (where $0 \le j \le m$ and $0 \le k \le n$), construct a relation of the form $F(a_j b_k) = 0$ in which $F(X) = X^N + p_1 X^{N-1} + p_2 X^{N-2} + \cdots + p_N$ is a monic polynomial whose coefficients p_1, p_2, ..., p_N are elements of R that are polynomials in c_0, c_1, ..., c_{m+n} with integer coefficients.

Proof. Let f, g, and h be the polynomials with coefficients in R defined by $f(x) = a_0 x^m + a_1 x^{m-1} + \cdots + a_m$, $g(x) = b_0 x^n + b_1 x^{n-1} + \cdots + b_n$, $h(x) = f(x)g(x)$. Construct a splitting field for $h(x)$, and let $\xi_1, \xi_2, \ldots, \xi_{m+n}$ be the negatives of the roots of $h(x)$ in this field. Then $h(x) = c_0 x^{m+n} + c_1 x^{m+n-1} + \cdots + c_{m+n} = c_0 \prod (x + \xi_i)$, so $c_i = c_0 \Sigma_i$, where Σ_i is the ith elementary symmetric polynomial in $\xi_1, \xi_2, \ldots, \xi_{m+n}$. The equation $h(x) = f(x)g(x)$ partitions the factors $x + \xi_i$ of $h(x)$ into two subsets, say $f(x) = a_0(x+\xi_1)(x+\xi_2)\cdots(x+\xi_m)$ and $g(x) = b_0(x+\xi_{m+1})(x+\xi_{m+2})\cdots(x+\xi_{m+n})$. Then $a_j b_k = a_0 b_0 \sigma_j \tau_k = c_0 \sigma_j \tau_k$, where σ_j is the jth elementary symmetric polynomial in $\xi_1, \xi_2, \ldots, \xi_m$ and τ_k is the kth elementary symmetric polynomial in ξ_{m+1}, $\xi_{m+2}, \ldots, \xi_{m+n}$. Let $F(X) = \prod_S (X - c_0 S(\sigma_j \tau_k))$, where S runs over all

$(m+n)!$ permutations S of $\xi_1, \xi_2, \ldots, \xi_{m+n}$. Since $a_j b_k$ is a root of the monic polynomial $F(X)$ (it is the root of the factor in which S is the identity), the theorem will be proved if the coefficients of F are shown to be polynomials in $c_0, c_1, \ldots, c_{m+n}$ with integer coefficients.

The coefficient of $X^{(m+n)!-\rho}$ in F is $(-c_0)^\rho$ times the sum of all products of ρ distinct conjugates of $\sigma_j \tau_k$ under permutation of $\xi_1, \xi_2, \ldots, \xi_{m+n}$. The degree of any conjugate of $\sigma_j \tau_k$ in ξ_1 is 1, so this coefficient is c_0^ρ times a symmetric polynomial in $\xi_1, \xi_2, \ldots, \xi_{m+n}$ whose degree in ξ_1—and therefore in any one ξ_i—is at most ρ. Because the polynomial is symmetric, this coefficient of F has the form $c_0^\rho C(\Sigma_1, \Sigma_2, \ldots, \Sigma_{m+n})$, where C is a polynomial with integer coefficients, by Corollary 2 of Essay 2.4. Let each term $\gamma \Sigma_1^{d_1} \Sigma_2^{d_2} \cdots \Sigma_{m+n}^{d_{m+n}}$ of C, where γ is the integer coefficient of the term, be expressed as a polynomial in $\xi_1, \xi_2, \ldots, \xi_{m+n}$, and let the terms be arranged in lexicographic order (terms with highest degree in ξ_1 come first, among terms with the same degree in ξ_1 the terms with highest degree in ξ_2 come first, and so forth). The leading term of the result is $\gamma \xi_1^{e_1} \xi_2^{e_2} \cdots \xi_{m+n}^{e_{m+n}}$, where $e_1 = d_1 + d_2 + \cdots + d_{m+n}$, $e_2 = d_2 + d_3 + \cdots + d_{m+n}$, $\ldots$, $e_{m+n-1} = d_{m+n-1} + d_{m+n}$, $e_{m+n} = d_{m+n}$, because the leading term of a product is the product of the leading terms and the leading term of Σ_k is $\xi_1 \xi_2 \cdots \xi_k$. No two of these leading terms contain the same exponents (the e's determine the d's as differences of successive e's), so the leading term of C, when it is expressed in terms of $\xi_1, \xi_2, \ldots, \xi_{m+n}$ and written in lexicographic order, is one of these leading terms $\gamma \xi_1^{e_1} \xi_2^{e_2} \cdots \xi_{m+n}^{e_{m+n}}$. In particular, its degree in ξ_1 is $d_1 + d_2 + \cdots + d_n$ for some term, so its degree in ξ_1 is the total degree of C in the Σ's. Therefore, $c_0^\rho C(\Sigma_1, \Sigma_2, \ldots, \Sigma_{m+n}) = c_0^\rho C(c_1/c_0, c_2/c_0, \ldots, c_{m+n}/c_0)$ can be expressed as a polynomial in $c_0, c_1, \ldots, c_{m+n}$ with integer coefficients, as was to be shown.

See Part 0 of [19] and [23, Nr. 20] for fuller accounts and other references.

3

Some Quadratic Problems

Essay 3.1 The Problem $A\square + B = \square$ and "Hypernumbers"

The problem that motivates the study of 'hypernumbers' in the next few essays comes from the prehistory of mathematics.

In "The Measurement of the Circle," Archimedes states that $\frac{265}{153} < \sqrt{3}$ and $\frac{1351}{780} > \sqrt{3}$ without giving any derivation. The closeness of these approximations becomes clear when one compares $265^2 = 70225$ to $3 \cdot 153^2 = 70227$ in the case of the first and $1351^2 = 1825201$ to $3 \cdot 780^2 = 1825200$ in the case of the second to find that $265^2 + 2 = 3 \cdot 153^2$ and $1351^2 = 3 \cdot 780^2 + 1$. There have been many attempts to guess how Archimedes might have derived these estimates. One can be certain that they were not found by trial and error; very probably, they involve some analogue of what is today called the continued fractions algorithm, but there is no documentary evidence on which to base such speculations.

A similar problem is treated in earlier Greek mathematics. As early as the time of Pythagoras, Greek mathematicians are said to have derived[*] an entire sequence of approximations to $\sqrt{2}$ in the form of "side and diagonal" numbers. If d is the length of the diagonal of a square and s is the length of its side, then $d^2 = 2s^2$ by the Pythagorean theorem. The followers of Pythagoras are thought to have discovered that there are no whole-number solutions (d, s) of this equation—and to have been very dismayed to learn that numbers, in the simplest sense, are not sufficient for the description of this simple geometrical construction. But their study of the problem probably went well beyond the impossibility of $d^2 = 2s^2$ in whole numbers to the following sequence of *approximate* solutions $d_n^2 = 2s_n^2 \pm 1$. A solution (d_n, s_n) of $d_n^2 = 2s_n^2 \pm 1$ implies a solution (d_{n+1}, s_{n+1}) of $d_{n+1}^2 = 2s_{n+1}^2 \mp 1$ via the formulas $s_{n+1} = d_n + s_n$ and $d_{n+1} = s_n + s_{n+1}$, as is easy for us, with our simple

---

[*] See Dickson [15, vol. 2, p. 341], where the reference is to a work of Proclus.

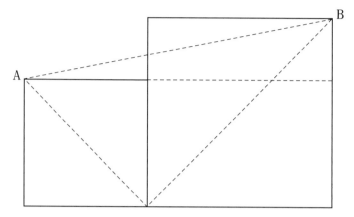

Fig. 3.1. The line segment AB is the hypotenuse of two different right triangles.

algebraic notation, to verify.[†] Let $(d_0, s_0) = (1, 1)$. Then $d_0^2 = 2s_0^2 - 1$, and the formula generates the sequence $(1, 1)$, $(3, 2)$, $(7, 5)$, $(17, 12)$, $(41, 29)$, $(99, 70)$, $(239, 169)$, $(577, 408)$, ..., which alternates between solutions of $d^2 + 1 = 2s^2$ and solutions of $d^2 = 2s^2 + 1$; that is, $1^2 + 1 = 2 \cdot 1^2$, $3^2 = 2 \cdot 2^2 + 1$, $7^2 + 1 = 2 \cdot 5^2$, and so forth. In this way one derives, for example, $577^2 = 2 \cdot 408^2 + 1$, which shows that $\frac{577}{408}$ is a very good approximation to $\sqrt{2}$, because $(\frac{577}{408})^2 = 2 + \frac{1}{408^2}$, but that it is a bit too large.

Plato's reference in *Theaetetus* to the study of the irrationality of square roots up to $\sqrt{17}$ probably indicates that mathematicians of his time studied rational approximations of these other square roots as well. There are other indications that techniques of finding approximate solutions of $y^2 = Ax^2$ were studied by ancient mathematicians whose works are lost. In India, Brahmagupta, in the 7th century, stated the formula that will be studied later in

---

[†] The statement that $d_n^2 = 2s_n^2 \pm 1$ implies $d_{n+1}^2 = 2s_{n+1}^2 \mp 1$ can be deduced from the formula $(2s + d)^2 + d^2 = 2s^2 + 2(s + d)^2$, which is Proposition 10 of Book 2 of Euclid's *Elements* [25], except that Euclid had no algebraic symbolism and expressed the equation in words rather than a formula. This proposition can be proved algebraically by noting that both sides are equal to $4s^2 + 4sd + 2d^2$ or can be deduced from a diagram consisting of two squares, one with side s and one with side $s + d$, resting side by side on a horizontal line with the smaller square on the left. Let A be the upper left vertex of the small square and let B be the upper right vertex of the large square. The desired identity comes from two ways of describing the square on the line segment AB. On the one hand, it is $2s^2 + 2(s + d)^2$, because AB is the hypotenuse of the right triangle whose third vertex is diagonally opposite the vertices A and B in their respective squares. On the other hand, it is $(s + s + d)^2 + d^2$, because AB is the hypotenuse of the right triangle whose third vertex is the point where the extension of the top side of the small square intersects the right side of the large square. Thus, $d^2 \geq 2s^2$ implies $(2s + d)^2 \leq 2(s + d)^2$ and $2(s + d)^2 - (2s + d)^2 = d^2 - 2s^2$; in the same way, $d^2 \leq 2s^2$ implies $(2s + d)^2 - 2(s + d)^2 = 2s^2 - d^2$.

this essay, and Bhascára Achárya, in the 12th century, mentioned the spectacular fact that the smallest number x for which $61x^2 + 1$ is a square is $x = 226153980$. Problems similar to this one are connected with the famous "cattle problem" of Archimedes [49], causing scholars to believe that Archimedes knew far more about such number-theoretic problems than our usual view of Greek mathematics as being primarily geometrical would lead us to expect.

These problems will be studied in this group of essays in the form of the problem that will be indicated by the symbolic equation $A\square + B = \square$, which is to say the problem "Given numbers A and B, find numbers x for which $Ax^2 + B$ is a square," say $Ax^2 + B = y^2$. (At first glance, Archimedes' solution $265^2 + 2 = 3 \cdot 153^2$ does not appear to be an instance of this problem, because $A = 3$ and $B = 2$ are on opposite sides of the equation, but if the equation is multiplied by 3, it becomes $3 \cdot 265^2 + 6 = (3 \cdot 153)^2$, which is a solution of $3\square + 6 = \square$. Conversely, in any solution (x, y) of $3x^2 + 6 = y^2$, y must be divisible by 3, say $z = y/3$, and division of $3x^2 + 6 = 9z^2$ by 3 gives a solution of $x^2 + 2 = 3z^2$.) Because the solution of this problem is easy when A is a square,* the case in which A is a square will be ignored.

Brahmagupta stated (but in words, not as an algebraic formula) the crucial tool that is used in the solution of $A\square + B = \square$. It is the observation[†] that a solution of $A\square + B = \square$ can be combined with a solution of $A\square + C = \square$ to find a solution of $A\square + BC = \square$. Specifically, if $Ax^2 + B = y^2$ and $Au^2 + C = v^2$, then $A(xv + yu)^2 + BC = (Axu + yv)^2$. It seems likely that some version of this remarkable fact was known was known in Greek times and that it was involved in the calculation of approximations to square roots. Archimedes' approximations to $\sqrt{3}$ can be derived using Brahmagupta's formula in the following way: Combine the simple equation $3 \cdot 1^2 + 1 = 2^2$ with itself to obtain $3 \cdot 4^2 + 1 = 7^2$, then combine this new equation with $3 \cdot 1^2 + 1 = 2^2$ to obtain $3 \cdot 15^2 + 1 = 26^2$, and so forth, to obtain the infinite sequence

$$3 \cdot 1^2 + 1 = 2^2,$$
$$3 \cdot 4^2 + 1 = 7^2,$$

---

* When A is a square, the problem is to write the given B in the form $s^2 - t^2 = (s - t)(s + t)$, where t is a multiple of the square root of A. Thus $s - t = B_1$ and $s + t = B_2$, where $B = B_1 B_2$ is one of the finite set of factorizations of B in which $B_1 \leq B_2$. Since $B_1 + B_2 = 2s$ is even, the problem is thus to find all factorizations $B_1 B_2 = B$, if any, in which $B_1 \leq B_2$, $B_1 \equiv B_2 \bmod 2$, and $(B_2 - B_1)/2$ is a multiple of the square root of A. For each of them, $\left(\frac{B_1 + B_2}{2}\right)^2 = \left(\frac{B_1 - B_2}{2}\right)^2 + B$ is a solution, and there are no others.

† See [10, p. 363]. The proof, using modern algebraic notation, is a simple calculation. How Brahmagupta might have proved it without algebraic notation—or how he might have known it is true—is a mystery. Certainly Euclid's Proposition 10 of Book 2 (see note above) indicates a Greek awareness of a similar phenomenon many centuries earlier, but there is no reason to suppose that the Greeks were the first.

$$3 \cdot 15^2 + 1 = 26^2,$$
$$3 \cdot 56^2 + 1 = 97^2,$$
$$3 \cdot 209^2 + 1 = 362^2,$$
$$3 \cdot 780^2 + 1 = 1351^2,$$

$$\cdots$$

of solutions of $3\square + 1 = \square$ that includes the one Archimedes used.

Each of these equations can be combined with $3 \cdot 1^2 + 6 = 3^2$ to obtain

$$3 \cdot 5^2 + 6 = (3 \cdot 3)^2,$$
$$3 \cdot 19^2 + 6 = (3 \cdot 11)^2,$$
$$3 \cdot 71^2 + 6 = (3 \cdot 41)^2,$$
$$3 \cdot 265^2 + 6 = (3 \cdot 153)^2,$$
$$3 \cdot 989^2 + 6 = (3 \cdot 571)^2,$$
$$3 \cdot 3691^2 + 6 = (3 \cdot 2131)^2,$$

$$\cdots$$

and the results divided by 3 to obtain an infinite sequence

$$5^2 + 2 = 3 \cdot 3^2,$$
$$19^2 + 2 = 3 \cdot 11^2,$$
$$71^2 + 2 = 3 \cdot 41^2,$$
$$265^2 + 2 = 3 \cdot 153^2,$$
$$989^2 + 2 = 3 \cdot 571^2,$$
$$3691^2 + 2 = 3 \cdot 2131^2,$$

$$\cdots$$

of solutions of $\square + 2 = 3\square$ that includes, of course, the other solution Archimedes used. (One naturally wonders whether $\square + 1 = 3\square$ is possible; it is not, because -1 is not a square mod 3.)

In modern terminology, Brahmagupta's formula has become the statement that *for expressions of the form* $y + x\sqrt{A}$, *the product of the norms is the norm of the product*. Here the **norm** of $y + x\sqrt{A}$ is by definition its product with its conjugate $y - x\sqrt{A}$, which is to say that it is $(y + x\sqrt{A})(y - x\sqrt{A}) = y^2 - Ax^2$. When one computes with these expressions using the normal rules of algebra—the *Buchstabenrechnung* of Essay 1.1—one finds that the conjugate of a product is the product of the conjugates, because

$$\left(y + x\sqrt{A}\right)\left(v + u\sqrt{A}\right) = (yv + Axu) + (yu + xv)\sqrt{A},$$

whereas

$$\left(y - x\sqrt{A}\right)\left(v - u\sqrt{A}\right) = (yv + Axu) - (yu + xv)\sqrt{A},$$

so the norm of a product can be computed in either of two ways: the product can be expanded $\left(y + x\sqrt{A}\right)\left(v + u\sqrt{A}\right) = (yv + Axu) + (yu + xv)\sqrt{A}$ to find that the norm of the product is $(yv + Axu)^2 - A(yu + xv)^2$, or the norm of the product can be computed by multiplying the product $\left(y + x\sqrt{A}\right)\left(v + u\sqrt{A}\right)$ by its conjugate $\left(y - x\sqrt{A}\right)\left(v - u\sqrt{A}\right)$. Thus

$$(yv + Axu)^2 - A(yu + xv)^2 = (y^2 - Ax^2)(v^2 - Au^2).$$

With $B = y^2 - Ax^2$ and $C = v^2 - Au^2$, this is Brahmagupta's formula rewritten as $(yv + Axu)^2 - A(yu + xv)^2 = BC$.

A solution of $A\square + B = \square$ is an expression $y + x\sqrt{A}$, in which x and y are numbers, whose norm is B. For want of a better term, I will call such an expression $y + x\sqrt{A}$ a **hypernumber** for A, so that the problem "find all solutions of $A\square + B = \square$" for given numbers A and B, with A not a square, becomes "find all hypernumbers for A whose norms are B."

More precisely, the hypernumbers for a given number A not a square can be described in the following way: As in Essay 1.1, a number is a term in the sequence $0, 1, 2, \ldots$. For a given number A not a square, a hypernumber is an expression $y + x\sqrt{A}$ in which x and y are numbers and $\sqrt{A}$ is a mere symbol. Hypernumbers for the same A are added in the obvious way,

$$\left(y_1 + x_2\sqrt{A}\right) + \left(y_2 + x_2\sqrt{A}\right) = (y_1 + y_2) + (x_1 + x_2)\sqrt{A},$$

and they are multiplied using the rule $\left(\sqrt{A}\right)^2 = A$ to obtain

$$\left(y_1 + x_2\sqrt{A}\right)\left(y_2 + x_2\sqrt{A}\right) = (y_1y_2 + Ax_1x_2) + (y_1x_2 + y_2x_1)\sqrt{A}.$$

Otherwise stated, the hypernumbers for A are $\mathbf{N}(X)$ mod $(X^2 - A)$, the set of all polynomials in X whose coefficients are numbers, when two such polynomials are considered to be equal if they are congruent mod $(X^2 - A)$. Every polynomial in X whose coefficients are numbers is congruent mod $(X^2 - A)$ to one and only one polynomial of degree less than 2 (replace X^2 with A, X^3 with AX, X^4 with A^2, and so forth), and the defining relation $X^2 = A$ justifies writing $\sqrt{A}$ in place of X. This definition of course implies the rules of addition and multiplication of hypernumbers just stated.

(The assumption that A is not a square guarantees that nonzero factors can be canceled in the arithmetic of hypernumbers $y + x\sqrt{A}$, because it guarantees that $X^2 - A$ is irreducible, from which it follows that $\mathbf{Z}[X]$ mod $(X^2 - A)$ is an integral domain; then for *integers* r, s, x, y, u, v the congruence $(s + rX)(y + xX) \equiv (s + rX)(v + uX)$ mod $(X^2 - A)$ implies $s + rX = 0$ or $y + xX \equiv v + uX$ mod $(X^2 - A)$ and therefore implies $s + rX = 0$ or $y + xX = v + uX$. On the other hand, if $A = r^2$ then, $(r + X)r \equiv (r + X)X$ mod $(X^2 - A)$ even though $r + X \neq 0$ and $r \not\equiv X$ mod $(X^2 - A)$.)

The exclusion of negative numbers is a bit inconvenient—the norm $y^2 - Ax^2$ of a hypernumber $y + x\sqrt{A}$ may not be a number in this strict sense because Ax^2 may be larger than y^2, and the conjugate $y - x\sqrt{A}$ of $y + x\sqrt{A}$ will be a hypernumber only when $x = 0$—but insistence on the narrow definition of "number" can be maintained with very little real difficulty and gives the theory a pleasing economy of structure. All of the results in the first five essays of this section, including the law of quadratic reciprocity in Essay 3.5, are deduced using only the arithmetic of numbers 0, 1, 2, ... in the narrowest sense.

Essay 3.2 Modules

The notion of a module of hypernumbers that is introduced in this essay is used in the next essay to solve $A\square + B = \square$ and in the following essays to deal with other questions in number theory. Very simply put, a module of hypernumbers for a given A is a list of hypernumbers for that A, written between square brackets to indicate that the list is to be used to define a congruence relation. The concept is motivated by the following reexamination of the Euclidean algorithm.

Gauss's notion of what it means to say that $a \equiv b \bmod m$—that is, two numbers a and b are congruent modulo a third number m—was generalized by Kronecker* as follows: Given a list of numbers $m_1, m_2, \ldots, m_\mu$, two numbers a and b are **congruent modulo** $[m_1, m_2, \ldots, m_\mu]$, written

$$a \equiv b \bmod [m_1, m_2, \ldots, m_\mu],$$

if there are numbers $i_1, i_2, \ldots, i_\mu$ and $j_1, j_2, \ldots, j_\mu$ such that $a + \sum_{\alpha=1}^{\mu} i_\alpha m_\alpha = b + \sum_{\alpha=1}^{\mu} j_\alpha m_\alpha$. A module is a (nonempty, finite) list of numbers $[m_1, m_2, \ldots, m_\mu]$ written between square brackets to indicate that they are to be used to define a congruence relation in this way. Two modules are **equal** if they define the same congruence relation.

Clearly, two modules are equal if the lists of numbers they contain can be obtained from one another by a sequence of steps in which (1) terms are rearranged, or (2) a zero is omitted from the list or annexed to it, or (3) a term is added to or subtracted from another term. (A subtraction assumes, of course, that the term being subtracted is less than or equal to the term from which it is being subtracted.) In the case of operations of types (1) or (2) the assertion is obvious. In the case of an operation of type (3), it follows from the observations that $a + i_1(m_1 + m_2) + i_2 m_2 + i_3 m_3 + \cdots = b + j_1(m_1 + m_2) + j_2 m_2 + j_3 m_3 + \cdots$ implies $a + i_1 m_1 + (i_1 + i_2)m_2 + i_3 m_3 + \cdots = b + j_1 m_1 + (j_1 + j_2)m_2 + j_3 m_3 + \ldots$ and, conversely, $a + i_1' m_1 + i_2' m_2 + i_3' m_3 + \cdots = b + j_1' m_1 + j_2' m_2 + j_3' m_3 + \cdots$ implies $a + i_1'(m_1 + m_2) + (j_1' + i_2')m_2 + i_3' m_3 + \cdots = b + j_1'(m_1 + m_2) + (i_1' + j_2')m_2 + j_3' m_3 + \cdots$ when $i_1' m_2 + j_1' m_2$ is added to both sides. These simple observations lead to a version of the Euclidean algorithm:

The Euclidean Algorithm.

Input: A list of numbers describing a module.

Algorithm: While the list contains more than one number
   If the first entry is zero, drop it from the list.
   If the first entry is greater than the second entry, interchange the first two entries.
   Otherwise, subtract the first entry from the second.
End

---

* See, for example, [44, p. 144]. Kronecker did not go to the extreme that I have of insisting that the multipliers all be natural numbers, so he did not need to put sums of multiples of m's on both sides of the equation.

Output: The list with one entry that remains.

For example, $[21, 15, 6] = [15, 21, 6] = [15, 6, 6] = [6, 15, 6] = [6, 9, 6] = [6, 3, 6] = [3, 6, 6] = [3, 3, 6] = [3, 0, 6] = [0, 3, 6] = [3, 6] = [3, 3] = [3, 0] = [0, 3] = [3]$.

Each step results in a new module equal to the preceding one; it either reduces the length of the list (the first alternative holds), or it reduces the sum of the entries (the third alternative), or it is followed by a step in which one of these two types of reduction occurs (the second alternative). Therefore, the algorithm eventually terminates, and reduces the module to a very simple form:

Theorem 1. *Given any module* $[m_1, m_2, \ldots, m_\mu]$, *there is a number n for which* $[m_1, m_2, \ldots, m_\mu] = [n]$.

This theorem gives a canonical form for modules, because $[n_1] = [n_2]$ only if $n_1 = n_2$. (If $[n_1] = [n_2]$, then each of the numbers n_1 and n_2 is a multiple of the other, which implies $n_1 = n_2$, because it implies that if one is zero, then both are, and otherwise, each is less than or equal to the other.) One can determine whether two given modules are equal by putting them both in canonical form; they are equal if and only if the canonical forms are identical. The number n is obviously the greatest common divisor of $m_1, m_2, \ldots, m_\mu$, except when the numbers m_i are all zero, in which case there is no greatest common divisor because all numbers are common divisors.

Corollary. *If two lists determine the same module, they can be transformed into one another by a sequence of steps of types (1), (2), and (3) described above.*

Deduction. One can pass from either of them to their common canonical form and back by a sequence of such steps, so one can pass from either of them to the other.

The "Euclidean algorithm" of Essay 1.4 shows that if $[m_1, m_2] = [n]$, there are *integers* ϕ and ψ for which $\phi m_1 + \psi m_2 = n$. Without using integers, this fact can be stated and generalized as follows:

Proposition 1. *If* $[m_1, m_2, \ldots, m_\mu] = [n]$ *and* $m_\mu \neq 0$, *then there are numbers* $k_1, k_2, \ldots, k_\mu$ *for which* $k_1 m_1 + k_2 m_2 + \cdots + k_{\mu-1} m_{\mu-1} + n = k_\mu m_\mu$.

Proof. Because $n \equiv 0 \bmod [n]$ and therefore $n \equiv 0 \bmod [m_1, m_2, \ldots, m_\mu]$, there are numbers $i_1, i_2, \ldots, i_\mu$ and $j_1, j_2, \ldots, j_\mu$ such that $n + \sum i_\alpha m_\alpha = \sum j_\alpha m_\alpha$. What is to be shown is that there is an equation of this form in which $i_\alpha \geq j_\alpha$ for $\alpha = 1, 2, \ldots, \mu - 1$. If $i_\alpha < j_\alpha$ for some α, one can add $m_\alpha m_\mu$ to both sides by adding m_μ to i_α and m_α to j_μ, increasing i_α without changing any other i and without changing any j other than j_μ. Repetition of this step enough times makes $i_\alpha \geq j_\alpha$ without changing the relation between i_β and j_β for any $\beta < \mu$ other than α. Since this can be done for each $\alpha < \mu$, the desired conclusion follows.

Two numbers m_1 and m_2 are **relatively prime** if $[m_1, m_2] = [1]$. Proposition 1 implies that if m_1 and m_2 are relatively prime and nonzero then each is invertible mod the other. It also implies an important theorem of elementary number theory:

The Chinese remainder theorem. *If $f > 0$ and $F > 0$ are relatively prime and g and G are given numbers, the congruences $x \equiv g \bmod f$ and $x \equiv G \bmod F$ determine a number $x \bmod fF$ in the sense that there is a solution x of the congruences and any two solutions are congruent mod fF.*

Proof. By the proposition, there are numbers k_1, k_2, l_1, and l_2 for which $k_1 f + 1 = k_2 F$ and $l_1 F + 1 = l_2 f$. Then $x = g \cdot k_2 F + G \cdot l_2 f$ satisfies $x \equiv g \cdot k_2 F = g \cdot (k_1 f + 1) \equiv g \bmod f$ and $x \equiv G \cdot l_2 f = G \cdot (l_1 F + 1) \equiv G \bmod F$, as required. The uniqueness of $x \bmod fF$ follows from simple counting: Since one of the fF numbers x less than fF solves each of the fF possible problems $x \equiv g \bmod f$ and $x \equiv G \bmod F$ in which $g < f$ and $G < F$, no two of them solve the same problem.

There is a natural way to multiply modules: The product of $[m_1, m_2, \ldots, m_\mu]$ and $[n_1, n_2, \ldots, n_\nu]$ is by definition the module described by a list $[\ldots, m_\alpha n_\beta, \ldots]$ made up of all products of one m and one n, arranged in some order. Multiplication is well defined for *modules* in the sense that if one list is replaced by another list describing the same module, then the product list may change, but the module it describes will not. (This statement is clear if the passage from one factor to an equal factor involves rearranging the list or omitting or annexing zeros. If the passage involves adding a term to or subtracting a term from another, it is only slightly less obvious.) Multiplication of modules is obviously commutative and associative.

All of the same ideas apply without change to modules of hypernumbers, except that there is no Euclidean algorithm in the case of hypernumbers, and the problem of establishing a canonical form for modules of hypernumbers is more challenging.

Let a number A, not a square, be fixed throughout the discussion. A **module of hypernumbers** is simply a list $[m_1, m_2, \ldots, m_\mu]$ of hypernumbers (for the given A) enclosed in square brackets. Two hypernumbers a and b are **congruent** modulo a given module, written $a \equiv b \bmod [m_1, m_2, \ldots, m_\mu]$, if there are hypernumbers $i_1, i_2, \ldots, i_\mu$ and $j_1, j_2, \ldots, j_\mu$ such that $a + \sum i_\alpha m_\alpha = b + \sum j_\alpha m_\alpha$. Two modules are **equal** if they determine the same congruence relation.

Again it is easy to see that two modules are equal if the lists of numbers they contain can be obtained from one another by a sequence of steps in which (1) terms are rearranged, or (2) a zero is omitted from the list or annexed to it, or (3) a term is added to or subtracted from another term. (A subtraction is of course possible only when the coefficients of the term being subtracted are no larger than the corresponding coefficients of the term from which it is being subtracted.) In the hypernumber case, there is another elementary operation that does not change the module, namely, (4) $\sqrt{A}$ times a term is added to or subtracted from another term.

This set of four types of transformations that change a module into an equal module are sufficient to establish a method for determining whether two given modules are equal:

Theorem 2. *Let A be a fixed number, not a square. Every module of hypernumbers for A that is not* equal to $[0]$ is equal to a module of the form $[ef, eg + e\sqrt{A}]$, where e, f, and g are numbers for which $ef \neq 0$, $g < f$, and $g^2 \equiv A$ mod f. Two modules of this form are equal only if they are identical.*

A module $[ef, eg + e\sqrt{A}]$ in which $ef \neq 0$, $g < f$, and $g^2 \equiv A$ mod f will be said to be in **canonical form**.

Proof. The following elaboration of the Euclidean algorithm puts any module that is not equal to $[0]$ in canonical form after a finite number of steps.

By assumption, the list that presents the given module contains at least one nonzero entry, call it $y + x\sqrt{A}$. Because the number $|y^2 - Ax^2|$ can be annexed to the list, one can assume without loss of generality that the list that presents the given module contains a nonzero number. (The annexed number is $y(y+x\sqrt{A}) - x\sqrt{A}(y+x\sqrt{A})$ if $y^2 > Ax^2$ and $x\sqrt{A}(y+x\sqrt{A}) - y(y+x\sqrt{A})$ if $y^2 < Ax^2$. Therefore, annexing it to the list does not change the module. It cannot be zero because A is not a square, so Ax^2 cannot be a square[†] unless $x = 0$.)

Therefore, provided the given module is not $[0]$, one can assume without loss of generality that the list representing the given module has a nonzero *number* as its first entry. Moreover, because the first entry times $\sqrt{A}$ can be annexed to the list if necessary, one can also assume without loss of generality that the list contains at least one hypernumber that is not a number.

Reduction to Canonical Form

Input: A presentation of a module in which the first entry is a nonzero number, and at least one entry is a hypernumber that is not a number.

---

* The module $[0]$ is equal only to modules of the form $[0, 0, \ldots, 0]$. It is a very trivial sort of module—congruence mod it is simply equality—which for the most part will be ignored. A module in canonical form is not $[0]$.

† If $y^2 = Ax^2$ and $x \neq 0$, then A must be a square, as can be seen as follows: Let $[x, y] = [d] \neq [0]$. If $y = 0$, then $A = 0$ is a square. Otherwise, by Proposition 1, there are numbers α and β for which $\alpha x + d = \beta y$, from which it follows that $[xd] = [xd, Ax^2] = [xd, y^2] = [xd, y^2, \beta^2 y^2] = [xd, y^2, (\alpha x + d)^2] = [xd, y^2, d^2] = [d^2]$. Thus, $xd = d^2$, $x = d$, $y^2 = Ad^2$, and $(y/d)^2 = A$. Or if one is willing to take the unique factorization of numbers as known, one can simply observe that some prime factor of A divides A an odd number of times, and therefore divides Ax^2 an odd number of times, so Ax^2 is not a square.

Algorithm: While the module is not in canonical form

If any number in the list is preceded by an entry that is not a number, interchange the two.

Otherwise, if the second entry is a number, use the Euclidean algorithm to replace the first two entries with their greatest common divisor.

Otherwise, if there is a third term (in which case the first term is a number and the second and third terms are not numbers), make use of the first term to perform the Euclidean algorithm on the coefficients of $\sqrt{A}$ in the second and third terms. Specifically, if the coefficient of $\sqrt{A}$ in the second term is less than or equal to the coefficient of $\sqrt{A}$ in the third term, add the first term to the third term as many times as necessary and then subtract the second term from the third; otherwise, interchange the second and third terms.

Otherwise (in which case there are just two entries, the first a number and the second not), if the coefficient of $\sqrt{A}$ in the second entry does not divide the other coefficient of the second entry, annex $\sqrt{A}$ times the second entry to the list as a third entry.

Otherwise, if the coefficient of $\sqrt{A}$ in the second entry does not divide the first entry, annex $\sqrt{A}$ times the first entry to the list as a third entry.

Otherwise (in which case the module has the form $[ef, eg + e\sqrt{A}]$ but is not in canonical form), subtract the first term from the second if possible.

Otherwise, annex the difference of the numbers eA and eg^2, which is the difference of the hypernumbers $\sqrt{A}(eg + e\sqrt{A})$ and $g(eg + e\sqrt{A})$, to the list as a third entry.

End

Output: The module in canonical form with which the algorithm terminates.

Example. To apply the algorithm to $[7+5\sqrt{3}]$ one must first annex $|7^2 - 3 \cdot 5^2| = 26$. The succeeding steps are

$$\begin{aligned}
[26, 7 + 5\sqrt{3}] &= [26, 7 + 5\sqrt{3}, 15 + 7\sqrt{3}] = [26, 7 + 5\sqrt{3}, 8 + 2\sqrt{3}] \\
&= [26, 8 + 2\sqrt{3}, 7 + 5\sqrt{3}] = [26, 8 + 2\sqrt{3}, 25 + 3\sqrt{3}] \\
&= [26, 8 + 2\sqrt{3}, 17 + \sqrt{3}] = [26, 17 + \sqrt{3}, 8 + 2\sqrt{3}] \\
&= [26, 17 + \sqrt{3}, 17 + \sqrt{3}] = [26, 17 + \sqrt{3}, 0] \\
&= [26, 0, 17 + \sqrt{3}] = [26, 17 + \sqrt{3}].
\end{aligned}$$

The final module is in canonical form because 1 divides both 17 and 26, 17 is less than 26, and $17^2 \equiv 3 \bmod 26$.

This algorithm terminates for the following reasons:

By assumption, the input list contains both a number and a hypernumber that is not a number. A step that changes a hypernumber into a number leaves a hypernumber in the list unchanged, and the only step that reduces the number of numbers in the list replaces two numbers with a single one (the greatest common divisor of the first two entries). Therefore, at each step there is at least one number and at least one hypernumber not a number. They are arranged by the first step to put all numbers first, and the second step eventually reduces the number of numbers in the list to one.

Steps of the first three types do not change the greatest common divisor of the coefficients of $\sqrt{A}$ that occur in entries of the list. Since they reduce the total of the numbers in the list or the total of the coefficients of $\sqrt{A}$ (except for the finitely many steps that rearrange terms), eventually a step beyond the first three must be reached.

Each such step reduces either the greatest common divisor of the coefficients of $\sqrt{A}$ or the greatest common divisor of the numbers in the list (except for finitely many steps that reduce the first coefficient of the second term), so only a finite number of them can occur before canonical form is achieved.

Lemma. *Let* $[ef, eg + e\sqrt{A}]$ *be a module in canonical form. The congruence* $y + x\sqrt{A} \equiv 0 \bmod [ef, eg + e\sqrt{A}]$ *is equivalent to the pair of congruences* $y \equiv gx \bmod ef$ *and* $x \equiv 0 \bmod e$.

Proof. Because

$$ef\sqrt{A} + efg = f\left(eg + e\sqrt{A}\right)$$

and

$$\sqrt{A}\left(eg + e\sqrt{A}\right) \equiv g\left(eg + e\sqrt{A}\right) \bmod ef,$$

an equation of the form

$$y + x\sqrt{A} + i_1 ef + i_2\left(eg + e\sqrt{A}\right) = y' + x'\sqrt{A} + j_1 ef + j_2\left(eg + e\sqrt{A}\right)$$

in which i_1, i_2, j_1, j_2 are *hypernumbers* implies another equation of the same form in which i_1, i_2, j_1, j_2 are *numbers*. (For example, if $i_1 = \alpha + \beta\sqrt{A}$, one can add βefg to both sides and replace $\beta\sqrt{A}ef + \beta efg$ with $\beta f(eg + e\sqrt{A})$ to obtain another equation of the same form in which i_1 is α, i_2 is increased by βf, and j_1 is increased by βg. If $i_2 = \alpha + \beta\sqrt{A}$, one can use the fact that $\sqrt{A}(eg + e\sqrt{A}) + \gamma ef = g(eg + e\sqrt{A}) + \delta ef$ for suitable γ and δ to add $\beta\gamma ef$ to both sides and replace $\beta\sqrt{A}(eg + e\sqrt{A}) + \beta\gamma ef$ with $\beta g(eg + e\sqrt{A}) + \beta\delta ef$ to obtain another equation of the same form in which i_2 is $\alpha + \beta g$, while $\beta\delta$ is added to i_1 and $\beta\gamma$ is added to j_1, and so forth.) Therefore, $y + x\sqrt{A} \equiv 0 \bmod [ef, eg + e\sqrt{A}]$ if and only if $y + x\sqrt{A} + i_1 ef + i_2(eg + e\sqrt{A}) = j_1 ef + j_2(eg + e\sqrt{A})$ for some numbers i_1, i_2, j_1, j_2. Comparison of the coefficients of $\sqrt{A}$ shows not only that $x \equiv 0 \bmod e$ but also that $x + i_2 e = j_2 e$; then comparison of the other terms shows that $y + i_1 ef + i_2 eg = j_1 ef + j_2 eg = j_1 ef + (x + i_2 e)g$ and therefore that $y \equiv gx \bmod ef$. Conversely, if $x + ie = je$ and $y + kef = gx + lef$ for numbers i, j, k, l, then $y + x\sqrt{A} + kef + i(eg + e\sqrt{A}) = gx + lef + je\sqrt{A} + ieg = gje + lef + je\sqrt{A} = j(eg + e\sqrt{A}) + lef$, so $y + x\sqrt{A} \equiv 0 \bmod [ef, eg + e\sqrt{A}]$.

Completion of the Proof of Theorem 2. Thus, if $[ef, eg + e\sqrt{A}] = [e'f', e'g' + e'\sqrt{A}]$, then $e'g' + e'\sqrt{A} \equiv 0 \bmod [ef, eg + e\sqrt{A}]$, which implies in particular that $e' \equiv 0 \bmod e$. By symmetry, $e \equiv 0 \bmod e'$, so $e = e'$. Then $e'f' \equiv g \cdot 0 \bmod ef$ implies $f' \equiv 0 \bmod f$, and $f = f'$ follows as before by symmetry. Finally, $eg' + e\sqrt{A} \equiv 0 \bmod [ef, eg + e\sqrt{A}]$ implies $eg' \equiv ge \bmod ef$, which is to say $g' \equiv g \bmod f$. Since both g and g' are less than $f = f'$, $g = g'$ follows.

Corollary. *If two modules are equal, each can be transformed into the other by a sequence of steps of types (1)–(4).*

Deduction. Such steps suffice to transform a module into its canonical form and vice versa.

The **product** of two modules of hypernumbers can be defined, exactly as in the case of modules of numbers, to be the module described by the list containing all products in which one factor is from a list describing the first module and the other factor is from a list describing the second. Products are easily shown to be well defined for modules; that is, if a factor is replaced by an equal module, the new product is equal to the old one.

The product operation defined in this way is commutative and associative, which is to say that it makes the set of modules of hypernumbers for a given A into a **commutative semigroup**. In this semigroup, $[1]$ is an identity.

The "modules" described here are closely related to Dedekind's "ideals" in the ring $\mathbf{Z}[\sqrt{A}]$, but the underlying attitude is opposite to Dedekind's. His goal was to divorce the theory as much as possible from algorithmic techniques, and he felt that he had achieved his goal by considering *the infinite set of all ring elements that are zero mod* $[ef, eg+e\sqrt{A}]$ to be a mathematical entity. To me, it borders on the absurd to believe that a mathematical idea is made "concrete" [23, Remark 21, p. 60] by describing it as an infinite set whose elements are themselves abstractions. Modules of hypernumbers for a given A are made concrete by specifying how they are to be described (as finite, nonempty lists of hypernumbers between square brackets) and how to compute with them (they are multiplied by the familiar rule, and one determines whether two given modules are equal by reducing them both to canonical form).

Examples. When $A = 3$, some modules in canonical form are $[2, 1 + \sqrt{3}]$, $[3, \sqrt{3}]$, $[11, 5 + \sqrt{3}]$, $[11, 6 + \sqrt{3}]$. Some products of such modules are

$$[2, 1+\sqrt{3}][2, 1+\sqrt{3}] = [4, 2(1+\sqrt{3}), 4 + 2\sqrt{3}] = [2][2, 1+\sqrt{3}, 2+\sqrt{3}]$$
$$= [2][2, 1+\sqrt{3}, 1] = [2],$$
$$[11, 5+\sqrt{3}][11, 5+\sqrt{3}] = [11^2, 11(5+\sqrt{3}), 28 + 10\sqrt{3}]$$
$$= [121, 55 - 28 + (11-10)\sqrt{3}, 28 + 10\sqrt{3}]$$
$$= [121, 27 + \sqrt{3}, 28 + 10\sqrt{3} + 2 \cdot 121 - 10(27 + \sqrt{3})]$$

$$= \left[121, 27 + \sqrt{3}\right],$$

$$\left[11, 5 + \sqrt{3}\right]\left[11, 6 + \sqrt{3}\right] = \left[121, 11(5 + \sqrt{3}), 11(6 + \sqrt{3}), 33 + 11\sqrt{3}\right]$$

$$= [11]\left[11, 5 + \sqrt{3}, 6 + \sqrt{3}, 3 + \sqrt{3}\right]$$

$$= [11]\left[11, 2, 3, 3 + \sqrt{3}\right] = [11]\left[11, 2, 1, 3 + \sqrt{3}\right] = [11].$$

Essay 3.3 The Class Semigroup. Solution of $A\square + B = \square$.

... die schwierigste Frage ... nämlich die, ob zwei reducirte Formen dersel-
ben Determinante, welche verschiedenen Perioden angehören, äquivalent
sein können oder nicht. (...the most difficult question ..., namely, whether
two reduced forms with the same determinant that belong to different periods can be equivalent.)—P. G. Lejeune Dirichlet [16, §80]

Again let A be a fixed number, not a square. As was seen in the previous essay,
the modules of hypernumbers for A form a *commutative semigroup* under
multiplication, or, to put it more simply, the operation of multiplication of
modules is commutative and associative. Computations in the semigroup of
modules will be used in this essay to solve $A\square + B = \square$. A key role will be
played by the following notion of equivalence of modules.

A module will be called **principal*** if it can be expressed in the form
$[y + x\sqrt{A}]$ for some hypernumber $y + x\sqrt{A}$ that satisfies $y^2 > Ax^2$. The principal modules form a subsemigroup—in other words, a product of principal modules is principal—by virtue of Brahmagupta's formula $(y^2 - Ax^2)(v^2 - Au^2) =
(yv + Axu)^2 - A(yu + xv)^2$, because this formula shows that the product
$[(y + x\sqrt{A})(v + u\sqrt{A})] = [(yv + Axu) + (yu + xv)\sqrt{A}]$ of $[y + x\sqrt{A}]$ and
$[v + u\sqrt{A}]$ satisfies $(yv + Axu)^2 > A(yu + xv)^2$ when $y^2 > Ax^2$ and $v^2 > Au^2$.
Two modules M_1 and M_2 will be called **equivalent**, written $M_1 \sim M_2$, if
there are principal modules P_1 and P_2 for which $M_1 P_1 = M_2 P_2$. This is
an equivalence relation (transitivity follows from the fact that a product of
principal modules is principal, because $M_1 P_1 = M_2 P_2$ and $M_2 P_1' = M_3 P_2'$ imply $M_1 P_1 P_1' = M_2 P_2 P_1' = M_3 P_2 P_2'$) that is consistent with multiplication of
modules ($M_1 \sim M_2$ implies $M_1 M_3 \sim M_2 M_3$ for any module M_3). The **class
semigroup** is simply the set of equivalence classes, multiplied by multiplying
representatives. Otherwise stated, the class semigroup is the *quotient* semigroup of the semigroup of modules relative to the subsemigroup of principal
modules.

Computations in the class semigroup depend on solving the problem of
determining whether two given modules are equivalent; this problem, which I
will call the equivalence problem, is solved by the theorem of this essay. It is
the main step in the solution of $A\square + B = \square$.

The equivalence problem cannot be solved by giving a *canonical form* that
picks one representative out of each equivalence class, because there is no
natural canonical form for this particular equivalence relation. Instead, the
solution of the equivalence problem will follow a procedure like the one Gauss
used in Section 5 of *Disquisitiones Arithmeticae* to determine whether two
given binary quadratic forms are equivalent; it consists of two parts, the first
establishing that every module is equivalent to one in a certain finite set of

---

* This term derives from the fact that the module is in the *principal class* of the
class group. It has always seemed to me peculiar to apply the adjective "principal"
to the module itself, but the usage is universal among mathematicians.

Fig. 3.2. Gauss.

stable[†] modules, and the second giving a method of determining whether two *stable* modules are equivalent. Specifically, an algorithm—the "comparison algorithm"—will be given for generating a sequence of modules equivalent to a given one. A sequence of equivalent modules generated by the comparison algorithm eventually begins to cycle, as will be shown; a module will be called **stable** if the sequence of modules obtained by applying the comparison algorithm to it cycles back to this module itself. The equivalence problem will be solved by showing that the obvious sufficient condition for the equivalence of two modules—namely, that application of the comparison algorithm to them leads to the same cycle of stable modules—is also necessary. Thus, the answer to Dirichlet's "most difficult question" is no: Reduced forms in different periods are *not* equivalent, or, in the present formulation, stable modules in different cycles are not equivalent.

By the definition of equivalence, a module $[e][f, g + \sqrt{A}]$ in canonical form is equivalent to $[f, g + \sqrt{A}]$. Therefore, in solving the equivalence problem one can assume without loss of generality that the given modules in canonical form have $e = 1$.

[†] I have avoided Gauss's term "reduced" because it conflicts with my term "reduction algorithm," an algorithm for reducing a coefficient of $\sqrt{A}$, not for reducing the module.

Comparison Algorithm.

Input: A module $[f, g + \sqrt{A}]$ in canonical form with $e = 1$.

Algorithm: Let r be the smallest solution of $r + g \equiv 0 \bmod f$ for which $r^2 > A$.
   Let $f_1 = (r^2 - A)/f$.
   Let g_1 be the smallest solution of $g_1 \equiv r \bmod f_1$.

Output: A module $[f_1, g_1 + \sqrt{A}]$ in canonical form with $e = 1$ and an equation $[r + \sqrt{A}][f, g + \sqrt{A}] = [f][f_1, g_1 + \sqrt{A}]$ showing that it is equivalent to the input module.

   That the definition of f_1 makes sense—that is, that $r^2 \equiv A \bmod f$—follows from $r \equiv -g \bmod f$ and the fact that g is a square root of $A \bmod f$. That g_1 is a square root of $A \bmod f_1$ follows from $r \equiv g_1 \bmod f_1$ and $r^2 - A = f f_1$. Of course $g_1 < f_1$, because g_1 is the smallest number in its class mod f_1. Finally, when q is defined by $qf = r + g$ one obtains the output equation

$$
\begin{aligned}
[r + \sqrt{A}][f, g + \sqrt{A}] &= [f(r + \sqrt{A}), rg + A + (r + g)\sqrt{A}] \\
&= [f(r + \sqrt{A}), fq(r + \sqrt{A}), rg + r^2 - f f_1 + (r + g)\sqrt{A}] \\
&= [f(r + \sqrt{A}), fq(r + \sqrt{A}), rfq - f f_1 + fq\sqrt{A}] \\
&= [f(r + \sqrt{A}), f f_1, rfq - f f_1 + fq\sqrt{A}] \\
&= [f(r + \sqrt{A}), f f_1] = [f][f_1, r + \sqrt{A}] = [f][f_1, g_1 + \sqrt{A}].
\end{aligned}
$$

   Let the output $[f_1, g_1 + \sqrt{A}]$ of the comparison algorithm be called the **immediate successor** of the input $[f, g + \sqrt{A}]$, and let the **successors** of $[f, g + \sqrt{A}]$ be the modules in the sequence generated by repeated application of the comparison algorithm. Not only is each successor of $[f, g + \sqrt{A}]$ equivalent to $[f, g + \sqrt{A}]$, but the algorithm gives an explicit equivalence

$$
(1) \qquad \left[\prod_{i=1}^{k} (r_i + \sqrt{A})\right][f, g + \sqrt{A}] = \left[\prod_{i=0}^{k-1} f_i\right][f_k, g_k + \sqrt{A}],
$$

where $[f_i, g_i + \sqrt{A}]$ is the ith successor of $[f, g + \sqrt{A}] = [f_0, g_0 + \sqrt{A}]$ and where r_i is the value of r used by the comparison algorithm to go from $[f_{i-1}, g_{i-1} + \sqrt{A}]$ to $[f_i, g_i + \sqrt{A}]$.

Theorem. *Let $[f, g + \sqrt{A}]$ and $[F, G + \sqrt{A}]$ be modules in canonical form with $e = 1$, and let $[F, G + \sqrt{A}]$ be stable. Formula (1) describes all equivalences between $[f, g + \sqrt{A}]$ and $[F, G + \sqrt{A}]$ in the sense that any equivalence $[Y + X\sqrt{A}][f, g + \sqrt{A}] = [V + U\sqrt{A}][F, G + \sqrt{A}]$ in which $Y^2 > AX^2$ and $V^2 > AU^2$ must satisfy*

$$
\left(V + U\sqrt{A}\right)\prod_{i=1}^{k}(r_i + \sqrt{A}) = \left(Y + X\sqrt{A}\right)\prod_{i=0}^{k-1} f_i,
$$

where k is a number for which $[F, G + \sqrt{A}]$ is the kth successor of $[f, g + \sqrt{A}]$. In particular, there are no equivalences when $[F, G + \sqrt{A}]$ is not a successor of $[f, g + \sqrt{A}]$.

Equation (1) implies that both coefficients of the hypernumber $f \prod_{i=1}^{k}(r_i + \sqrt{A})$ are divisible by $\prod_{i=0}^{k-1} f_i$. Therefore both coefficients of $(r_1 + \sqrt{A})(r_2 + \sqrt{A}) \cdots (r_k + \sqrt{A})$ are divisible by $f_1 f_2 \cdots f_{k-1}$. Thus, (1) can be divided by $f_1 f_2 \cdots f_{k-1}$, which will normally be a very large number, to put it in the form

$$(2) \qquad [y + x\sqrt{A}][f, g + \sqrt{A}] = [f][f_k, g_k + \sqrt{A}]$$

$$\text{where} \quad y + x\sqrt{A} = \frac{\prod_{i=1}^{k}(r_i + \sqrt{A})}{\prod_{i=1}^{k-1} f_i}.$$

The theorem shows that the most general equivalence

$$[Y + X\sqrt{A}][f, g + \sqrt{A}] = [V + U\sqrt{A}][F, G + \sqrt{A}]$$

is obtained from (2) by multiplying by $[(V + U\sqrt{A}) \prod_{i=1}^{k-1} f_i]$ and dividing by $[\prod_{i=0}^{k-1} f_i]$.

Proposition 1. *The number of stable modules is finite, and every module has a stable successor.*

Proof. As above, let $[f_i, g_i + \sqrt{A}]$ be the ith successor of the given module, and let r_i be the number used in passing from $[f_{i-1}, g_{i-1} + \sqrt{A}]$ to $[f_i, g_i + \sqrt{A}]$. In other words, let r_i be the least solution of $r_i + g_{i-1} \equiv 0 \bmod f_{i-1}$ whose square is greater than A. It will be shown that if $|r_i - f_{i-1}|^2 > A$, then $|r_i - f_{i-1}|^2 > |r_{i+1} - f_i|^2$.

Note first that $|r_i - f_{i-1}|^2 < A$ if and only if $|r_i - f_i|^2 < A$, because both are equivalent to $f_{i-1} + f_i < 2r_i$, as one sees when one writes them as $r_i^2 + f_{i-1}^2 < 2r_i f_{i-1} + A$ and $r_i^2 + f_i^2 < 2r_i f_i + A$, respectively, subtracts A from both sides, and uses $r_i^2 - A = f_{i-1}f_i$ to obtain $f_{i-1}f_i + f_{i-1}^2 < 2r_i f_{i-1}$ and $f_{i-1}f_i + f_i^2 < 2r_i f_i$, respectively. In the same way, the three inequalities $|r_i - f_{i-1}|^2 > A$, $|r_i - f_i|^2 > A$, and $f_{i-1} + f_i > 2r_i$ all imply one another.

Also, on successive steps, the inequality $r_i + r_{i+1} \geq 2f_i$ holds, as can be seen as follows: Because $r_i + r_{i+1} \equiv g_i + r_{i+1} \equiv 0 \bmod f_i$, it will suffice to prove that $r_i + r_{i+1} > f_i$. This is true if $r_i > f_i$. It is also true if $r_i < f_{i-1}$, because then $r_i^2 > r_i^2 - A = f_i f_{i-1} > f_i r_i$, so $r_i > f_i$. Otherwise, $f_{i-1} \leq r_i \leq f_i$, in which case $(r_i - f_{i-1})^2 < A$ by the definition of r_i (A is not a square, so $(r_i - f_{i-1})^2 \neq A$), which implies $|r_i - f_i|^2 < A$, as was just seen. Thus, $(f_i - r_i)^2 < r_{i+1}^2$ and $f_i - r_i < r_{i+1}$ in this case as well.

Suppose now that $|r_i - f_{i-1}|^2 > A$. If $f_i \leq r_{i+1}$, the definition of r_{i+1} implies $|r_{i+1} - f_i|^2 < A$, so of course $|r_i - f_{i-1}|^2 > |r_{i+1} - f_i|^2$ in that case. Otherwise, $r_{i+1} < f_i$, in which case the inequality of the last paragraph implies that $r_i - f_i \geq f_i - r_{i+1} > 0$. On the other hand, the assumption $|r_i - f_{i-1}|^2 > A$

implies $f_{i-1} + f_i > 2r_i$, as was seen above. Therefore $f_{i-1} - r_i > r_i - f_i$, which combines with the previous inequality to give $f_{i-1} - r_i > f_i - r_{i+1} > 0$, from which the desired inequality $|r_i - f_{i-1}|^2 > |r_{i+1} - f_i|^2$ follows.

Therefore $|r_i - f_{i-1}|^2$ decreases as long as it is greater than A, so a step must be reached at which $|r_i - f_{i-1}|^2 < A$. That the same inequality holds on all subsequent steps—which is to say that $|r_i - f_{i-1}|^2 < A$ implies $|r_{i+1} - f_i|^2 < A$—can be proved as follows:

If $|r_i - f_{i-1}|^2 < A$, then $|r_i - f_i|^2 < A$, as was seen. If $|r_{i+1} - f_i|^2$ were greater than A, then f_i would be greater than r_{i+1} (r_{i+1} is the least number in its class mod f_i whose square is greater than A), in which case the above inequality $r_i + r_{i+1} \geq 2f_i$ would imply $r_i - f_i \geq f_i - r_{i+1} > 0$, from which $|r_i - f_i|^2 \geq |f_i - r_{i+1}|^2 > A$ would follow. Therefore, $|f_i - r_{i+1}|^2$ must be less than A.

Thus, the sequence of successors of any module eventually reaches a module $[f, g + \sqrt{A}]$ in canonical form for which $|r - f|^2 < A$, where r is the least solution of $r + g \equiv 0 \bmod f$ for which $r^2 > A$. Let $\mathcal{M}$ denote the set of such modules. The set $\mathcal{M}$ is *finite*, as one sees when one sets $\phi = |r - f|$ and notes that then $\phi^2 < A$ and $\phi \equiv \pm r \equiv \pm g \bmod f$, so f divides $A - \phi^2$. In particular, $f \leq A$. Since canonical form requires that g be less than f, $\mathcal{M}$ is therefore finite.

The comparison algorithm defines a function from $\mathcal{M}$ to itself, as was shown above. Since it carries $[f, g + \sqrt{A}]$ in $\mathcal{M}$ to a module $[f_1, g_1 + \sqrt{A}]$ for which $|r - f_1|^2 < A$, f_1 and g_1 determine r as the least number in the class of $g_1 \bmod f_1$ whose square is greater than A. (If $g_1^2 > A$, then $r = g_1$; otherwise, $r = g_1 + \mu f_1$ for $\mu > 0$.) Therefore, $[f_1, g_1 + \sqrt{A}]$ determines r and determines $[f, g + \sqrt{A}]$ by the rules $f = (r^2 - A)/f_1$ and $g \equiv -r \bmod f$. In short, the function from $\mathcal{M}$ to itself defined by the comparison algorithm is one-to-one.

Therefore, the comparison algorithm permutes the finite set $\mathcal{M}$, which implies that every module in $\mathcal{M}$ is *stable*—application of the comparison algorithm to it cycles back to this module itself—and the proof of the proposition is complete. Moreover, it has been shown that the stable modules are precisely those in $\mathcal{M}$. (It is not difficult to show that these are the modules $[f, g + \sqrt{A}]$ in canonical form in which f divides a number of the form $A - \phi^2$ and the square root g of $A \bmod f$ satisfies either $g^2 < A$ or $(f - g)^2 < A$.)

See the table at the end of the essay for a list of the stable modules for a few values of A and the cycles into which they are partitioned by the comparison algorithm.

The first step in finding all equivalences between $[f, g + \sqrt{A}]$ and $[F, G + \sqrt{A}]$, where $[F, G + \sqrt{A}]$ is stable, will be to find all equivalences of the special form $[y + x\sqrt{A}][f, g + \sqrt{A}] = [n][F, G + \sqrt{A}]$ in which $y + x\sqrt{A} \equiv 0 \bmod [F, G + \sqrt{A}]$. The solution of this problem will use the following algorithm:

Reduction Algorithm.

Input: An equation $[y + x\sqrt{A}][f, g + \sqrt{A}] = [n][F, G + \sqrt{A}]$ in which $x > 0$, $y^2 > Ax^2$, the modules $[f, g + \sqrt{A}]$ and $[F, G + \sqrt{A}]$ are in canonical form, and $y + x\sqrt{A} \equiv 0 \bmod [F, G + \sqrt{A}]$. ($[F, G + \sqrt{A}]$ need not be stable.)

Algorithm: Determine ρ as the least number congruent to G mod F for which $y \leq \rho x$.

Define $y_1 + x_1\sqrt{A}$ to be $(\rho - \sqrt{A})(y + x\sqrt{A})/F$.

Define F_1 to be $(\rho^2 - A)/F$ and G_1 to be the least solution of $\rho + G_1 \equiv 0$ mod F_1.

Output: A new equation $[y_1 + x_1\sqrt{A}][f, g + \sqrt{A}] = [n][F_1, G_1 + \sqrt{A}]$, with $x_1 < x$, which can be used as a new input equation—that is, $[F_1, G_1 + \sqrt{A}]$ is in canonical form, $y_1^2 > Ax_1^2$, and $y_1 + x_1\sqrt{A} \equiv 0$ mod $[F_1, G_1 + \sqrt{A}]$—unless $x_1 = 0$.

Justification. By the choice of ρ, $\rho x \geq y$ and $\rho \equiv G$ mod F, so the definition of x_1 as $(\rho x - y)/F$ is valid by virtue of $y \equiv Gx \equiv \rho x$ mod F (because $y + x\sqrt{A} \equiv 0$ mod $[F, G + \sqrt{A}]$). Moreover, $Fx_1 = \rho x - y < Fx$ because $\rho x - y \geq Fx$ would imply $\rho \geq F$ and $(\rho - F)x \geq y$, contrary to the definition of ρ. Thus, $x_1 < x$. Since $\rho^2 x^2 \geq y^2 > Ax^2$ implies $\rho^2 > A$ (because $x > 0$), it follows that $(\rho y)^2 = \rho^2 \cdot y^2 > A \cdot Ax^2 = (Ax)^2$ and $\rho y > Ax$; at the same time, $\rho y \equiv Gy \equiv G^2 x \equiv Ax$ mod F, which shows that the definition of y_1 as $(\rho y - Ax)/F$ is valid. That $y_1^2 > Ax_1^2$ follows from $(\rho^2 - A)(y^2 - Ax^2) > 0$ when one rewrites this inequality first as $\rho^2 y^2 + A^2 x^2 > Ax^2\rho^2 + Ay^2$, then as $(\rho y - Ax)^2 > A(x\rho - y)^2$, and divides by F^2. Since $\rho^2 \equiv G^2 \equiv A$ mod F, the definition of F_1 as $(\rho^2 - A)/F$ is valid and $F_1 > 0$. Also, $G_1^2 \equiv (-\rho)^2 \equiv \rho^2 \equiv A$ mod F_1 by virtue of $\rho^2 - A = FF_1$, so $[F_1, G_1 + \sqrt{A}]$ is in canonical form. When q is defined by $\rho + G_1 = qF_1$, one deduces $[F_1][F, G + \sqrt{A}] = [F_1][F, \rho + \sqrt{A}] = [(\rho - \sqrt{A})(\rho + \sqrt{A}), F_1(\rho + \sqrt{A})] = [(qF_1 - G_1 - \sqrt{A})(\rho + \sqrt{A}), F_1(\rho + \sqrt{A})] = [(qF_1 - G_1 - \sqrt{A})(\rho + \sqrt{A}), F_1(\rho + \sqrt{A}), qF_1(\rho + \sqrt{A}) - (qF_1 - G_1 - \sqrt{A})(\rho + \sqrt{A})]$ (the third entry is q times the second minus the first) $= [(qF_1 - G_1 - \sqrt{A})(\rho + \sqrt{A}), F_1(\rho + \sqrt{A}), (G_1 + \sqrt{A})(\rho + \sqrt{A})] = [F_1(\rho + \sqrt{A}), (G_1 + \sqrt{A})(\rho + \sqrt{A})] = [\rho + \sqrt{A}][F_1, G_1 + \sqrt{A}]$. Since $(\rho + \sqrt{A})(y_1 + x_1\sqrt{A}) = (\rho^2 - A)(y + x\sqrt{A})/F = F_1(y + x\sqrt{A}) \equiv 0$ mod $[F_1][F, G + \sqrt{A}]$, the equation $[F_1][F, G + \sqrt{A}] = [\rho + \sqrt{A}][F_1, G_1 + \sqrt{A}]$ implies $(\rho + \sqrt{A})(y_1 + x_1\sqrt{A}) \equiv 0$ mod $[\rho + \sqrt{A}][F_1, G_1 + \sqrt{A}]$ and therefore implies* $y_1 + x_1\sqrt{A} \equiv 0$ mod $[F_1, G_1 + \sqrt{A}]$. Finally, multiplication of the input equation by $[F_1]$ gives $[F_1][y + x\sqrt{A}][f, g + \sqrt{A}] = [n][\rho + \sqrt{A}][F_1, G_1 + \sqrt{A}]$; multiply by $[\rho - \sqrt{A}]$—which is valid even though $\rho - \sqrt{A}$ is not a hypernumber because the hypernumbers $y + x\sqrt{A}$ and $\rho + \sqrt{A}$ can both be multiplied by $\rho - \sqrt{A}$—to put this equation in the form $[F_1][F(y_1 + x_1\sqrt{A})][f, g + \sqrt{A}] = [n][FF_1][F_1, G_1 + \sqrt{A}]$ and divide by $[FF_1]$ to conclude that the output equation holds.

The theorem will be proved by proving that if $[F, G + \sqrt{A}]$ is *stable* and if the reduction algorithm is applied iteratively until x is reduced to zero, then (1) the terminal equation is obvious from the original equation and (2) the steps of the algorithm can be retraced using the comparison algorithm

---

* The definitions imply—when use is made of the fact that if a, b and c are hypernumbers with $c \neq 0$ then $ac = bc$ implies $a = b$—that, for any nonzero hypernumber c and any module M, a congruence $ac \equiv bc$ mod $[c]M$ implies $a \equiv b$ mod M.

to go from the terminal equation back to the original, thereby determining the possible original equations and showing that $[F, G + \sqrt{A}]$ is a successor of $[f, g + \sqrt{A}]$. For example, the input equation $[236 + 89\sqrt{7}][83, 16 + \sqrt{7}] = [83][3, 1 + \sqrt{7}]$ leads to

$$
\begin{array}{ll}
[236 + 89\sqrt{7}][83, 16 + \sqrt{7}] = [83][3, 1 + \sqrt{7}] & (\rho_1 = 4), \\
[107 + 40\sqrt{7}][83, 16 + \sqrt{7}] = [83][3, 2 + \sqrt{7}] & (\rho_2 = 5), \\
[85 + 31\sqrt{7}][83, 16 + \sqrt{7}] = [83][6, 1 + \sqrt{7}] & (\rho_3 = 7), \\
[63 + 22\sqrt{7}][83, 16 + \sqrt{7}] = [83][7, \sqrt{7}] & (\rho_4 = 7), \\
[41 + 13\sqrt{7}][83, 16 + \sqrt{7}] = [83][6, 5 + \sqrt{7}] & (\rho_5 = 5), \\
[19 + 4\sqrt{7}][83, 16 + \sqrt{7}] = [83][3, 1 + \sqrt{7}] & (\rho_6 = 7), \\
[35 + 3\sqrt{7}][83, 16 + \sqrt{7}] = [83][14, 7 + \sqrt{7}] & (\rho_7 = 21), \\
[51 + 2\sqrt{7}][83, 16 + \sqrt{7}] = [83][31, 10 + \sqrt{7}] & (\rho_8 = 41), \\
[67 + \sqrt{7}][83, 16 + \sqrt{7}] = [83][54, 13 + \sqrt{7}] & (\rho_9 = 67), \\
[83][83, 16 + \sqrt{7}] = [83][83, 16 + \sqrt{7}]. &
\end{array}
$$

Each step leaves the second factor on the left and the first factor on the right unchanged. At the last step, the uniqueness of canonical form implies that the two sides are identical. Therefore, the terminal equation $[83][83, 16 + \sqrt{7}] = [83][83, 16 + \sqrt{7}]$ is determined without computation by the original one. Moreover, at each step the module on the right will be seen to be the immediate successor of the module below it. In fact, the number ρ_i used to go from equation $i - 1$ to equation i is the number r used by the comparison algorithm to go from the module in equation i to the one in equation $i - 1$, which implies that the input equation at the top of the list can be obtained by starting with the identity at the bottom and successively multiplying by $67 + \sqrt{7}$, dividing by 83, multiplying by $41 + \sqrt{7}$, dividing by 54, and so forth, applying the operations to the hypernumbers in the first factors on the left and to the modules in the second factors on the right.

As this example indicates, the key fact used to determine the possible input equations is that application of the reduction algorithm to an equation $[y + x\sqrt{A}][f, g + \sqrt{A}] = [n][F, G + \sqrt{A}]$ in which $[F, G + \sqrt{A}]$ is stable produces a sequence of equations $[y_i + x_i\sqrt{A}][f, g + \sqrt{A}] = [n][F_i, G_i + \sqrt{A}]$ in which the immediate successor of $[F_i, G_i + \sqrt{A}]$ is $[F_{i-1}, G_{i-1} + \sqrt{A}]$ and the number ρ_i used by the reduction algorithm to go from equation $i - 1$ to equation i is the number used by the comparison algorithm to go from $[F_i, G_i + \sqrt{A}]$ to its immediate successor. Let a step of the reduction algorithm be called **traceable** if the number ρ used to perform it is equal to the number r used by the comparison algorithm to determine the immediate successor of $[F_1, G_1 + \sqrt{A}]$.

Lemma. *A step of the reduction algorithm is traceable if $[F, G + \sqrt{A}]$ is stable or if it follows a traceable step.*

Proof. Let $[y + x\sqrt{A}][f, g + \sqrt{A}] = [n][F, G + \sqrt{A}]$ be an input to the reduction algorithm. To say that the resulting step of the reduction algorithm is traceable is to say that $\rho = r$ where ρ is the number used by the reduction algorithm and r is the number used by the comparison algorithm to determine the immediate successor of $[F_1, G_1 + \sqrt{A}]$. Let s be the number used by the comparison algorithm to determine the immediate successor of $[F, G + \sqrt{A}]$. The first step of the proof will be to show that if the step is not traceable, then $\rho + s = F$.

If the step is not traceable, then because $\rho^2 > A$ and $\rho + G_1 \equiv 0 \bmod F_1$, and because r is by definition the smallest number for which $r^2 > A$ and $r + G_1 \equiv 0 \bmod F_1$, ρ must be at least as great as F_1, and the square of $\rho - F_1$ must be greater than A. Then $(\rho - F_1)^2 > A$ or $\rho^2 + F_1^2 > 2\rho F_1 + A$, from which it follows (subtract A and divide by F_1) that $F + F_1 > 2\rho$ and $F > \rho + (\rho - F_1) \geq \rho$, so $F - \rho > \rho - F_1 \geq 0$. If $\rho + s$ were greater than or equal to $2F$, it would follow that $s \geq F + (F - \rho) > F$ and $(s - F)^2 \geq (F - \rho)^2 > (\rho - F_1)^2 > A$, contrary to the definition of s. Thus, $\rho + s < 2F$. Since $\rho + s \equiv G + s \equiv 0 \bmod F$, the desired conclusion $\rho + s = F$ follows from the assumption that the step is not traceable.

Since $\rho + s = F$ implies $|s - F|^2 = \rho^2 > A$, which in turn implies that $[F, G + \sqrt{A}]$ is not stable (see the proof of Proposition 1), the first statement of the lemma, that *if $[F, G + \sqrt{A}]$ is stable then the step is traceable*, follows.

Suppose, finally, that the step follows a traceable step of the reduction algorithm. Then the step that it follows is retraced by multiplying by $s + \sqrt{A}$ and dividing by F. Thus, the previous x is $(y + sx)/F$. Since the reduction algorithm reduces x, it follows that $x < \frac{y + sx}{F}$, which is to say $Fx < y + sx$. If the step were not traceable, F would be $s + \rho$, so $\rho x + sx$ would be less than $y + sx$, and ρx would be less than y, contrary to the definition of ρ. Therefore, the proof of the lemma is complete.

Proposition 2 (Solution of Pell's equation).   *The only solutions of Pell's equation $Ax^2 + 1 = y^2$ are those given by the reduction algorithm, namely, the pairs (x, y) given by formula (2) when $[f, g + \sqrt{A}] = [f_k, g_k + \sqrt{A}] = [1]$.*

Proof. Putting $[y + x\sqrt{A}]$ in canonical form when y and x are relatively prime and $y^2 > Ax^2$ easily gives $[y + x\sqrt{A}] = [y^2 - Ax^2, g + \sqrt{A}]$, where g is determined by $y \equiv gx \bmod (y^2 - Ax^2)$. Therefore, $Ax^2 + 1 = y^2$ implies $[y + x\sqrt{A}] = [1]$. Conversely, if $y^2 > Ax^2$ and $[y + x\sqrt{A}] = [1]$, then x and y are relatively prime and $y^2 - Ax^2 = 1$. In short, solutions of Pell's equation correspond one-to-one to hypernumbers $y + x\sqrt{A}$ for which $y^2 > Ax^2$ and $[y + x\sqrt{A}] = [1]$.

Since $[1]$ is stable, infinitely many of its successors are $[1]$. Each such successor implies a solution (x_k, y_k) of Pell's equation given by the formula

$$y_k + x_k\sqrt{A} = \frac{\prod_{i=1}^{k}(r_i + \sqrt{A})}{\prod_{i=0}^{k-1} f_i}.$$

What is to be shown is that there are no others.

But a solution of Pell's equation is, as was just shown, a hypernumber $y + x\sqrt{A}$ that satisfies $[y + x\sqrt{A}][1] = [1]$. This equation is an input to the reduction algorithm (write the right side as $[1][1, \sqrt{A}])$, and repeated application of the reduction algorithm reduces it to $[1][1, \sqrt{A}] = [1][1, \sqrt{A}]$. The input $y + x\sqrt{A} = 1 + 0 \cdot \sqrt{A}$ of course is already reduced and corresponds to the trivial solution $A \cdot 0^2 + 1 = 1^2$ of Pell's equation. Otherwise, the reduction requires $k \geq 1$ steps, all steps are traceable, and $y + x\sqrt{A}$ is obtained when 1 is multiplied by $\rho_k + \sqrt{A}$ and divided by 1, multiplied by $\rho_{k-1} + \sqrt{A}$ and divided by f_{k-1}, and so forth. Since the sequence of ρ's is the sequence—in reverse order—of r's obtained by applying the comparison algorithm to $[1]$, Proposition 2 follows.

For example, when $A = 13$, the cycle of $[1]$ is $[1]$, $[3, 1 + \sqrt{13}]$, $[4, 1 + \sqrt{13}]$, $[9, 7 + \sqrt{13}]$, $[12, 11 + \sqrt{13}]$, $[13, \sqrt{13}]$, $[12, 1 + \sqrt{13}]$, $[9, 2 + \sqrt{13}]$, $[4, 3 + \sqrt{13}]$, $[3, 2 + \sqrt{13}]$, after which the sequence returns to $[1]$ and repeats. The r's used at the successive steps are 4, 5, 7, 11, 13, 13, 11, 7, 5, 4, after which they repeat. Thus the smallest solution of $13x^2 + 1 = y^2$ other than the trivial one is given by the coefficients of

$$\frac{(4 + \sqrt{13})^2(5 + \sqrt{13})^2(7 + \sqrt{13})^2(11 + \sqrt{13})^2(13 + \sqrt{13})^2}{3^2 \cdot 4^2 \cdot 9^2 \cdot 12^2 \cdot 13},$$

which is easily found to be $649 + 180\sqrt{13}$. That is, the smallest solution of Pell's equation when $A = 13$ is $13 \cdot 180^2 + 1 = 649^2$. Since $(649 + 180\sqrt{13})^2 = 842401 + 233640\sqrt{13}$, the next smallest solution is $13 \cdot 233640^2 + 1 = 842401^2$, and so forth.

(For any A, as for $A = 13$, the sequence of r's is in fact a palindrome, so the sequence of ρ's is identical to the sequence of r's.)

Proposition 3. *If $[f, g + \sqrt{A}]$ is principal, then $[f, f - g + \sqrt{A}]$ is also principal, and the product of these modules is $[f]$.*

Proof. Suppose $[f, g + \sqrt{A}] = [y + x\sqrt{A}]$, where $y^2 > Ax^2$. Since any common divisor of x and y divides the coefficient of $\sqrt{A}$ in $g + \sqrt{A}$, x and y are relatively prime. Therefore, the number $y^2 - Ax^2$, call it N, is relatively prime to x, and x has a reciprocal, call it r, mod N. Then $[y + x\sqrt{A}] = [N, y + x\sqrt{A}, r(y + x\sqrt{A})] = [N, G + \sqrt{A}]$, where $G \equiv ry$ mod N, so $f = N$, $g = G$, and $y \equiv xry \equiv gx$ mod N.

The solutions (X, Y) of Pell's equation obviously grow without bound, so there is a solution $Y^2 = AX^2 + 1$ of Pell's equation in which $X > x$. Since $y^2 - Ax^2 = f > 0$, it follows that $X^2y^2 = AX^2x^2 + fX^2 > AX^2x^2 + x^2 = Y^2x^2$, which implies $Xy > Yx$. Also, $Y^2y^2 = A^2x^2X^2 + Ax^2 + fAX^2 + f > A^2X^2x^2$, so $Yy > AXx$. Therefore, the formula $z + w\sqrt{A} = (Y + X\sqrt{A})(y - x\sqrt{A})$ defines a hypernumber $z + w\sqrt{A}$ (even though, by the strict definition being used here, $y - x\sqrt{A}$ is not a hypernumber). This hypernumber satisfies $(z + w\sqrt{A})(y + x\sqrt{A}) = (Y + X\sqrt{A})(y - x\sqrt{A})(y + x\sqrt{A}) = (Y + X\sqrt{A})f$. Thus,

$[z + w\sqrt{A}][y + x\sqrt{A}] = [Y + X\sqrt{A}][f] = [f]$, and what is to be shown is that $[z + w\sqrt{A}] = [f, f - g + \sqrt{A}]$.

Now, $z^2 = (Yy - AxX)^2 = Y^2y^2 - 2AxyXY + A^2x^2X^2 = Y^2y^2 - AY^2x^2 + AY^2x^2 - 2AxyXY + AX^2y^2 - AX^2y^2 + A^2x^2X^2 = (Y^2 - AX^2)(y^2 - Ax^2) + A(Xy - Yx)^2 = Aw^2 + f$. Thus $z^2 - Aw^2 = f$, and it remains only to show that $z \equiv (f - g)w \bmod f$. But equating coefficients of $\sqrt{A}$ in $(z + w\sqrt{A})(y + x\sqrt{A}) = (Y + X\sqrt{A})f$ gives $wy + zx = fX$, so $0 \equiv wy + zx \equiv wgx + zx \bmod f$, which implies, because x is relatively prime to f, that $wg + z \equiv 0 \bmod f$, or $z \equiv (f - g)w \bmod f$, as was to be shown.

Corollary.   *A module that is equivalent to [1] is principal.*

Deduction. To say that M is equivalent to [1] means that there are principal modules P_1 and P_2 for which $MP_1 = P_2$. By Proposition 3, there is a principal module P_3 such that $P_1P_3 = [n]$ for some number n. Thus $M[n] = P_2P_3$, which implies $M[n] = [z + w\sqrt{A}]$, where $z^2 > Aw^2$. This equation implies that n divides both z and w, so $M = [\frac{z}{n} + \frac{w}{n}\sqrt{A}]$ is principal.

Proof of the Theorem. Suppose that $[f, g + \sqrt{A}]$ and $[F, G + \sqrt{A}]$ are equivalent—say $[y + x\sqrt{A}][f, g + \sqrt{A}] = [v + u\sqrt{A}][F, G + \sqrt{A}]$ where $y^2 > Ax^2$ and $v^2 > Au^2$—and that $[F, G + \sqrt{A}]$ is stable. By Proposition 3, there is a hypernumber $z + w\sqrt{A}$ for which $z^2 > Aw^2$ and $[z + w\sqrt{A}][v + u\sqrt{A}] = [n]$ for some number n. Let the given equivalence between $[f, g + \sqrt{A}]$ and $[F, G + \sqrt{A}]$ be multiplied by $[F(z + w\sqrt{A})]$ to yield an equation of the form $[Y + X\sqrt{A}][f, g + \sqrt{A}] = [N][F, G + \sqrt{A}]$, where $N = Fn$, that is an input to the reduction algorithm. Application of the reduction algorithm reduces this equation to $[N][f, g + \sqrt{A}] = [N][f, g + \sqrt{A}]$. Since $[F, G + \sqrt{A}]$ is stable, the steps of the algorithm can be retraced by applying the comparison algorithm to $[f, g + \sqrt{A}]$, from which it follows that $Y + X\sqrt{A}$ can be obtained by multiplying N by $r_1 + \sqrt{A}$, dividing by f, multiplying by $r_2 + \sqrt{A}$, dividing by f_1, and so forth, stopping with the kth step, where $[F, G + \sqrt{A}]$ is the kth successor of $[f, g + \sqrt{A}]$. In short,

$$Y + X\sqrt{A} = N \cdot \frac{\prod_{i=1}^{k}(r_i + \sqrt{A})}{\prod_{i=0}^{k-1} f_i}.$$

Since $Y + X\sqrt{A} = F(z + w\sqrt{A})(y + x\sqrt{A})$ and $N = Fn$, the equation $F(z + w\sqrt{A})(y + x\sqrt{A}) \prod_{i=0}^{k-1} f_i = Fn \prod_{i=1}^{k}(r_i + \sqrt{A})$ follows. The equation

$$(y + x\sqrt{A}) \prod_{i=0}^{k-1} f_i = (v + u\sqrt{A}) \prod_{i=1}^{k}(r_i + \sqrt{A})$$

of the theorem follows when one multiplies by $v + u\sqrt{A}$ and divides by Fn.

Corollary (Solution of the equivalence problem).   *Stable modules in different cycles are not equivalent.*

Deduction. If two stable modules are equivalent, the theorem implies that each is a successor of the other, so they are in the same cycle.

Solution of $A\square + B = \square$.   *A solution of $Ax^2 + B = y^2$ is called* **primitive** *if x and y are relatively prime. The primitive solutions are found in the following way:*

For each square root ρ of A mod B, use formula (2) to find all solutions

$$
(3) \qquad y + x\sqrt{A} = \frac{\prod_{i=1}^{k}(r_i + \sqrt{A})}{\prod_{i=1}^{k-1} f_i}
$$

of the problem $[y + x\sqrt{A}][B, \rho + \sqrt{A}] = [B]$. Each pair (x, y) found in this way is a primitive solution of $Ax^2 + B = y^2$, and there are no others.

The solutions (x, y) that are not primitive are of the form (ud, vd), where $d^2 > 1$ is a square factor of B and (u, v) is a primitive solution of $Au^2 + \frac{B}{d^2} = v^2$. Therefore, they can be found by finding all square factors d^2 of B that are greater than 1 and, for each of them, using the method just described to find all primitive solutions (u, v) of $Au^2 + \frac{B}{d^2} = v^2$.

Proof. The hypernumbers (3) are the solutions of $[y + x\sqrt{A}][B, \rho + \sqrt{A}] = [B][1, \sqrt{A}]$ found by the construction of the theorem. Multiplication by $[B, B - \rho + \sqrt{A}]$ then gives $[y + x\sqrt{A}][B] = [B][B, B - \rho + \sqrt{A}]$, which implies $[y + x\sqrt{A}] = [B, B - \rho + \sqrt{A}]$, so x and y are relatively prime and satisfy $y^2 - Ax^2 = B$. Conversely, if x and y satisfy these conditions, then reduction of $[y + x\sqrt{A}]$ to canonical form gives $[B, g + \sqrt{A}]$ for some square root g of A mod B, so $[y + x\sqrt{A}][B, B - g + \sqrt{A}] = [B]$ for another square root $B - g$ of A mod B, and $y + x\sqrt{A}$ is among the solutions given by (3).

For example, the solution of $79\square + 21 = \square$ requires finding the square roots of 79 mod 21. (Note that 21 has no square factors, so all solutions are primitive solutions.) These are easily found by finding the square roots ± 1 of 79 mod 3 and the square roots ± 3 of 79 mod 7 and putting them together using the Chinese remainder theorem to find the four square roots 4, 10, 11, and 17 of 79 mod 21. The module $[21, 4 + \sqrt{79}]$ is stable, and its cycle under the comparison algorithm contains 8 stable modules; since $[1]$ is not among them, the square root 4 of 79 mod 21 gives rise to no solutions of $79\square + 21 = \square$. Similarly, the module $[21, 17 + \sqrt{79}]$ is stable. Its cycle—which contains the conjugates of the modules in the cycle of $[21, 4 + \sqrt{79}]$, as is shown in the table below—also has length 8 and does not contain $[1]$, so this square root does not give rise to any solutions of $79\square + 21 = \square$ either. Application of the comparison algorithm to $[21, 10 + \sqrt{79}]$, on the other hand, reaches $[1]$ in two steps from $[21, 10 + \sqrt{79}]$ to $[2, 1 + \sqrt{79}]$ to $[1]$; the values of r are first 11 and then 9, so $y + x\sqrt{79} = \frac{(11+\sqrt{79})(9+\sqrt{79})}{2} = 89 + 10\sqrt{79}$, and the smallest solution of $79\square + 21 = \square$ in the sequence of solutions corresponding to $[21, 10 + \sqrt{79}]$ is $79 \cdot 10^2 + 21 = 89^2$. The next solution is found by taking the comparison algorithm two steps further, which multiplies $89 + 10\sqrt{79}$ by $\frac{(9+\sqrt{79})(9+\sqrt{79})}{2} = 80 + 9\sqrt{79}$, which of course describes the smallest solution $79 \cdot 9^2 + 1 = 80^2$ of Pell's equation in the case $A = 79$. Since $(89 + 10\sqrt{79})(80 + 9\sqrt{79}) = 14230 + 1601\sqrt{79}$, the next solution of $79\square + 21 = \square$ in this sequence is $79 \cdot 1601^2 + 21 = 14230^2$. More generally, the nth solution in the sequence is contained in the coefficients of $(89 + 10\sqrt{79})(80 + 9\sqrt{79})^{n-1}$. In the same way, the fact that $[21, 11 + \sqrt{79}] \sim [1]$ leads to an infinite sequence of solutions of $79x^2 + 21 = y^2$, namely, the solutions in which x is the coefficient of $\sqrt{79}$ in $(10 + \sqrt{79})(80 + 9\sqrt{79})^{n-1}$. *All* solutions are contained in these two infinite sequences.

Orbits of Stable Modules for Various Values of A

$A = 2$. (2 modules, 1 cycle)
$$[1] \sim [2, \sqrt{2}];$$

$A = 3$. (3 modules, 2 cycles)
$$[1] \quad \text{(Cycle contains just one module.)}$$
$$[2, 1 + \sqrt{3}] \sim [3, \sqrt{3}];$$

$A = 5$. (5 modules, 2 cycles)
$$[1] \sim [4, 3 + \sqrt{5}] \sim [5, \sqrt{5}] \sim [4, 1 + \sqrt{5}],$$
$$[2, 1 + \sqrt{5}];$$

$A = 6$. (6 modules, 2 cycles)
$$[1] \sim [3, \sqrt{6}],$$
$$[2, \sqrt{6}] \sim [5, 4 + \sqrt{6}] \sim [6, \sqrt{6}] \sim [5, 1 + \sqrt{6}];$$

$A = 7$. (7 modules, 2 cycles)
$$[1] \sim [2, 1 + \sqrt{7}],$$
$$[3, 1 + \sqrt{7}] \sim [6, 5 + \sqrt{7}] \sim [7, \sqrt{7}] \sim [6, 1 + \sqrt{7}] \sim [3, 2 + \sqrt{7}];$$

$A = 8$. (7 modules, 3 cycles)
$$[1],$$
$$[2, \sqrt{8}] \sim [4, \sqrt{8}],$$
$$[7, 1 + \sqrt{8}] \sim [4, 2 + \sqrt{8}] \sim [7, 6 + \sqrt{8}] \sim [8, \sqrt{8}];$$

$A = 10$. (10 modules, 2 cycles)
$$[1] \sim [6, 4 + \sqrt{10}] \sim [9, 8 + \sqrt{10}] \sim [10, \sqrt{10}] \sim [9, 1 + \sqrt{10}] \sim [6, 2 + \sqrt{10}],$$
$$[2, \sqrt{10}] \sim [3, 1 + \sqrt{10}] \sim [5, \sqrt{10}] \sim [3, 2 + \sqrt{10}];$$

$A = 11$. (9 modules, 2 cycles)
$$[1] \sim [5, 4 + \sqrt{11}] \sim [5, 1 + \sqrt{11}],$$
$$[2, 1 + \sqrt{11}] \sim [7, 5 + \sqrt{11}] \sim [10, 9 + \sqrt{11}] \sim [11, \sqrt{11}] \sim [10, 1 + \sqrt{11}] \sim$$
$$[7, 2 + \sqrt{11}];$$

$A = 12$. (11 modules, 4 cycles)
$$[1] \sim [4, \sqrt{12}],$$
$$[2, \sqrt{12}],$$
$$[3, \sqrt{12}] \sim [8, 6 + \sqrt{12}] \sim [11, 10 + \sqrt{12}] \sim [12, \sqrt{12}] \sim [11, 1 + \sqrt{12}] \sim [8, 2 + \sqrt{12}],$$

$[6, \sqrt{12}] \sim [4, 2 + \sqrt{12}];$

$A = 13.$ (13 modules, 2 cycles)
$\quad [1] \sim [3, 1 + \sqrt{13}] \sim [4, 1 + \sqrt{13}] \sim [9, 7 + \sqrt{13}] \sim [12, 11 + \sqrt{13}] \sim [13, \sqrt{13}] \sim$
$[12, 1 + \sqrt{13}] \sim [9, 2 + \sqrt{13}] \sim [4, 3 + \sqrt{13}] \sim [3, 2 + \sqrt{13}],$
$\quad [2, 1 + \sqrt{13}] \sim [6, 5 + \sqrt{13}] \sim [6, 1 + \sqrt{13}];$

$A = 14.$ (10 modules, 2 cycles)
$\quad [1] \sim [2, \sqrt{14}],$
$\quad [13, 1 + \sqrt{14}] \sim [10, 2 + \sqrt{14}] \sim [5, 3 + \sqrt{14}] \sim [7, \sqrt{14}] \sim [5, 2 + \sqrt{14}] \sim [10, 8 +$
$\sqrt{14}] \sim [13, 12 + \sqrt{14}] \sim [14, \sqrt{14}];$

$A = 15.$ (12 modules, 4 cycles)
$\quad [1],$
$\quad [5, \sqrt{15}] \sim [2, 1 + \sqrt{15}],$
$\quad [3, \sqrt{15}] \sim [7, 6 + \sqrt{15}] \sim [7, 1 + \sqrt{15}],$
$\quad [15, \sqrt{15}] \sim [14, 1 + \sqrt{15}] \sim [11, 2 + \sqrt{15}] \sim [6, 3 + \sqrt{15}] \sim [11, 9 + \sqrt{15}] \sim$
$[14, 13 + \sqrt{15}];$

$A = 17.$ (13 modules, 2 cycles)
$\quad [1] \sim [8, 5 + \sqrt{17}] \sim [13, 11 + \sqrt{17}] \sim [16, 15 + \sqrt{17}] \sim [17, \sqrt{17}] \sim [16, 1 + \sqrt{17}] \sim$
$[13, 2 + \sqrt{17}] \sim [8, 3 + \sqrt{17}],$
$\quad [2, 1 + \sqrt{17}] \sim [4, 1 + \sqrt{17}] \sim [8, 7 + \sqrt{17}] \sim [8, 1 + \sqrt{17}] \sim [4, 3 + \sqrt{17}];$

$A = 18.$ (12 modules, 2 cycles)
$\quad [1] \sim [7, 5 + \sqrt{18}] \sim [9, \sqrt{18}] \sim [7, 2 + \sqrt{18}],$
$\quad [2, \sqrt{18}] \sim [9, 6 + \sqrt{18}] \sim [14, 12 + \sqrt{18}] \sim [17, 16 + \sqrt{18}] \sim [18, \sqrt{18}] \sim [17, 1 +$
$\sqrt{18}] \sim [14, 2 + \sqrt{18}] \sim [9, 3 + \sqrt{18}];$

$A = 19.$ (17 modules, 2 cycles)
$\quad [1] \sim [6, 5 + \sqrt{19}] \sim [5, 2 + \sqrt{19}] \sim [9, 8 + \sqrt{19}] \sim [9, 1 + \sqrt{19}] \sim [5, 3 + \sqrt{19}] \sim$
$[6, 1 + \sqrt{19}],$
$\quad [2, 1 + \sqrt{19}] \sim [3, 2 + \sqrt{19}] \sim [10, 7 + \sqrt{19}] \sim [15, 13 + \sqrt{19}] \sim [18, 17 + \sqrt{19}] \sim$
$[19, \sqrt{19}] \sim [18, 1 + \sqrt{19}] \sim [15, 2 + \sqrt{19}] \sim [10, 3 + \sqrt{19}] \sim [3, 1 + \sqrt{19}];$

$A = 20.$ (14 modules, 3 cycles)
$\quad [1] \sim [5, \sqrt{20}],$
$\quad [2, \sqrt{20}] \sim [8, 6 + \sqrt{20}] \sim [10, \sqrt{20}] \sim [8, 2 + \sqrt{20}],$
$\quad [20, \sqrt{20}] \sim [19, 1 + \sqrt{20}] \sim [16, 2 + \sqrt{20}] \sim [11, 3 + \sqrt{20}] \sim [4, \sqrt{20}] \sim [11, 8 +$
$\sqrt{20}] \sim [16, 14 + \sqrt{20}] \sim [19, 18 + \sqrt{20}];$

$A = 21.$ (18 modules, 4 cycles)
$\quad [1] \sim [4, 1 + \sqrt{21}] \sim [7, \sqrt{21}] \sim [4, 3 + \sqrt{21}],$
$\quad [3, \sqrt{21}] \sim [5, 1 + \sqrt{21}] \sim [12, 9 + \sqrt{21}] \sim [17, 15 + \sqrt{21}] \sim [20, 19 + \sqrt{21}] \sim$
$[21, \sqrt{21}] \sim [20, 1 + \sqrt{21}] \sim [17, 2 + \sqrt{21}] \sim [12, 3 + \sqrt{21}] \sim [5, 4 + \sqrt{21}],$
$\quad [2, 1 + \sqrt{21}],$
$\quad [6, 3 + \sqrt{21}] \sim [10, 9 + \sqrt{21}] \sim [10, 1 + \sqrt{21}].$

Finally, an example that is frequently cited by Gauss.*
$A = 79$. (51 modules, 6 cycles)

$[1] \sim [2, 1 + \sqrt{79}]$,

$[3, 1 + \sqrt{79}] \sim [14, 11 + \sqrt{79}] \sim [15, 2 + \sqrt{79}] \sim [6, 1 + \sqrt{79}] \sim [7, 4 + \sqrt{79}]$,

$[9, 4 + \sqrt{79}] \sim [13, 1 + \sqrt{79}] \sim [5, 2 + \sqrt{79}] \sim [18, 13 + \sqrt{79}] \sim [25, 23 + \sqrt{79}] \sim$
$[26, 1 + \sqrt{79}] \sim [21, 4 + \sqrt{79}] \sim [10, 7 + \sqrt{79}]$,

$[27, 22 + \sqrt{79}] \sim [35, 32 + \sqrt{79}] \sim [39, 38 + \sqrt{79}] \sim [39, 1 + \sqrt{79}] \sim [35, 3 + \sqrt{79}] \sim$
$[27, 5 + \sqrt{79}] \sim [15, 7 + \sqrt{79}] \sim [30, 23 + \sqrt{79}] \sim [43, 37 + \sqrt{79}] \sim [54, 49 + \sqrt{79}] \sim$
$[63, 59 + \sqrt{79}] \sim [70, 67 + \sqrt{79}] \sim [75, 73 + \sqrt{79}] \sim [78, 77 + \sqrt{79}] \sim [79, \sqrt{79}] \sim$
$[78, 1 + \sqrt{79}] \sim [75, 2 + \sqrt{79}] \sim [70, 3 + \sqrt{79}] \sim [63, 4 + \sqrt{79}] \sim [54, 5 + \sqrt{79}] \sim$
$[43, 6 + \sqrt{79}] \sim [30, 7 + \sqrt{79}] \sim [15, 8 + \sqrt{79}]$,

$[9, 5 + \sqrt{79}] \sim [10, 3 + \sqrt{79}] \sim [21, 17 + \sqrt{79}] \sim [26, 25 + \sqrt{79}] \sim [25, 2 + \sqrt{79}] \sim$
$[18, 5 + \sqrt{79}] \sim [5, 3 + \sqrt{79}] \sim [13, 12 + \sqrt{79}]$,

$[3, 2 + \sqrt{79}] \sim [7, 3 + \sqrt{79}] \sim [6, 5 + \sqrt{79}] \sim [15, 13 + \sqrt{79}] \sim [14, 3 + \sqrt{79}]$.

---

* *Disquisitiones Arithmeticae*, §§185, 186, 187, 195, 196, 198, 223. The reason 79 is
of interest is that it is the smallest value of A for which the class group contains
a square that is not the identity—for example, $[3, 1 + \sqrt{79}]^2 \sim [9, 4 + \sqrt{79}] \not\sim [1]$.
Perhaps Gauss's attention was drawn to this case by the fact that it occurs in a
counterexample Lagrange gave to a conjecture of Euler [48, Article 84]. Lagrange
notes that the problem $79 \cdot \square + 733 = \square$ has a solution (the comparison algorithm
gives $[733, 476 + \sqrt{79}] \sim [90, 77 + \sqrt{79}] \sim [1]$) but the problem $79 \cdot \square + 101 = \square$
does not $([101, 33 + \sqrt{79}] \sim [45, 23 + \sqrt{79}] \sim [9, 4 + \sqrt{79}] \not\sim [1])$, contrary to a
conjecture of Euler that would have implied that the answer to "Does $A\square + B = \square$
have a solution?" might, for prime B, depend only on the class of $B \bmod 4A$.

Essay 3.4 Multiplication of Modules and Module Classes

In this essay, the semigroup of modules and the class semigroup are examined more closely. In the semigroup of modules, modules can be decomposed as products of their "p-parts" where p ranges over the primes, and in this way the semigroup can be described quite fully, except for the crucial problem of *determining the primes p mod which A is a square*, which are the primes for which $[p]$ is a product in which neither factor is $[1]$. This question is the subject of Essay 3.5.

The structure of the *class* semigroup depends on more subtle considerations, and general statements are harder to come by. The theorem that is proved in this essay and that is used in the next to prove the law of quadratic reciprocity merely describes the subgroups of index 2 of the class group in a few simple cases (namely, the cases in which A is an odd prime or twice an odd prime or a product of two primes that are congruent mod 4).

The computation of products of modules comes down to the computation of products $[f, g + \sqrt{A}][F, G + \sqrt{A}]$ (where $g^2 \equiv A$ mod f and $G^2 \equiv A$ mod F), because multiplication of a module by $[e]$ is easy. The following theorem determines these products when f and F are relatively prime:

Theorem 1. *If f and F are relatively prime, then $[f, g+\sqrt{A}][F, G+\sqrt{A}] = [fF, z + \sqrt{A}]$, where z is determined by $z \equiv g$ mod f and $z \equiv G$ mod F.*

Proof. By the Chinese remainder theorem,* the congruences $z \equiv g$ mod f and $z \equiv G$ mod F determine a unique z mod fF, so the formula $[fF, z+\sqrt{A}]$ determines a module. The desired product is $[fF, f(G+\sqrt{A}), F(g+\sqrt{A}), (g+\sqrt{A})(G + \sqrt{A})]$, which is $[fF, f(z + \sqrt{A}), F(z + \sqrt{A}), gG + A + (g + G)\sqrt{A}]$, because $fG \equiv fz$ mod fF and $Fg \equiv Fz$ mod fF. If $\beta f = \alpha F + 1$, then $z + \sqrt{A}$ is the difference of $\beta f(z + \sqrt{A})$ and $\alpha F(z + \sqrt{A})$, so the desired product is $[fF, z + \sqrt{A}, gG + A + (g + G)\sqrt{A}]$. Since $A \equiv z^2$ mod f and $A \equiv z^2$ mod F and since f and F are relatively prime, $A \equiv z^2$ mod fF; moreover, $(z - g)(z - G) \equiv 0$ mod fF, so $A + gG \equiv z(g + G)$ mod fF, which implies that $gG + A$ in the last term can be replaced by $z(g + G)$, and the desired product becomes $[fF, z + \sqrt{A}, (g + G)(z + \sqrt{A})] = [fF, z + \sqrt{A}]$, as was to be shown.

Given a prime p and a module $[f, g + \sqrt{A}]$, let the p-**part** of $[f, g + \sqrt{A}]$ be the module $[p^n, g + \sqrt{A}]$, where n is the number of times p divides f. (If p does not divide f, the p-part of $[f, g + \sqrt{A}]$ is $[1]$ by this definition.) By Theorem 1, every module is the product of its p-parts, and one can find the product of $[f, g+\sqrt{A}]$ and $[F, G+\sqrt{A}]$ by finding the product of their p-parts for each prime divisor p of fF, which reduces to finding the product of the p-parts for the primes p that divide both f and F. In short, the computation of $[f, g + \sqrt{A}][F, G + \sqrt{A}]$ can be done prime by prime. For all but a very few primes, the needed products are given by the three propositions that follow.

---

* See Essay 3.2.

Proposition 1. *Let p be an odd prime that does not divide A. If A is not a square mod p, there are no modules $[f, g + \sqrt{A}]$ in canonical form other than $[1]$ in which f is a power of p. If A is a square mod p, there are exactly two modules $[p^n, g + \sqrt{A}]$ in canonical form for each $n > 0$, and they are the nth powers of those in which $n = 1$. The product of the two in which $n = 1$ is $[p]$.*

Thus, a product of the form $[p^m, g + \sqrt{A}][p^n, G + \sqrt{A}]$ can be found by writing it as $[p, g + \sqrt{A}]^m [p, G + \sqrt{A}]^n$ and observing that the theorem implies that their product is $[p, g + \sqrt{A}]^{m+n}$ if $[p, g + \sqrt{A}] = [p, G + \sqrt{A}]$, and is $[p]^m [p, G + \sqrt{A}]^{n-m}$ or $[p]^n [p, g + \sqrt{A}]^{m-n}$ in the obvious way if $[p, g + \sqrt{A}] \neq [p, G + \sqrt{A}]$.

Proof. Because a polynomial of degree n with integer coefficients that is not zero mod p has at most n roots mod p, A has at most two square roots mod p. Because $-g$ is a square root of A mod p whenever g is, and because g and $-g$ are different mod p when this is the case (if they were the same, then $2g$ would be zero mod p so $4A \equiv (2g)^2 \equiv 0 \bmod p$, contrary to hypothesis), A has either no square roots mod p or exactly two. When it has two, and when $g < p$ is one of them, the product $[p, g + \sqrt{A}][p, p - g + \sqrt{A}]$ is $[p^2, p(p - g + \sqrt{A}), p(g + \sqrt{A}), pg - g^2 + A + p\sqrt{A}]$. The third term minus the second is $2pg \bmod p^2$ and $[p^2, 2pg] = [p]$ (again, because the square of $2g$ is not divisible by p and p is prime), so the first term p^2 can be replaced by p, and the module is equal to $[p]$.

If, for some $n > 1$, g_{n-1} is a square root of A mod p^{n-1}, then there is a unique square root of A mod p^n that is congruent to g_{n-1} mod p^{n-1}, because the formula $(g_{n-1} + \beta p^{n-1})^2 \equiv A \bmod p^n$ implies $2\beta g_{n-1} \equiv (A - g_{n-1}^2)/p^{n-1} \bmod p$, which determines $\beta \bmod p$ because $2g_{n-1}$ is relatively prime to p. Therefore, for each value of n and for each square root g of $A \bmod p$ there is exactly one square root of $A \bmod p^n$ that is $g \bmod p$; call it g_n. If $[p, g + \sqrt{A}]^{n-1} = [p^{n-1}, g_{n-1} + \sqrt{A}]$, then the same formula holds with n in place of $n - 1$, because $[p, g + \sqrt{A}]^n = [p^{n-1}, g_{n-1} + \sqrt{A}][p, g + \sqrt{A}] = [p^{n-1}, g_n + \sqrt{A}][p, g_n + \sqrt{A}] = [p^n, p(g_n + \sqrt{A}), g_n^2 + A + 2g_n\sqrt{A}] = [p^n, p(g_n + \sqrt{A}), 2g_n(g_n + \sqrt{A})] = [p^n, g_n + \sqrt{A}]$ (because $[p, 2g_n] = [1]$). Thus $[p, g + \sqrt{A}]^n$ is the unique module $[p^n, G + \sqrt{A}]$ in canonical form in which $G \equiv g \bmod p$, as was to be shown.

Proposition 2. *If p is a prime that divides A once but not twice, the only module $[p^n, g + \sqrt{A}]$ in canonical form in which $n > 0$ is $[p, \sqrt{A}]$; the square of $[p, \sqrt{A}]$ is $[p]$.*

Proof. If g were a square root of A mod p^n for $n > 1$, then p would divide g^2 and therefore would divide g itself, so p^2 would divide $g^2 \equiv A \bmod p^2$, contrary to the assumption that $A \not\equiv 0 \bmod p^2$. Therefore, the only module $[p^n, g + \sqrt{A}]$ in canonical form in which $n > 0$ is $[p, \sqrt{A}]$ because $g^2 \equiv A \equiv 0 \bmod p$ implies $g \equiv 0 \bmod p$. The square of this module is $[p^2, p\sqrt{A}, A] = [p]$, because $[p^2, A] = [p]$ by assumption.

Proposition 3. *If A is odd, the modules in canonical form $[2^n, g + \sqrt{A}]$ with $n > 0$ are as follows:*

(i) When $A \equiv 3 \bmod 4$, $[2, 1 + \sqrt{A}]$ is the only one; its square is $[2]$.

(ii) When $A \equiv 5 \bmod 8$, there are three: $\alpha = [2, 1 + \sqrt{A}]$, $\beta = [4, 1 + \sqrt{A}]$, and $\overline{\beta} = [4, 3 + \sqrt{A}]$. The product of α with any of the three is $[2]\alpha$, whereas $\beta^2 = [2]\overline{\beta}$, $\overline{\beta}^2 = [2]\beta$ and $\beta\overline{\beta} = [4]$.

(iii) When $A \equiv 1 \bmod 16$, let $\alpha = [2, 1 + \sqrt{A}]$, $\beta = [8, 5 + \sqrt{A}]$, and $\overline{\beta} = [8, 3 + \sqrt{A}]$. The modules in question are α and four infinite sequences of modules $\beta^n / [2^{n-1}]$, $\overline{\beta}^n / [2^{n-1}]$, $\alpha\beta^n / [2^n]$, and $\alpha\overline{\beta}^n / [2^n]$ for $n > 0$, these modules being distinct from one another. They can be multiplied using $\alpha^2 = [2]\alpha$ and $\beta\overline{\beta} = [8]$.

(iv) When $A \equiv 9 \bmod 16$, the answer is given by (iii) when β is changed to $[8, 1 + \sqrt{A}]$ and $\overline{\beta}$ to $[8, 7 + \sqrt{A}]$.

Proof. (i) The square of any odd number is $1 \bmod 4$, so $A \equiv 3 \bmod 4$ implies that no module $[2^n, g + \sqrt{A}]$ is in canonical form when $n > 1$. The square of $[2, 1 + \sqrt{A}]$ is $[4, 2 + 2\sqrt{A}, A + 1 + 2\sqrt{A}] = [4, 2 + 2\sqrt{A}, A - 1]$, which is $[2]$ because $[4, A - 1] = [2]$.

(ii) Similarly, the square of any odd number is $1 \bmod 8$, so $A \equiv 5 \bmod 8$ implies that there are no modules $[2^n, g + \sqrt{A}]$ in canonical form in which $n > 2$. The multiplication formulas that are given are easy to verify. For example, $[4, 1 + \sqrt{A}]^2 = [16, 4(1 + \sqrt{A}), 1 + A + 2\sqrt{A}]$; twice the third term minus the second is $2 + 2A - 4 = 2(A-1) \equiv 8 \bmod 16$, so the 16 in the first term can be replaced by 8, after which the second term can be dropped because it can be expressed in terms of the first and third, resulting in $[8, 1 + A + 2\sqrt{A}] = [2][4, 3 + \sqrt{A}]$.

(iii) and (iv) Let A be $1 \bmod 8$. *For each $n \geq 1$, there is a unique solution $g_n < 2^{n+3}$ of $g_n^2 \equiv A + 2^{n+2} \bmod 2^{n+3}$ for which $g_n \equiv 1 \bmod 4$.* In fact, $g_1 = 5$ is determined by these conditions when $A \equiv 1 \bmod 16$, and $g_1 = 1$ is determined by them when $A \equiv 9 \bmod 16$. For $n > 1$, knowledge of g_{n-1} enables one to find g_n in the following way. The congruence $h^2 \equiv A + 2^{n+2} \bmod 2^{n+3}$ of which g_n is a root implies, because $g_{n-1}^2 \equiv A + 2^{n+1} \bmod 2^{n+2}$, that $h^2 - g_{n-1}^2 \equiv 2^{n+1} \bmod 2^{n+2}$. If $h \equiv 1 \equiv g_{n-1} \bmod 4$, then $h + g_{n-1}$ is divisible by 2 but not 4, so the congruence $(h + g_{n-1})(h - g_{n-1}) \equiv 2^{n+1} \bmod 2^{n+2}$ implies that $h - g_{n-1}$ is divisible by 2^n but not 2^{n+1}. Thus $h \equiv g_{n-1} + 2^n \bmod 2^{n+1}$. In short, if the conditions on g_n can be met, then $g_n \equiv g_{n-1} + 2^n \bmod 2^{n+1}$. The two possible values $g_{n-1} \pm 2^n$ of $g_n \bmod 2^{n+2}$ determined in this way have squares that differ by $2^{n+2} \bmod 2^{n+3}$, so the condition $g_n^2 \equiv A + 2^{n+2} \bmod 2^{n+3}$ is satisfied by exactly one of them.

The sequence $g_1, g_2, g_3, \ldots$ determined in this way describes the modules $\beta^n / [2^{n-1}] = [2^{n+2}, g_n + \sqrt{A}]$, as can be seen as follows: By the very definition of g_1, $\beta = [8, g_1 + \sqrt{A}]$ in both (iii) and (iv). What is to be shown, then, is that $[8, g_1 + \sqrt{A}][2^{n+1}, g_{n-1} + \sqrt{A}] = [2][2^{n+2}, g_n + \sqrt{A}]$. The product on the left side is $[2^{n+4}, 8(g_{n-1} + \sqrt{A}), 2^{n+1}(g_1 + \sqrt{A}), g_1 g_{n-1} + A + (g_1 + g_{n-1})\sqrt{A}]$. Because $g_{n-1} \equiv g_n + 2^n \bmod 2^{n+1}$, one has $8g_{n-1} \equiv 8g_n + 2^{n+3} \bmod 2^{n+4}$, so the second term can be changed to $8(g_n + \sqrt{A}) + 2^{n+3}$. Similarly, the third term can be changed to $2^{n+1}(g_n + \sqrt{A}) + 2^{n+3}$. (Use $g_1 \equiv g_2 + 4 \bmod 8$ and $g_2 \equiv g_n \bmod 8$.) Finally, the congruences $g_n \equiv g_1 + 4 \bmod 8$ and $g_n \equiv g_{n-1} + 2^n \bmod 2^{n+1}$ imply $(g_n - g_1)(g_n - g_{n-1}) \equiv 2^{n+2} \bmod 2^{n+3}$, which combines with $A \equiv g_n^2 + 2^{n+2} \bmod 2^{n+3}$ to give $A + g_1 g_{n-1} \equiv g_n(g_{n-1} + g_1) \bmod 2^{n+3}$, and the product $[8, g_1 + \sqrt{A}][2^{n+1}, g_{n-1} + \sqrt{A}]$ can be written as $[2^{n+4}, 8(g_n + \sqrt{A}) + 2^{n+3}, 2^{n+1}(g_n + \sqrt{A}) + 2^{n+3}, (g_1 + g_{n-1})(g_n + \sqrt{A}) + 2^{n+3}\sigma]$, where $\sigma = 0$ or 1. The difference between $\frac{g_1 + g_{n-1}}{2}$ times the second term and 4 times the last term is $2^{n+3} \bmod 2^{n+4}$, because $\frac{g_1 + g_{n-1}}{2}$ is odd. Thus the first term can be changed to 2^{n+3}, and the desired product can be expressed as $[2^{n+3}, 8(g_n + \sqrt{A}), 2^{n+1}(g_n + \sqrt{A}), (g_1 + g_{n-1})(g_n + \sqrt{A})]$. The third term is a multiple of the second and can be dropped. Because $[8, g_1 + g_n] = [2]$, this brings $[8, g_1 + \sqrt{A}][2^{n+1}, g_{n-1} + \sqrt{A}]$ to the desired form $[2 \cdot 2^{n+2}, 2(g_n + \sqrt{A})]$.

The sequence $\bar{g}_1, \bar{g}_2, \bar{g}_2, \ldots$ defined by $\bar{g}_n \equiv -g_n \bmod 2^{n+2}$ satisfies the same conditions as the sequence $g_1, g_2, g_3 \ldots$, except that the condition $g_n \equiv 1 \bmod 4$ is replaced by $\bar{g}_n \equiv 3 \bmod 4$. Therefore, the same argument gives $\bar{\beta}^n / [2^{n-1}] = [2^{n+2}, -g_n + \sqrt{A}] = [2^{n+2}, \bar{g}_n + \sqrt{A}]$. For $n \geq 1$ there are four modules of the form $[2^{n+2}, g + \sqrt{A}]$, one for each of the four square roots of $A \bmod 2^{n+2}$, which are $\pm g_n$ and $\pm g_n + 2^{n+1}$. The first two have been accounted for. The remaining two are $\alpha\beta^{n+1}/[2^{n+1}]$ and $\alpha\bar{\beta}^{n+1}/[2^{n+1}]$, as follows from the observation that the last term in $[2, 1 + \sqrt{A}][2^{n+3}, \pm g_{n+1} + \sqrt{A}] = [2^{n+4}, 2(\pm g_{n+1} + \sqrt{A}), 2^{n+3}(1 + \sqrt{A}), \pm g_{n+1} + A + (1 \pm g_{n+1})\sqrt{A}]$ can be changed to $(1 \pm g_{n+1})(\pm g_{n+1}) + 2^{n+3} + (1 \pm g_{n+1})\sqrt{A}$, which is 2^{n+3} plus a multiple of the second term; therefore, the first term can be replaced with 2^{n+3}, so that the module becomes $[2^{n+3}, 2(\pm g_{n+1} + \sqrt{A})] = [2][2^{n+2}, \pm g_n + 2^{n+1} + \sqrt{A}]$, as was to be shown. When $n = 0$ the result is $[2][4, \pm 1 + \sqrt{A}]$, which accounts for the two modules $\alpha\beta/[2] = [4, 1 + \sqrt{A}]$ and $\alpha\bar{\beta}/[2] = [4, 3 + \sqrt{A}]$ and completes the proof.

These three propositions cover all products $[f, g + \sqrt{A}][F, G + \sqrt{A}]$ *except* those in which some prime p divides both f and F *and* divides A twice. In particular, if A is *square-free,* it describes all products.

The description of products of equivalence classes of modules is in principle much easier, because the number of equivalence classes is *finite,* so one can simply compile a multiplication table showing all products of all pairs of classes. However, multiplication tables are rarely very enlightening. More insight into the class semigroup is obtained by considering specific features. In particular, as Gauss's work in Section 5 of *Disquisitiones Arithmeticae* showed, enough information about the class semigroup on which to base a proof of quadratic reciprocity is provided by analyzing the classes that are *ambiguous* in the sense defined below.

Lemma. *If $[f, g + \sqrt{A}]$ is stable and if $[f_1, g_1 + \sqrt{A}]$ is its successor in the comparison algorithm, then $[f, f - g + \sqrt{A}]$ is the successor of $[f_1, f_1 - g_1 + \sqrt{A}]$ in the comparison algorithm.*

Proof. By definition, $[f_1, g_1 + \sqrt{A}]$ is determined by $ff_1 = r^2 - A$ and $g_1 \equiv r \bmod f_1$ where r is the least solution of $r + g \equiv 0 \bmod f$ whose square is greater than A. Similarly, the successor of $[f_1, f_1 - g_1 + \sqrt{A}]$ is $[f', g' + \sqrt{A}]$ where $f'f_1 = r_1^2 - A$, $g' \equiv r_1 \bmod f'$, and r_1 is the least solution of $r_1 \equiv g_1 \bmod f_1$ whose square is greater than A. It was shown in Essay 3.3 in the proof that the comparison algorithm permutes stable modules that r is the least number in its class mod f_1 whose square is greater than A. Thus, since r_1 and r are both $g_1 \bmod f_1$, $r_1 = r$, which implies $f' = f$ and $g' \equiv r_1 \equiv -g \bmod f$, as was to be shown.

Thus, if $[f, g + \sqrt{A}]$ is stable, the cycle of $[f, f - g + \sqrt{A}]$ contains the modules obtained by changing g to $f - g$ in the modules in the cycle of $[f, g + \sqrt{A}]$, but they are traversed in the opposite direction. In particular, if $[f, f - g + \sqrt{A}] \sim [f, g + \sqrt{A}]$ holds for *one* module in a cycle it holds for *all*. When this is the case, the cycle is called **ambiguous**.

Let M_1, M_2, ..., M_ν be the modules in an ambiguous cycle in the order given them by the comparison algorithm. Let the definition of M_i be extended to all integers i by setting $M_i = M_j$ whenever $i \equiv j \bmod \nu$, and, for $M_i = [f_i, g_i + \sqrt{A}]$, let $\overline{M}_i$ denote $[f_i, f_i - g_i + \sqrt{A}]$. Call a stable module M_i **pivotal** if (1) $\overline{M}_i = M_i$ or (2) $\overline{M}_i = M_{i-1}$. An ambiguous cycle contains *exactly two* pivotal modules—unless it contains just one module—as can be seen as follows:

Let the given cycle be ambiguous and let $\mu \geq 0$ satisfy $M_0 = \overline{M}_\mu$. By the lemma, $M_1 = \overline{M}_{\mu-1}$, $M_2 = \overline{M}_{\mu-2}$, and so forth. If $\mu > 1$ renumber the modules in the cycle by setting $M_i' = M_{i+1}$. Then $M_i' = M_{i+1} = \overline{M}_{\mu-i-1} = \overline{M}_{\mu-i-2}'$, so the renumbering of the modules has the effect of reducing μ by 2. In this way, μ can be successively reduced until it is 0 or 1.

If $\mu = 0$, then $M_i = \overline{M}_{(-i)} = \overline{M}_{\nu-i}$ for each i, which is to say $\overline{M}_i = M_{\nu-i}$, so M_i is pivotal of type (1) if and only if $i \equiv -i \bmod \nu$ and pivotal of type (2) if and only if $i \equiv 1 - i \bmod \nu$. Thus, when $\nu = 2\sigma$ for $\sigma > 0$ the only pivotal modules are M_0 and M_σ (because 0 and σ are the only numbers i less than ν for which $2i \equiv 0 \bmod 2\sigma$ and $2i \equiv 1 \bmod 2\sigma$ has no solutions at all) and when $\nu = 2\tau + 1$ the only pivotal modules are M_0 and $M_{\tau+1}$. Similarly, if $\mu = 1$, then $\overline{M}_i = M_{\nu+1-i}$, so M_i is pivotal of type (1) if and only if $i \equiv 1 - i \bmod \nu$ and pivotal of type (2) if and only if $i \equiv 2 - i \bmod \nu$, from which it follows that there are two pivotal modules when $\nu = 2\sigma$ for $\sigma > 0$, namely, M_1 and $M_{\sigma+1}$, and two pivotal modules when $\nu = 2\tau + 1$, namely, M_1 and $M_{\tau+1}$. That there are two pivotal modules unless $\nu = 1$ follows from the observation that the modules given by these formulas are distinct unless $\nu = 1$.

Theorem 2. *(a) If A is an odd prime, there are four pivotal modules, $[1, \sqrt{A}]$, $[2, 1 + \sqrt{A}]$, $[A, \sqrt{A}]$, and $[\frac{A-1}{2}, 1 + \sqrt{A}]$ (except that in the cases $A = 3$ and $A = 5$ the last of these coincides with one of the first two to form a cycle of length 1).*

(b) If A is twice an odd prime, say $A = 2p$, there are four pivotal modules, $[1, \sqrt{A}]$, $[2, \sqrt{A}]$, $[p, \sqrt{A}]$, and $[A, \sqrt{A}]$.

(c) If A is a product of two odd primes, say $A = pq$, where $p < q$, there are eight pivotal modules, $[1, \sqrt{A}]$, $[p, \sqrt{A}]$, $[q, \sqrt{A}]$, $[A, \sqrt{A}]$, $[2, 1 + \sqrt{A}]$, $[2p, p + \sqrt{A}]$, $[\frac{A-1}{2}, 1 + \sqrt{A}]$ and $[\frac{q-p}{2}, p + \sqrt{A}]$ (except that the last of these coincides with one of the others to form a cycle of length 1 if $q - p = 2$ or 4).

Thus, in cases (a) and (b) there are 2 ambiguous cycles, and in case (c) there are four.

Proof. (a) A module $[f, g + \sqrt{A}]$ in canonical form is pivotal of type (1) if and only if $g \equiv -g \bmod f$. When this is the case, $2g \equiv 0 \bmod f$ and $4A \equiv (2g)^2 \equiv 0 \bmod f$; that is, f is a factor of $4A$. Since A is an odd prime, $f = 1, 2, 4$, or A, because $f \leq A$ in a stable module. Since $f = 4$ would imply $2g \equiv 0 \bmod 4$ and $g \equiv 0 \bmod 2$, it is impossible in view of $g^2 \equiv A \bmod f$. Therefore, the only possible pivotal modules of type (1) in this case are $[1, \sqrt{A}]$, $[2, 1 + \sqrt{A}]$, and $[A, \sqrt{A}]$, all of which are clearly pivotal of type (1). If $[f, g + \sqrt{A}]$ is pivotal of type (2), then its predecessor in the comparison algorithm is $[f, f - g + \sqrt{A}]$, and the r that determines the step from $[f, f - g + \sqrt{A}]$ to $[f, g + \sqrt{A}]$ satisfies $f^2 = r^2 - A$. Thus, $A = (r - f)(r + f)$, which implies that $r - f = 1$ and $r + f = A$ and therefore that $f = \frac{A-1}{2}$, $r = \frac{A+1}{2}$. Since the new g is congruent to $r \bmod f$, and since $r \equiv 1 \bmod f$, the only possible pivotal module of type (2) is $\left[\frac{A-1}{2}, 1 + \sqrt{A}\right]$, which is in fact pivotal of type (2).

(b) As in the proof of (a), if $[f, g + \sqrt{A}]$ is pivotal of type (1), then $4A \equiv 0 \bmod f$ and $f \leq A$, so $f = 1, 2, 4, p$, or $2p$. Again, $f = 4$ is impossible because it would imply $A \equiv 0 \bmod 4$, so the only pivotal modules of type (1) are the ones listed in (b). There are no pivotal modules of type (2) because $A \equiv 2 \bmod 4$, so $r^2 - f^2 = A$ is impossible.

(c) If f is a factor of $4A$ less than or equal to A, then $f = 1, 2, 4, p, q, 2p, 2q$, or $A = pq$. Since $[2q, q + \sqrt{A}]$ is not stable (both q^2 and $(2q - q)^2$ are $q^2 > pq = A$), the only possible pivotal modules of type (1) are $[1]$, $[2, 1 + \sqrt{A}]$, $[p, \sqrt{A}]$, $[q, \sqrt{A}]$, $[2p, p + \sqrt{A}]$, and $[A, \sqrt{A}]$, all of which are indeed pivotal of type (1). Those that are pivotal of type (2) satisfy $A = (r - f)(r + f)$ as in the proof of (a), where r is the number used by the comparison algorithm to go to the pivotal module from its predecessor. Thus, either $r - f = 1$ and $r + f = A$ or $r - f = p$ and $r + f = q$. In the first case, $r \equiv 1 \bmod f$, so the module must be $\left[\frac{A-1}{2}, 1 + \sqrt{A}\right]$, and in the second case, $r \equiv p \bmod f$, so the module must be $\left[\frac{q-p}{2}, p + \sqrt{A}\right]$, which completes the list.

The complications that arise in the cases $A \equiv 1 \bmod 4$ of Proposition 3 stem from the fact that in these cases *the class semigroup is not a group*, which is to say that there are classes without inverses. For example, the square of $[2, 1 + \sqrt{5}]$ is $[2][2, 1 + \sqrt{5}]$, but $[2, 1 + \sqrt{5}]$ constitutes a cycle of length 1 and is therefore not equivalent to $[1]$. Therefore, the class of $\alpha = [2, 1 + \sqrt{5}]$ has no inverse because $\alpha\gamma \sim [1]$ would imply $\alpha = \alpha[1] \sim \alpha^2\gamma = [2]\alpha\gamma \sim \alpha\gamma \sim [1]$. A module whose class is invertible in the class semigroup is called **primitive**.

The **class group** for a number A, not a square, consists of the invertible elements of the class semigroup, those classes whose elements are primitive. For any given A, the class semigroup is found by finding the cycles of stable modules. One can then find the class group by using the following theorem to determine which cycles contain primitive modules:

Theorem 3. *Let $[e][f, g + \sqrt{A}]$ be a module in canonical form, and let d be the greatest common divisor of f, $2g$, and $\frac{|g^2 - A|}{f}$. If $d = 1$, then $[f, g + \sqrt{A}][f, f - g + \sqrt{A}] = [f]$; in particular, $[e][f, g + \sqrt{A}]$ is primitive, because the class of $[f, f - g + \sqrt{A}]$ is inverse to its class. If $d > 1$, $[e][f, g + \sqrt{A}]$ is not primitive.*

In particular, if $A \equiv 1 \bmod 4$, the module $[2, 1 + \sqrt{A}]$ is not primitive, because in this case $d = 2$.

Proof. By direct computation

$$[f, g + \sqrt{A}][f, f - g + \sqrt{A}] = [f^2, -fg + f\sqrt{A}, fg + f\sqrt{A}, fg - g^2 + A + f\sqrt{A}]$$
$$= [f^2, -2fg, fg + f\sqrt{A}, |g^2 - A|]$$
$$= [f]\left[f, 2g, \frac{|g^2 - A|}{f}, g + \sqrt{A}\right]$$
$$= [f][d, g + \sqrt{A}],$$

where use is made of the first term f^2 to compute with the coefficients of the other terms as numbers mod f^2. If $d = 1$, this product is $[f]$, which proves the first statement.

Now, $[d, g + \sqrt{A}][f, g + \sqrt{A}] = [df, d(g + \sqrt{A}), g^2 + A + 2g\sqrt{A}] = [df, d(g + \sqrt{A}), A - g^2]$ (subtract $2g/d$ times the second term from the third, computing mod df) $= [d][f, g + \sqrt{A}]$ (because $g^2 \equiv A \bmod df$). Therefore $[d, g + \sqrt{A}][f, g + \sqrt{A}] = [d][f, g + \sqrt{A}] \sim [f, g + \sqrt{A}]$, which shows that if $[f, g + \sqrt{A}]$ is primitive, then $[d, g + \sqrt{A}] \sim [1]$, which is to say that *repeated application of the comparison algorithm to $[d, g + \sqrt{A}]$ must eventually reach* $[1]$. But if d divides f, $2g$, and $\frac{|g^2 - A|}{f}$ for any module $[f, g + \sqrt{A}]$ in canonical form—as is the case with the module $[d, g_1 + \sqrt{A}]$ when g_1 is the smallest number congruent to $g \bmod d$ because $g_1 + g$ and $g_1 - g$ are both zero mod d, so $g_1^2 - A \equiv g^2 - A + (g_1 + g)(g_1 - g) \equiv 0 \bmod d^2$—then d divides F, $2G$, and $\frac{|G^2 - A|}{F}$, where $[F, G + \sqrt{A}]$ is the successor of $[f, g + \sqrt{A}]$ in the comparison algorithm, as can be seen as follows: Let $r = \nu f - g$ be the number used to find $[F, G + \sqrt{A}]$. Then $fF = r^2 - A = \nu^2 f^2 - 2g\nu f + g^2 - A \equiv 0 \bmod df$, which implies $F \equiv 0 \bmod d$. Moreover, $G = \mu F + r$ for some μ, which gives $2G \equiv 2\mu F + 2\nu f - 2g \equiv 0 \bmod d$. Finally, $G^2 - A \equiv r^2 - A + 2\mu Fr + \mu^2 F^2 \equiv Ff + 2\mu Fr + \mu^2 F^2 \equiv Ff + 2\mu FG \equiv 0 \bmod dF$, so d divides $\frac{|G^2 - A|}{F}$. (Note that μ may be negative in these congruences, so that the argument is still valid when $G < r$.)

Therefore, if $[f, g + \sqrt{A}]$ is primitive, d must divide 1, as was to be shown.

Corollary. *Elements of the class group whose squares are the identity correspond one-to-one to ambiguous cycles whose modules are primitive.*

Deduction. An equivalence class whose square is the identity contains modules whose squares are equivalent to $[1]$; in particular, the modules it contains are all primitive. By the theorem just proved, if $[f, g + \sqrt{A}]$ is a stable module in canonical form that is in such a class, then both $[f, g + \sqrt{A}]$ and $[f, f - g + \sqrt{A}]$ are in the class inverse to the class of $[f, g + \sqrt{A}]$, which shows that the cycle of $[f, g + \sqrt{A}]$ is ambiguous. Conversely, if $[f, g + \sqrt{A}]$ is both primitive and ambiguous, then $[f, g + \sqrt{A}]^2 \sim [f, g + \sqrt{A}][f, f - g + \sqrt{A}] = [f] \sim [1]$.

When this corollary is combined with Theorem 2, it determines the elements of the class group of order 2 in a few cases:

Theorem 4. *If A is prime and congruent to 1 mod 4, the class group has no element of order 2. If A is prime and congruent to 3 mod 4, or if A is a product of two primes $A = pq$ for which $p + q \not\equiv 0$ mod 4, the class group has a unique element of order 2.*

Proof. When A is prime and 1 mod 4, $[2, 1 + \sqrt{A}]$ and $[\frac{A-1}{2}, 1 + \sqrt{A}]$ are not primitive, because the greatest common divisor of f, $2g$, and $\frac{|A - g^2|}{f}$ is 2 for each of them, so at most two pivotal modules are primitive. The cycle of $[1]$ therefore is the only one that represents an element of the class group whose square is the identity. (The few cases in which there are cycles of length 1 are enumerated in Theorem 2.) Therefore, the class group contains no elements of order 2.

On the other hand, when A is prime and 3 mod 4, or when $A = 2p$ where p is an odd prime, all four of the pivotal modules identified in Theorem 2 are primitive. Therefore, there are two primitive, ambiguous classes. The case $A = pq$ in which p and q are odd primes that satisfy $p \equiv q$ mod 4 is similar, because the four pivotal modules other than $[1]$, $[p, \sqrt{A}]$, $[q, \sqrt{A}]$, and $[A, \sqrt{A}]$ are not primitive (for all of them, f, $2g$, and $\frac{|g^2 - A|}{f}$ are all even). Thus, in these cases, the class group has just one element of order 2.

The Class Semigroup for Various Values of A
(Compare to the table of Essay 3.3)

Value of A	Class group	Representatives of imprimitive classes
2	Trivial group	none
3	Group of order 2	none
5	Trivial group	$[2, 1 + \sqrt{5}]$
6	Group of order 2	none
7	Group of order 2	none
8	Group of order 2	$[2, \sqrt{8}]$
10	Group of order 2	none
11	Group of order 2	none
12	Group of order 2	$[2, \sqrt{12}], [6, \sqrt{12}]$
13	Trivial group	$[2, 1 + \sqrt{13}]$
14	Group of order 2	none
15	Four-group	none
17	Trivial group	$[2, 1 + \sqrt{17}]$
18	Group of order 2	none
19	Group of order 2	none
20	Group of order 2	$[2, \sqrt{20}]$
21	Group of order 2	$[2, 1 + \sqrt{21}], [6, 3 + \sqrt{21}]$
79	Cyclic of order 6	none

Essay 3.5 Is A a Square Mod p?

...eine noch höhere Bedeutung haben sie [die Reciprocitätsgesetze] in der geschichtlichen Entwickelung dieser mathematischen Disciplin [Zahlentheorie] dadurch erlangt, daß die Beweise derselben, so weit sie überhaupt gefunden sind, fast durchgängig aus neuen, bis dahin noch unerforschten Gebieten haben geschöpft werden müssen, welche so der Wissenschaft aufgeschlossen worden sind. (...the reciprocity laws attained an even greater significance in the historical development of number theory by the fact that their proofs, insofar as proofs have been found, had to be sought in areas that were hitherto almost completely unexplored and that in this way were opened to science.)—E. E. Kummer [47, Introduction]

Essay 3.4 gives a description of the semigroup of modules for a given A (A not a square) that is virtually complete except that it leaves untouched the obvious question raised by its Proposition 1: For which odd primes p not dividing A are there modules $[p, g + \sqrt{A}]$ in canonical form? This is the question

$$\text{"What is the value of } \chi_p(A)?\text{"}$$

where, for a given odd prime p, χ_p is defined to be the **quadratic character** of numbers mod p, which is to say that χ_p is the function that assigns to a number A the value* -1 if the congruence $A \equiv x^2$ mod p has no solution x, the value 0 if $A \equiv 0$ mod p, and the value 1 otherwise.

   In this way, the evaluation of $\chi_p(A)$ is essential to computation in the semigroup of modules of hypernumbers for a given A. The problem of evaluating $\chi_p(A)$ engaged Euler's interest rather early in his career (see [17]), when he discovered *empirically* the amazing fact that *if p and q are primes that satisfy $p \equiv q$ mod $4A$, then $\chi_p(A) = \chi_q(A)$.* In other words, the answer to the question "Is A a square mod p?" depends only on the class of the prime p mod $4A$. Euler made many attempts to prove what he had found empirically; in the process, he found refinements and generalizations of the phenomenon, thereby setting much of the agenda for number theory for the next hundred years, but the hoped-for proof eluded him.

   What Euler knew but couldn't prove about the values of $\chi_p(A)$ implies the law of quadratic reciprocity fairly easily. This law, which is stated below, was put in its usual form by Legendre and was first proved by the young Gauss, who gave *two* proofs in *Disquisitiones Arithmeticae*, published in 1801 when he was 24 years old. The second of these uses his theory of composition of binary quadratic forms. The proof that will be given in this essay is inspired by Gauss's second proof, but it will use *modules* and their *multiplication* instead of *quadratic forms* and their *composition*.

   The law of quadratic reciprocity is one case of a general formula for $\chi_p(A)$ of the form

---

* Use of the "negative number" -1 can be avoided by treating the values of χ_p as numbers mod 4, so that $-1 \equiv 3$, $(-1)^2 \equiv 1$.

$$(1) \qquad \chi_p(A) = \sigma_A(p) \prod_{A_i} \chi_{A_i}(p)$$

in which p is an odd prime, A is a square-free number that is not divisible by p, the product on the right is a product over all odd* prime factors A_i of A, and $\sigma_A(p)$ depends only on the classes of p and A mod 8; in fact $\sigma_A(p)$ depends only on the classes of p and A mod 4 when A is odd. The formula for $\sigma_A(p)$ is given at the end of the essay.

Euler's observation that $\chi_p(A) = \chi_q(A)$ when $p \equiv q \bmod 4A$ follows immediately from (1), because $p \equiv q \bmod 4A$ implies $p \equiv q \bmod A_i$ for all odd prime factors A_i of A and implies $p \equiv q \bmod 4$ when A is odd, $p \equiv q \bmod 8$ when A is even.

The derivation of formula (1) is the subject of the present essay. The law of quadratic reciprocity is simply the case in which A is an odd prime. However, the derivation will begin with this special case. It will use two lemmas:

Lemma 1. *If p is an odd prime, $\chi_p(p-1)$ is 1 when $p \equiv 1 \bmod 4$ and -1 when $p \equiv 3 \bmod 4$.*

Proof. I will use without proof the fundamental fact that *for any prime p there is a primitive root γ mod p*, which is to say a number γ with the property that every number not divisible by p is congruent to a power of γ mod p. (See Section 3 of *Disquisitiones Arithmeticae*, which contains two proofs, the first in Articles 39 and 54, the second in Article 55.)

If γ is a primitive root mod p, then each of the $p-1$ numbers between 0 and p is congruent mod p to a unique power of γ in which the exponent is less than p. Since $\gamma^{2\mu} \equiv \gamma^\lambda \bmod p$ if and only if $2\mu \equiv \lambda \bmod (p-1)$, a power γ^λ of γ is a square mod p for $\lambda < p$ if and only if λ is even. Since the roots γ^0 and $\gamma^{(p-1)/2}$ of $x^2 - 1 \bmod p$ coincide with $\pm 1 \bmod p$, $p - 1 \equiv \gamma^{(p-1)/2} \bmod p$, which is a square mod p if and only if $(p-1)/2$ is even, or in other words, $\chi_p(p-1) = 1$ if and only if $p \equiv 1 \bmod 4$, as was to be shown.

Let χ_4 denote the function that assigns the value 0 to even numbers, the value 1 to numbers congruent to 1 mod 4, and the value -1 to numbers congruent to 3 mod 4. Then Lemma 1 can be stated as

$$\chi_p(p-1) = \chi_4(p) \quad \text{for any odd prime } p.$$

As is easily proved, $\chi_p(mn) = \chi_p(m)\chi_p(n)$ for all numbers m and n whenever p is prime and also when $p = 4$.

Lemma 2. *Given a primitive module and given a number N, construct a module $[f, g + \sqrt{A}]$ in canonical form for which (1) f is relatively prime to N and (2) the product of $[f, g + \sqrt{A}]$ and the given module is equivalent to $[1]$.*

---

* Since $\chi_2(p) = 1$ for all odd primes p, it makes no difference whether $A_i = 2$ is included in the product in formula (1) when A is even.

Proof. Let $[E][F, G+\sqrt{A}]$ be the given primitive module. Since it is equivalent to $[F, G + \sqrt{A}]$, one may as well assume $E = 1$. Choose μ large enough that $(\mu F + G)^2 > A$. From $G^2 \equiv A \bmod F$ it follows that $(\mu F + G)^2 - A \equiv 0 \bmod F$, say $HF = (\mu F + G)^2 - A$. For all numbers ν, $(\nu F + \mu F + G)^2 - A$ can be written in the form $F \cdot q(\nu)$, where $q(\nu)$ is a polynomial of degree 2 with coefficients that are numbers, namely,

$$q(\nu) = F\nu^2 + 2(\mu F + G)\nu + (\mu^2 F + 2\mu G + H).$$

A common divisor d of the coefficients of $q(\nu)$ divides F, $2G$, and H, so $(\mu F + G)^2 \equiv A \bmod dF$ and $G^2 \equiv A \bmod dF$; thus, d divides F, $2G$, and $\frac{|G^2 - A|}{F}$, which implies $d = 1$ because $[F, G + \sqrt{A}]$ is primitive by assumption. Thus, for any prime p, $q(\nu)$ is a nonzero polynomial mod p whose degree is 2 at most. Therefore, $q(\nu)$ has at most 2 roots mod p that are less than p. Moreover, since F and H cannot both be even (because at least one of F, $2G$, and H is odd), $q(\nu)$ is either odd when ν is even (when H is odd) or odd when ν is odd (when H is even). Let $p_1, p_2, \ldots, p_\sigma$ list the distinct prime divisors of FN. For each p_i, choose a number $\nu_i < p_i$ for which $q(\nu_i) \not\equiv 0 \bmod p_i$. Use the Chinese remainder theorem to construct a number $\nu < \prod p_i$ such that $\nu \equiv \nu_i \bmod p_i$ for each i. Then $q(\nu) \equiv q(\nu_i) \not\equiv 0 \bmod p_i$ for each i. In other words, $q(\nu)$ is relatively prime to FN.

Let $\rho = (\nu + \mu)F + G$ for the ν chosen in this way. Then $[\rho + \sqrt{A}] = [\rho^2 - A, \rho + \sqrt{A}] = [F \cdot q(\nu), \rho + \sqrt{A}] = [F, \rho + \sqrt{A}][q(\nu), \rho + \sqrt{A}]$ (because $q(\nu)$ and F are relatively prime) $= [F, G + \sqrt{A}][q(\nu), \rho + \sqrt{A}]$ (because $\rho \equiv G \bmod F$). Thus, because $[\rho + \sqrt{A}] \sim [1]$, the module $[q(\nu), \rho + \sqrt{A}]$ has the required properties.

Proposition. *If p is an odd prime divisor of A, the value of $\chi_p(f)$ is the same for any two equivalent primitive modules $[f, g + \sqrt{A}]$ in which $f \not\equiv 0 \bmod p$. In this way, χ_p determines a homomorphism from the class group to the group with two elements ± 1. Similarly, if $A \equiv 3 \bmod 4$, the value of $\chi_4(f)$ is the same for any two equivalent primitive modules $[f, g + \sqrt{A}]$ in which f is odd, and the function from the class group to ± 1 that χ_4 determines in this way is a homomorphism.*

Proof. Let $[f, g + \sqrt{A}]$ be a given primitive module in canonical form. Use Lemma 2 to find a module $[F, G+\sqrt{A}]$ in canonical form with $[f, g+\sqrt{A}][F, G+\sqrt{A}] \sim [1]$ and F relatively prime to pf. Since Lemma 2 implies that there is a module in canonical form $[\mathcal{F}, \mathcal{G} + \sqrt{A}]$ with $[\mathcal{F}, \mathcal{F} + \sqrt{A}][F, G + \sqrt{A}] \sim [1]$ and $\mathcal{F}$ relatively prime to pF, the equivalence class of $[f, g+\sqrt{A}]$ contains a module $[\mathcal{F}, \mathcal{G} + \sqrt{A}]$ (because both $[f, g+\sqrt{A}]$ and $[\mathcal{F}, \mathcal{G}+\sqrt{A}]$ are in the class inverse to the class of $[F, G + \sqrt{A}]$) in canonical form in which $\mathcal{F}$ is relatively prime to pF. What is to be shown is that $\chi_p(\mathcal{F}) = \chi_p(f)$ whenever $f \not\equiv 0 \bmod p$. This will be done by proving that if $f \not\equiv 0 \bmod p$, then $\chi_p(f) = \chi_p(F)$.

Now $[fF, z + \sqrt{A}] = [f, g + \sqrt{A}][F, G + \sqrt{A}] \sim [1]$ for $z \equiv g \bmod f$ and $z \equiv G \bmod F$, so there are principal modules $[v + u\sqrt{A}]$ and $[y + x\sqrt{A}]$ such that $[v + u\sqrt{A}][fF, z + \sqrt{A}] = [y + x\sqrt{A}][1]$. Since there is a principal module $[t + s\sqrt{A}]$ for which $[t + s\sqrt{A}][v + u\sqrt{A}] = [N]$ for some number N (Proposition 3 of Essay 3.3), one can assume without loss of generality that $u = 0$; that is, $[v][fF, z + \sqrt{A}] = [y + x\sqrt{A}]$ where $y^2 > Ax^2$. Since this equation implies that $x \equiv 0 \bmod v$ and $y \equiv zx \bmod vfF$ (see Essay 3.2), the equation can be divided by v to give one of the form $[fF, z + \sqrt{A}] = [y + x\sqrt{A}]$. This equation implies that x is relatively prime to y (any common divisor divides the coefficient of $\sqrt{A}$ in $z + \sqrt{A}$) and therefore that x is relatively prime to $y^2 - Ax^2$, from which it follows that $[y + x\sqrt{A}] = [y^2 - Ax^2, y + x\sqrt{A}] = [y^2 - Ax^2, \rho + \sqrt{A}]$ for some ρ (x is invertible mod $y^2 - Ax^2$). Therefore $fF = y^2 - Ax^2$. Since $A \equiv 0 \bmod p$ and $fF \not\equiv 0 \bmod p$, it follows that $fF \equiv y^2 \not\equiv 0 \bmod p$. Thus $\chi_p(f) = \chi_p(fF^2) = \chi_p(y^2 F) = \chi_p(F)$, as was to be shown.

The proof of the analogous theorem for χ_4 in the case $A \equiv 3 \bmod 4$ follows the same steps, except that one needs to prove that if $fF = y^2 - Ax^2$ and fF is odd, then $\chi_4(f) = \chi_4(F)$. This follows easily from the observation that y and x must have opposite parity (because fF is odd), so one of the terms y^2 and $-Ax^2$ is 0 mod 4 and the other is 1 mod 4, resulting in $\chi_4(fF) = 1$, from which $\chi_4(f) = \chi_4(F)$ follows.

Theorem. *Let p and q be distinct odd primes. If $p \equiv 1 \bmod 4$, then $\chi_q(p) = 1$ implies $\chi_p(q) = 1$. If $p \equiv 3 \bmod 4$, then $\chi_q(p) = 1$ implies $\chi_p(q) = \chi_4(q)$. If $p \equiv q \equiv 3 \bmod 4$, then $\chi_p(q) = -\chi_q(p)$.*

Proof. If $\chi_q(p) = 1$, there is a module $[q, g + \sqrt{p}] \neq [1]$. By the proposition, the value of $\chi_p(q)$ depends only on the class of $[q, g + \sqrt{p}]$. The kernel of the homomorphism defined by χ_p from the class group to the group with two elements is either a subgroup of index two or it is the whole group.

When $p \equiv 1 \bmod 4$, Theorem 4 of the preceding essay states that the class group has no element of order two. Therefore, the group has odd order (see Essay 5.2), so it can have no subgroup of index two, which implies that the kernel is the whole group, so $\chi_q(p) = 1$ implies $\chi_p(q) = 1$, as was to be shown.

When $p \equiv 3 \bmod 4$, on the other hand, the class group contains a single element of order two, so the operation of squaring is a two-to-one homomorphism from the class group to itself whose image is a subgroup of index two. This subgroup is necessarily the kernel of the homomorphism determined by χ_p, because this kernel contains all squares but does not contain the class of $[p-1, 1 + \sqrt{p}]$, because $\chi_p(p-1) = -1$. The homomorphism determined by χ_4 also has the subgroup of squares as its kernel, because its value for the class of $[p, \sqrt{p}]$ is $\chi_4(p) = -1$. Therefore, these two homomorphisms are identical, and the statement to be proved—that $\chi_q(p) = 1$ implies $\chi_p(q) = \chi_4(q)$—follows.

When $p \equiv q \equiv 3 \bmod 4$ one finds in a similar way that if either $\chi_p(q)$ or $\chi_q(p)$ is 1, then the other is -1, but the possibility that both might be -1 remains. Consider the class group in the case $A = pq$. By Theorem 4 of the last essay, the class group in this case has a single element of order 2, so the squares form a subgroup of index two, as before, that is obviously contained in the kernel of the homomorphisms from the class group to the group with two elements that are determined by either χ_p or χ_q. In fact, since the class of $[pq-1, 1+\sqrt{pq}]$ is not in either kernel, these homomorphisms have the same kernel—the subgroup of squares—and are therefore identical. The stable modules $[1]$, $[p, \sqrt{pq}]$, $[q, \sqrt{pq}]$, and $[pq, \sqrt{pq}]$ are partitioned between the two ambiguous, primitive cycles. Since $[1]$ is in the principal cycle and $[pq, \sqrt{pq}]$ is not (one step of the comparison algorithm shows that $[pq, \sqrt{pq}] \sim [pq-1, 1+\sqrt{pq}]$), exactly one of $[p, \sqrt{pq}]$ and $[q, \sqrt{pq}]$ is in the kernel of the homomorphism in question, which is to say that exactly one of $\chi_q(p)$ and $\chi_p(q)$ is 1, as was to be shown.

The Law of Quadratic Reciprocity. *If p and q are distinct odd primes, then $\chi_p(q) = \chi_q(p)$ unless $p \equiv q \equiv 3 \bmod 4$, in which case $\chi_p(q) = -\chi_q(p)$.*

Proof. The last statement is of course part of the previous theorem.

Since $\chi_p(q) = -1$ is the negation of $\chi_p(q) = 1$, the statement that $\chi_p(q) = \chi_q(p)$ is the statement that $\chi_p(q) = 1$ if and only if $\chi_q(p) = 1$.

When $p \equiv q \equiv 1 \bmod 4$, the theorem proves that $\chi_q(p) = 1$ implies $\chi_p(q) = 1$, and the desired conclusion follows by symmetry.

When $p \equiv 1 \bmod 4$ and $q \equiv 3 \bmod 4$, the theorem proves that $\chi_q(p) = 1$ implies $\chi_p(q) = 1$ and that $\chi_p(q) = 1$ implies $\chi_q(p) = \chi_4(p) = 1$, as was to be shown.

Evaluation of $\chi_p(2)$. *If p is an odd prime, $\chi_p(2) = 1$ if and only if $p \equiv \pm 1 \bmod 8$.*

Proof. Consider the class group in the case $A = 8$. It has two elements, the class of $[7, 1+\sqrt{8}]$ and the principal class. An odd prime p satisfies $\chi_p(2) = 1$ if and only if it satisfies $\chi_p(8) = 1$, which is true if and only if there is a module $[p, g+\sqrt{8}] \neq [1]$ for some g. When this is the case, either $[p, g+\sqrt{8}]$ or $[p, q+\sqrt{8}][7, 1+\sqrt{8}]$ is principal, which implies (unless $p = 7$) that either p or $7p$ is of the form $y^2 - 8x^2$ and therefore that $p \equiv \pm 1 \bmod 8$.

If $p \equiv 1 \bmod 8$, then either $[8, 1+\sqrt{p}]$ or $[8, 5+\sqrt{p}]$ is primitive (because if $(p-1)/8$ is even, then $|p-25|/8$ is odd). The homomorphism from the class group to ± 1 determined by χ_p is trivial (since $p \equiv 1 \bmod 4$, the class group has no element of order 2), so $\chi_p(8) = 1$ and $\chi_p(2) = 1$.

Finally, if $p \equiv 7 \bmod 8$, then the last two of the four pivotal modules $[1, \sqrt{p}]$, $[2, 1+\sqrt{p}]$, $[p, \sqrt{p}]$, and $[\frac{p-1}{2}, 1+\sqrt{p}]$ are not principal—$[p, \sqrt{p}]$ because it is equivalent to $[p-1, 1+\sqrt{p}]$, for which $\chi_p(p-1) = -1$, and $[\frac{p-1}{2}, 1+\sqrt{p}]$ because $\chi_4(\frac{p-1}{2}) = -1$. Therefore $[2, 1+\sqrt{p}]$ must be principal, which implies that $[2, 1+\sqrt{p}] = [y+x\sqrt{p}]$, where $2 = y^2 - px^2$. Thus, $\chi_p(2) = \chi_p(y^2) = 1$, as was to be shown.

Let χ_8 denote the function that assigns the value 0 to even numbers, the value 1 to numbers congruent to $\pm 1 \bmod 8$, and the value -1 to numbers congruent to $\pm 3 \bmod 8$, so that $\chi_p(2)$ is $\chi_8(p)$ for all odd primes p.

Evaluation of $\chi_p(A)$.   *Let A be a square-free number and let p be an odd prime that does not divide A. The coefficient $\sigma_A(p)$ in formula (1) at the beginning of the essay is given by*

$$\sigma_A(p) = \begin{cases} 1 & \text{if } A \equiv 1 \bmod 4, \\ \chi_4(p) & \text{if } A \equiv 3 \bmod 4, \\ \chi_8(p) & \text{if } A \equiv 2 \bmod 8, \\ \chi_4(p)\chi_8(p) & \text{if } A \equiv 6 \bmod 8. \end{cases}$$

In particular, $\sigma_A(p)$ depends only on the classes of p and A mod 4 when A is odd and only on their classes mod 8 when A is even.

Proof. Since $\chi_p(A) = \prod \chi_p(A_i)$, where the product is over the prime factors A_i of A, the evaluations of $\chi_p(q)$ for prime q given above imply that

$$\sigma_A(p) = \chi_4(p)^\nu$$

if A is odd or

$$\sigma_A(p) = \chi_8(p)\chi_4(p)^\nu$$

if A is even, where ν is the number of prime factors A_i of A that are 3 mod 4. Because $\chi_4(p)^\nu$ is 1 when ν is even and $\chi_4(p)$ when ν is odd, the given formulas follow when one observes that an odd A is 3 mod 4 if and only if ν is odd, and an even A is 6 mod 8 if and only if ν is odd.

Essay 3.6 Gauss's Composition of Forms

The structure of Gauss's *Disquisitiones Arithmeticae* suggests that the original purpose of his theory of composition of forms was to put the law of quadratic reciprocity in a setting that would make it seem clearer and more natural. In the first three sections of the book he introduces the elementary theory of congruences and proves the important theorem that there is a primitive root mod p for every prime p. In Section 4 he goes on to the statement and proof of what he calls the "fundamental theorem," essentially the law of quadratic reciprocity. His proof in Section 4 was described by H. J. S. Smith [60, Part 1, Art. 18] as "repulsive to all but the most laborious students." Perhaps Gauss felt the same way about it, because he gave a second and altogether different proof in Section 5; in later years he gave other proofs, indicating that even then he was not satisfied that he had grasped the true basis of the phenomenon.

It is misleading to think of Section 5 as just one of seven sections of the book, because in number of pages it is more than *half* of the book. A large part of it is devoted to the theory of "composition" of binary quadratic forms, which is used (Article 262) to prove the "fundamental theorem," but which is also used in the study of ternary forms and is studied for its own sake. Surely "composition" represents an early step in Gauss's quest for the deeper secrets of number theory.

Section 5 had a profound effect on the development of number theory in the 19th century. Kummer's proof of his generalized reciprocity law in mid-century, a proof that was found only after years of intense effort, was directly inspired [47, p. 20 (700)] by Gauss's proof of quadratic reciprocity in Section 5. But beyond that, Section 5 was fundamental to the development of Dedekind's theory of "ideals" in the second half of the century, and in that way directly influenced the core ideas of modern abstract algebra. Moreover, the use of the structure of the group (without the name) of equivalence classes of binary quadratic forms in Section 5 (together with another implicit use of groups in Section 7) contributed to the development of the theory of groups. But perhaps the profoundest way in which Section 5 affected the development of mathematics lay in the challenge that it presented. Starting with Dirichlet, and continuing with Kummer, Dedekind, Kronecker, Hermite, and countless others, the unwieldy but fruitful theory of composition of forms called forth great efforts of study and theory-building that shaped modern mathematics.

The "forms" in Gauss's theory are **binary quadratic forms**, which is to say that they are homogeneous polynomials (forms) of degree 2 (quadratic) in 2 variables (binary) with integer coefficients. He used the notation $ax^2 + 2bxy + cy^2$ for such forms and, as this notation indicates, he only considered forms with *even* middle coefficients. Because I prefer not to impose this restriction, I have chosen different letters altogether and will denote binary quadratic forms by $rx^2 + sxy + ty^2$, where x and y are the variables of the form and r, s, and t are its integer coefficients. (In this essay and the next, I will do as Gauss did and use *integers* instead of the *numbers* in the narrowest sense that were used in the other essays of Part 3.) I will also write the form as (r, s, t) when the variables can remain unnamed.

The notion of "composition" generalizes Brahmagupta's formula

$$(x^2 - Dy^2)(u^2 - Dv^2) = X^2 - DY^2 \quad \text{when } X = xu + Dyv \text{ and } Y = xv + yu$$

(where D is a fixed integer). Other examples are

$$(x^2 + xy + y^2)(u^2 + uv + v^2) = X^2 + XY + Y^2$$
$$\text{when } X = xu - yv \text{ and } Y = xv + yu + yv$$

and

$$(16x^2 + 4xy - y^2)(4u^2 + 2uv - v^2) = 4X^2 + 6XY + Y^2$$
$$\text{when } X = 4xu - 2xv - yu \text{ and } Y = 4xv + 2yu + yv.$$

These formulas can be verified by the lengthy but simple process of performing the prescribed substitutions for X and Y on the right and expanding to find that the resulting polynomial in x, y, u, and v is indeed the product of the two polynomials on the left.

In general, given two binary quadratic forms $rx^2 + sxy + ty^2$ and $\rho u^2 + \sigma uv + \tau v^2$ with integer coefficients, a third form $RX^2 + SXY + TY^2$ is **transformable** into their product (see Gauss's §235) if one can define X and Y to be sums of integer multiples of the monomials xu, xv, yu, and yv in such a way that

$$(rx^2 + sxy + ty^2)(\rho u^2 + 2\sigma uv + \tau v^2) = RX^2 + SXY + TY^2.$$

A form that is transformable into the product of two forms **composes** those forms if the six 2×2 minors of the 2×4 matrix of coefficients of the expressions of X and Y in terms of xu, xv, yu, and yv that effect the transformation have no common divisor greater than 1. (It is easy to check that the three formulas above are compositions. For example, the last formula shows that $4X^2 + 6XY + Y^2$ composes $16x^2 + 4xy - y^2$ and $4u^2 + 2uv - v^2$ because the minors of the matrix of coefficients

$$\begin{bmatrix} 4 & -2 & -1 & 0 \\ 0 & 4 & 2 & 1 \end{bmatrix}$$

that effects the transformation are 16, 8, 4, 0, -2, and -1.)

To modern ears, the phrase "composition of forms" suggests a binary operation, assigning a *composed* form to each pair of given forms, but Gauss's compositions do not conform to this expectation. On the one hand, there may be *no* form that composes two given forms, while on the other hand, if some form *does* compose them, then *infinitely many others* also do,* because

---

* Both the English and the German translations of the *Disquisitiones* wrongly translate the theorems of §236 and §249, among others, when they use definite articles rather than indefinite ones; the original Latin of course has no articles.

infinitely many others can be obtained by a **unimodular change of variables** in the composed form. Specifically, a matrix $\begin{bmatrix} a & b \\ c & d \end{bmatrix}$ with determinant 1 can be used to define U and V as sums of multiples of xu, xv, yu, and yv by $U = aX + bY$ and $V = cX + dY$; then substitution of $X = dU - bV$ and $Y = -cU + aV$ in the known composition using X and Y produces a composition using U and V.

It seems fair to say that Gauss's theory in its full generality is largely forgotten. When André Weil writes [63, p. 334] that Dirichlet "restored its original simplicity" he is overlooking the fact that Dirichlet composes only *certain* pairs of forms—pairs that are concordant or *einig*—and that Dirichlet justified this limitation by shifting all emphasis from the composition of *forms* to the composition of *equivalence classes* of forms, thus disregarding Gauss's success in developing the theory in the greatest possible generality.

One could certainly argue that in this case Gauss's insistence on generality was excessive—that the purposes to which the theory is put are served just as well by the mere composition of equivalence classes, and that the classification of forms is so natural that the binary operation of composition of equivalence classes is a legitimate subject of study—but Gauss evidently disagreed.

The technical demands of developing the theory in the way Gauss does are indeed formidable. This becomes clear from Gauss's very statement of what amounts to the associative law, not to mention the difficulty of proving the statement that *if F composes f and f', if $\mathcal{F}$ composes F and f'', if F' composes f and f'', and if $\mathcal{F}'$ composes F' and f', then F and F' are properly equivalent* (§240, where it is assumed that all forms enter directly into all compositions in the sense defined in Essay 3.7).

The theory of multiplication of modules of hypernumbers resolves this conflict between the wish to preserve the full* generality of Gauss's theory and the wish to avoid its technical difficulties. Kummer's first paper [46, p. 324 (208)] on "ideal prime factors" mentions the possibility of applying his new theory to Gauss's—at least to justify Gauss's belief that the forms $ax^2 + 2bxy + cy^2$ and $ax^2 - 2bxy + cy^2$ should be considered to be inequivalent—but he never laid out the exact relation between the two theories. Similarly, Dedekind was well aware that Gauss's composition of forms was, in essence, the multiplication of modules—"ideals" in his terminology—but he did not develop the correspondence in detail. Nor, as far as I know, has anyone since Dedekind. In all probability, the reason is that it was felt that Gauss's approach was a false start that could be disregarded and replaced with Dedekind's. But such an attitude has the great disadvantage that it destroys the access of modern readers to Gauss's classic.

Essay 3.7 is meant to bridge the gap between the modern theory and Gauss's. On the one hand, modern readers will certainly see that the multiplication of modules of hypernumbers is the multiplication of ideals in quadratic

---

* In truth, Essay 3.7 does not preserve the *full* generality of Gauss's theory because it ignores forms (r, s, t) for which $s^2 - 4rt$ is a square.

number fields with the added technicality of dealing not with *all* integers in the field but with what Dedekind called *orders* (*Ordnungen*) of integers in the field. On the other hand, as Essay 3.7 shows, multiplication of modules makes possible the complete description of Gauss's compositions in the sense that it solves the problem, *Given two binary quadratic forms, determine whether they can be composed, and if so, find all possible compositions.* In brief, Gauss himself showed that two forms can be composed if and only if they pertain to the same square-free integer (see Essay 3.7) and showed that knowledge of *one* composition implies knowledge of *all,* because any two differ by a unimodular change of variables. Thus, the problem is solved by Proposition 3 of Essay 3.7, which describes how to use multiplication of modules to find an explicit composition of two given binary quadratic forms, provided they pertain to the same square-free integer.

Essay 3.7 The Construction of Compositions

Binary quadratic forms with integer coefficients will be called "forms" in this essay. For simplicity, all forms will be assumed to be irreducible; in other words, forms (r, s, t) whose discriminants $s^2 - 4rt$ are squares are excluded.*

Theorem.   *Given two forms, construct all forms that compose them.*

If (R, S, T) composes (r, s, t) and (ρ, σ, τ), then $(-R, -S, -T)$ composes $(-r, -s, -t)$ and (ρ, σ, τ). Therefore, one can assume without loss of generality that r is nonnegative. Since $s^2 - 4rt$ is not a square, r is positive in this case. Similarly, ρ can be assumed to be positive.

The required construction is accomplished by the four propositions that follow.

A form (r, s, t) will be said to **pertain** to a given square-free integer if its discriminant $s^2 - 4rt$ is a square times that integer. In this way, each form pertains to one and only one square-free integer. (An integer is square-free if it is not divisible by any square greater than 1.) The reducible forms that have been excluded from consideration are simply the forms that pertain to the square-free number 1 together with forms whose discriminant is zero.

Proposition 1.   *If a form composes two others, then all three pertain to the same square-free number.*

Corollary.   *If two given forms pertain to different square-free integers, then no form composes them.*

Proof. Given are three forms (r, s, t), (ρ, σ, τ), and (R, S, T) and a substitution

$$X = p_0 xu + p_1 xv + p_2 yu + p_3 yv,$$
$$Y = q_0 xu + q_1 xv + q_2 yu + q_3 yv,$$

in which the 2×2 minors of the matrix of coefficients have no common divisor greater than 1, whose substitution in $RX^2 + SXY + TY^2$ results in $(rx^2 + sxy + ty^2)(\rho u^2 + \sigma uv + \tau v^2)$. This last statement, when the coefficients of the various monomials x^2u^2, x^2v^2, $\ldots$, $xyuv$ are compared, amounts to nine equations:

(1)	$Rp_0^2 + Sp_0q_0 + Tq_0^2 = r\rho$
(2)	$Rp_1^2 + Sp_1q_1 + Tq_1^2 = r\tau,$
(3)	$Rp_2^2 + Sp_2q_2 + Tq_2^2 = t\rho,$
(4)	$Rp_3^2 + Sp_3q_3 + Tq_3^2 = t\tau,$
(5)	$2Rp_0p_1 + S(p_0q_1 + p_1q_0) + 2Tq_0q_1 = r\sigma,$

---

* Characteristically, Gauss does *not* exclude reducible forms, as is shown by the point he makes in §235 of avoiding the assumption that the first coefficients of his forms are nonzero.

$$(6) \qquad\qquad 2Rp_0p_2 + S(p_0q_2 + p_2q_0) + 2Tq_0q_2 = s\rho,$$

$$(7) \qquad\qquad 2Rp_1p_3 + S(p_1q_3 + p_3q_1) + 2Tq_1q_3 = s\tau,$$

$$(8) \qquad\qquad 2Rp_2p_3 + S(p_2q_3 + p_3q_2) + 2Tq_2q_3 = t\sigma,$$

$$(9)\ 2R(p_0p_3 + p_1p_2) + S(p_0q_3 + p_1q_2 + p_2q_1 + p_3q_0) + 2T(q_0q_3 + q_1q_2) = s\sigma.$$

Gauss's theory is based entirely on virtuosic algebraic deductions of general properties of transformations and compositions of forms from these nine equations.

His very first step is to subtract 4 times the product of equations (1) and (2) from the square of (5). In the notation above, the result is

$$(S^2 - 4RT)(p_0q_1 - p_1q_0)^2 = r^2(\sigma^2 - 4\rho\tau).$$

The right side is obvious; the left is not, but a little work with pencil and paper will confirm it.

Since $r > 0$ and $\sigma^2 - 4\rho\tau$ is not a square, this equation shows that $p_0q_1 \neq p_1q_0$, so (ρ, σ, τ) and (R, S, T) pertain to the same square-free integer, namely, the integer obtained by dividing $(S^2 - 4RT)(p_0q_1 - p_1q_0)^2 = r^2(\sigma^2 - 4\rho\tau)$ by its largest square factor. By the symmetry of the definition of composition, (r, s, t) and (R, S, T) must also pertain to the same square-free integer, and the proposition is proved.*

The **content** c of a form (r, s, t) is the greatest common divisor of its coefficients.

Proposition 2. *Let (r, s, t) and (ρ, σ, τ) be forms for which $s^2 - 4rt = \sigma^2 - 4\rho\tau$ and for which s and σ are even, say $s = 2s'$ and $\sigma = 2\sigma'$. In this case, $s^2 - 4rt = \sigma^2 - 4\rho\tau = 4A$, where $A = (s')^2 - rt = (\sigma')^2 - \rho\tau$ is not a square (by assumption). Let the module $[r, s' + \sqrt{A}][\rho, \sigma' + \sqrt{A}]$ be put in canonical form, say it is $[E][F, G + \sqrt{A}]$, and let H be defined to be $(G^2 - A)/F$. Then a multiple of $(F, 2G, H)$ composes (r, s, t) and (ρ, σ, τ). The specific multiplier is determined by the rule that the content of the composed form is the product of the contents of its "factors."*

(The theory of modules of hypernumbers was developed in the earlier essays only in the case in which A is a positive, nonsquare integer. Its extension to the case in which A is negative presents no difficulties at all. Therefore, forms with negative discriminants can be included in this proposition.)

Proof. The actual substitution that accomplishes the composition is given implicitly by the formula

$$(10)\ \left(rx + (s' + \sqrt{A})y\right)\left(\rho u + (\sigma' + \sqrt{A})v\right) = E\left(FX + (G + \sqrt{A})Y\right).$$

---

* By doing a good deal more algebraic work, Gauss proves Proposition 1—his "first consequence"—also in the case $r = 0$.

Multiplication of this equation by its conjugate (the equation obtained by changing $\sqrt{A}$ to $-\sqrt{A}$) gives

$$\left((rx + s'y)^2 - Ay^2\right)\left((\rho u + \sigma'v)^2 - Av^2\right) = E^2\left((FX + GY)^2 - AY^2\right),$$

or, what is the same,

$$r\rho(rx^2 + sxy + ty^2)(\rho u^2 + \sigma uv + \tau v^2) = E^2 F(FX^2 + 2GXY + HY^2).$$

Thus, the substitution implicitly defined by (10) transforms $FX^2 + 2GXY + HY^2$ times $E^2 F/r\rho$ into the product $(rx^2 + sxy + ty^2)(\rho u^2 + \sigma uv + \tau v^2)$. This is the multiple of $(F, 2G, H)$ described in the proposition. What is to be shown is that (1) the coefficients of the substitution implicitly defined by (10) are integers, that (2) $c\gamma = \frac{E^2 F}{r\rho} C$, where c, γ, and C are the contents of (r, s, t), (ρ, σ, τ), and $(F, 2G, H)$ respectively, and that (3) the greatest common divisor of the six 2×2 minors of the matrix of coefficients of the substitution is 1.

Comparison of the coefficients of $\sqrt{A}$ on the two sides of (10) shows that $rxv + \rho yu + (s' + \sigma')yv = EY$, after which a comparison of the remaining terms shows that $r\rho xu + r\sigma'xv + \rho s'yu + (s'\sigma' + A)yv = EFX + G(rxv + \rho yu + (s' + \sigma')yv)$. Therefore, the matrix of coefficients of the substitution determined by (10) is

$$\begin{bmatrix} \frac{r\rho}{EF} & \frac{r(\sigma' - G)}{EF} & \frac{\rho(s' - G)}{EF} & \frac{s'\sigma' + A - G(s' + \sigma')}{EF} \\ 0 & \frac{r}{E} & \frac{\rho}{E} & \frac{s' + \sigma'}{E} \end{bmatrix}.$$

That the entries of this matrix are integers can be seen as follows:

By definition, $[r, s' + \sqrt{A}][\rho, \sigma' + \sqrt{A}] = [E][F, G + \sqrt{A}]$. Thus, E, F, and G are found by putting $[r\rho, r\sigma' + r\sqrt{A}, \rho s' + \rho\sqrt{A}, s'\sigma' + A + (s' + \sigma')\sqrt{A}]$ in canonical form. Let $\mathcal{A}$, $\mathcal{B}$, and $\mathcal{C}$ be integers for which $\mathcal{A}r + \mathcal{B}\rho + \mathcal{C}(s' + \sigma')$ is the greatest common divisor of r, ρ, and $(s' + \sigma')$; call it d. Clearly, d divides both coefficients of all four numbers in the product module $[r\rho, r\sigma' + r\sqrt{A}, \rho s' + \rho\sqrt{A}, s'\sigma' + A + (s' + \sigma')\sqrt{A}]$ with the possible exception of $s'\sigma' + A$, and d divides this coefficient as well because $A \equiv (s')^2 \bmod r$, so $s'\sigma' + A \equiv (s'\sigma' + (s')^2) \equiv s'(\sigma' + s') \equiv 0 \bmod d$. Let G_0 be defined by the equation

$$d(G_0 + \sqrt{A}) = \mathcal{A}(r\sigma' + r\sqrt{A}) + \mathcal{B}(\rho s' + \rho\sqrt{A}) + \mathcal{C}(s'\sigma' + A + (s' + \sigma')\sqrt{A}).$$

Then $d(G_0 + \sqrt{A})$ is a sum of multiples of hypernumbers already in the product module, so it can be annexed to the list, and multiples of it can be subtracted from the previous entries to put the product module in the form $[r\rho, r\sigma' + r\sqrt{A}, \rho s' + \rho\sqrt{A}, s'\sigma' + A + (s' + \sigma')\sqrt{A}, d(G_0 + \sqrt{A})] = [r\rho, r\sigma' - rG_0, \rho s' - \rho G_0, s'\sigma' + A - (s' + \sigma')G_0, d(G_0 + \sqrt{A})]$. This module is $[dF_0, d(G_0 + \sqrt{A})]$, where by definition $[dF_0] = [r\rho, r\sigma' - rG_0, \rho s' - \rho G_0, s'\sigma' + A - (s' + \sigma')G_0]$. (Note that all entries in the module on the right are divisible by d.) In short, $[E][F, G + \sqrt{A}] = [d][F_0, G_0 + \sqrt{A}]$. By the uniqueness of canonical form, it follows that $E = d$, $F = F_0$, and $G \equiv G_0 \bmod F$, *provided* it is shown that $G_0^2 \equiv A \bmod F_0$.

Let a representation of a module be called **full** if $\sqrt{A}$ times any hypernumber in the list can be written as a sum of *integer* multiples of hypernumbers in the list. If $g^2 \equiv A \bmod f$, then $[f, g + \sqrt{A}]$ is full, because $f\sqrt{A} = -gf + f(g + \sqrt{A})$ and $(g + \sqrt{A})\sqrt{A} = qf + g(g + \sqrt{A})$, where q is defined by $q = (A - g^2)/f$. Thus, the representation of any module in canonical form is full. A product of full representations is full, and any representation obtained from a full representation by adding one hypernumber in the list to another or subtracting one from another is full, and the same is true for the operations of rearranging or of annexing to or deleting from the list a zero entry. Therefore, $[dF_0, dG_0 + d\sqrt{A}]$ is full, which implies that $(dG_0 + d\sqrt{A})\sqrt{A} = m \cdot dF_0 + n \cdot (dG_0 + d\sqrt{A})$ for integers m and n, which implies $dG_0 = nd$ and $dA = mdF_0 + ndG_0 = mdF_0 + dG_0^2$, so $A \equiv G_0^2 \bmod F_0$, as was to be shown.

The identity $E = d$ shows that the entries in the last row of the substitution matrix are integers. The identity $EF = dF_0$ shows that $[EF] = [r\rho, r\sigma' - rG_0, \rho s' - \rho G_0, s'\sigma' + A - (s' + \sigma')G_0]$ and therefore shows, because $dG_0 \equiv dG \bmod dF$, that the entries in the first row of the substitution matrix are integers.

By direct computation, the six 2×2 minors of this matrix of coefficients are seen to be $\frac{r\rho}{E^2 F}$ times r, ρ, $s' + \sigma'$, $\sigma' - s'$, τ, and t. Thus, the greatest common divisor of the six minors is 1 if and only if $[E^2 F] = [r\rho][r, \rho, s' + \sigma', \sigma' - s', \tau, t]$. That this is the case can be proved as follows:

It was seen above that $[EF] = [dF_0]$ is the greatest common divisor of $r\rho$, $r\sigma' - rG_0$, $\rho s' - \rho G_0$, and $s'\sigma' + A - (s' + \sigma')G_0$, so what is to be shown is that

$$[E][r\rho, r\sigma' - rG_0, \rho s' - \rho G_0, s'\sigma' + A - (s' + \sigma')G_0] = [r\rho][r, \rho, s' + \sigma', \sigma' - s', \tau, t].$$

The first three terms on the right—$r^2\rho$, $r\rho^2$ and $r\rho(s' + \sigma')$—are all divisible by $Er\rho$ and are therefore zero modulo the module on the left. That the remaining terms on the right are zero modulo the module on the left follows from the identities

$$r\rho(\sigma' - s') = \frac{\rho}{E} \cdot E(r\sigma' - rG_0) - \frac{r}{E} \cdot E(\rho s' - \rho G_0),$$

$$r\rho\tau = \frac{s' + \sigma'}{E} \cdot E(r\sigma' - rG_0) - \frac{r}{E} \cdot E(s'\sigma' + A - (s' + \sigma')G_0),$$

$$r\rho t = \frac{s' + \sigma'}{E} \cdot E(\rho s' - \rho G_0) - \frac{\rho}{E} \cdot E(s'\sigma' + A - (s' + \sigma')G_0).$$

On the other hand, the four terms in the module on the left are zero modulo the module on the right by virtue of

$$Er\rho = \mathcal{A}r^2\rho + \mathcal{B}r\rho^2 + \mathcal{C}r\rho(s' + \sigma'),$$

$$E(\sigma' - G_0) = \mathcal{B}\rho(\sigma' - s') + \mathcal{C}\rho\tau,$$

$$E(s' - G_0) = \mathcal{A}r(s' - \sigma') + \mathcal{C}rt,$$
$$E(s'\sigma' + A - (s' + \sigma')G_0) = -\mathcal{A}r\rho\tau - \mathcal{B}r\rho t.$$

(The last three equations are found by eliminating one of $\mathcal{A}$, $\mathcal{B}$, and $\mathcal{C}$ from $E = \mathcal{A}r + \mathcal{B}\rho + \mathcal{C}(s' + \sigma')$ and $EG_0 = \mathcal{A}r\sigma' + \mathcal{B}\rho s' + \mathcal{C}(s'\sigma' + A)$.)

Only the equation $c\gamma r\rho = E^2 FC$ of step (2) remains to be proved. As was shown in the proof of Theorem 3 of Essay 3.4, if $[f, g + \sqrt{A}]$ is in canonical form, then $[f, g+\sqrt{A}][f, f-g+\sqrt{A}] = [f][d, g+\sqrt{A}]$, where $[d] = [f, 2g, \frac{g^2-A}{f}]$. Thus, since $[c] = [r, 2s', t]$, the product of $[r, s' + \sqrt{A}]$ with its conjugate is $[r][c, s'+\sqrt{A}]$. In the same way, the product of $[\rho, \sigma'+\sqrt{A}]$ with its conjugate is $[\rho][\gamma, \sigma' + \sqrt{A}]$, and the product of $[F, G+\sqrt{A}]$ with its conjugate is $[F][C, G+\sqrt{A}]$. Therefore, multiplication of $[r, s' + \sqrt{A}][\rho, \sigma' + \sqrt{A}] = [E][F, G+\sqrt{A}]$ by its conjugate gives the equation $[r\rho][c, s' + \sqrt{A}][\gamma, \sigma' + \sqrt{A}] = [E^2F][C, G+\sqrt{A}]$. Since $s' \equiv -s'$ mod c, $\sigma' \equiv -\sigma'$ mod γ, and $G \equiv -G$ mod C, all modules in this last equation are self-conjugate, so the product of this equation with its conjugate is simply $[r^2\rho^2][c][c, s'+\sqrt{A}][\gamma][\gamma, \sigma'+\sqrt{A}] = [E^4F^2][C][C, G+\sqrt{A}]$. Replacement of $[r\rho][c, s'+\sqrt{A}][\gamma, \sigma' + \sqrt{A}]$ with $[E^2F][C, G+\sqrt{A}]$ on the left side of this equation gives $[r\rho c\gamma E^2 F][C, G + \sqrt{A}] = [E^4F^2C][C, G+\sqrt{A}]$. Reduction of G mod C puts both sides in canonical form, so $r\rho c\gamma E^2 F = E^4F^2C$, from which the desired equation $r\rho c\gamma = E^2 FC$ follows.

Proposition 3. *Given two forms pertaining to the same square-free integer, construct a form that composes them.*

Proof. Let (r, s, t) and (ρ, σ, τ) be the given forms. Because they pertain to the same square-free integer, positive integers m and μ can be chosen to make $m^2(s^2 - 4rt) = \mu^2(\sigma^2 - 4\rho\tau)$. Moreover, because m and μ can be doubled, if necessary, one can assume without loss of generality that ms and $\mu\sigma$ are even. Proposition 2 then constructs a form, call it $(F, 2G, H)$, that composes (mr, ms, mt) and $(\mu\rho, \mu\sigma, \mu\tau)$ whose content is the product of the contents of (mr, ms, mt) and $(\mu\rho, \mu\sigma, \mu\tau)$. Therefore, F, $2G$, and H are all divisible by $m\mu$, and the same substitution that gives $(F, 2G, H)$ as a composition of (mr, ms, mt) and $(\mu\rho, \mu\sigma, \mu\tau)$ gives $\left(\frac{F}{m\mu}, \frac{G}{m\mu}, \frac{H}{m\mu}\right)$ as a composition of (r, s, t) and (ρ, σ, τ), as was to be constructed.

It remains to find *all* compositions of two given forms that pertain to the same square-free number. The compositions constructed by Proposition 2—and therefore those constructed by Proposition 3—contain both forms **directly** in the sense Gauss defines that term, which is to say that the first two of the six minors, the minors $p_0q_1 - p_1q_0$ and $p_0q_2 - p_2q_0$, are both positive. (These minors are the positive integers $r^2\rho/E^2F$ and $r\rho^2/E^2F$.) More generally, the first form (r, s, t) enters **directly** if $p_0q_2 - p_2q_0$ is positive, **inversely** if it is negative. (As was seen in the proof of Proposition 1, it cannot be zero.) Whether (ρ, σ, τ) enters directly or inversely is similarly determined by the sign of $p_0q_1 - p_1q_0$. Since reversing the sign of y changes (r, s, t) to $(r, -s, t)$

and changes the signs in the last two columns of the matrix of coefficients of the substitution while leaving everything else unchanged, it changes a composition in which (r, s, t) enters directly into one in which it enters inversely and vice versa. Similarly, a composition in which (ρ, σ, τ) enters directly can be changed into one in which it enters inversely and vice versa. In this way, *all* compositions of (r, s, t) and (ρ, σ, τ) can be found by finding *one* composition of each of the four combinations of $(r, \pm s, t)$ and $(\rho, \pm \sigma, \tau)$ in which both forms enter directly and applying the final proposition:

Proposition 4. *Given one composition of a pair of forms in which both forms enter directly, every other such composition is obtained by a unimodular change of variables.*

Proof. Let the given forms be (r, s, t) and (ρ, σ, τ) and let the given composition $RX^2 + SXY + TY^2$ be effected by the substitution whose matrix of coefficients is

$$\begin{bmatrix} p_0 & p_1 & p_2 & p_3 \\ q_0 & q_1 & q_2 & q_3. \end{bmatrix}$$

Let Δ_{ij} be the six 2×2 minors of this matrix, where $0 \leq i < j \leq 3$ and $\Delta_{ij} = p_i q_j - p_j q_i$. It is given that Δ_{01} and Δ_{02} are positive. It is to be shown that any other composition matrix for the same two forms in which Δ_{01} and Δ_{02} are both positive can be obtained by multiplying the given matrix on the left by a 2×2 matrix of integers whose determinant is 1.

Every composition is accomplished, as was seen in the proof of Proposition 3, by a substitution that accomplishes the composition of two forms with the same discriminant. In this case, the formulas (1)–(9) above imply, as Gauss proves in §235, that $\Delta_{01}^2 = r^2$, $\Delta_{01}(\Delta_{03} - \Delta_{12}) = rs$, $\Delta_{01}\Delta_{23} = rt$, $\Delta_{02}^2 = \rho^2$, $\Delta_{02}(\Delta_{03} + \Delta_{12}) = \rho\sigma$, and $\Delta_{02}\Delta_{13} = \rho\tau$. (These are Gauss's equations (12)–(14) and (18)–(20) when use is made of his identity $\Delta^2 = dd'$.)

Since it is given that $\Delta_{01} > 0$, the first equation determines Δ_{01}, after which the next two equations determine $\Delta_{03} - \Delta_{12}$ and Δ_{23}. Similarly, the remaining three equations determine Δ_{02}, $\Delta_{03} + \Delta_{12}$, and Δ_{13}. In short, in any two compositions of the same forms (in which both forms enter directly), the six 2×2 minors of the matrix of coefficients of the substitution are *identical*. Then, as the elementary lemma Gauss proves in §234 shows, one matrix can be obtained from the other by multiplication on the left by a matrix of integers with determinant 1, as was to be shown.

Example: Compose $(16, 4, -1)$ and $(4, 2, -1)$. The discriminants of these forms are 80 and 20, respectively, so both pertain to 5. Multiply the second by 2 so that the problem is to compose $(16, 4, -1)$ and $(8, 4, -2)$ with discriminant 80. By Proposition 2, the composed form is computed by putting the product $[16, 2 + \sqrt{20}][8, 2 + \sqrt{20}]$ in canonical form. The computation of this canonical form is $[128, 32 + 16\sqrt{20}, 16 + 8\sqrt{20}, 24 + 4\sqrt{20}] = [4][32, 4 + 2\sqrt{20}, 6 + \sqrt{20}] = [4][32, 24, 6 + \sqrt{20}] = [4][8, 6 + \sqrt{20}]$. Thus, $E = 4$, $F = 8$, $G = 6$, so

$H = (6^2 - 20)/8 = 2$ and some multiple of $(8, 12, 2)$ composes $(16, 4, -1)$ and $(8, 4, -2)$. The product of the contents 1 and 2 of the "factors" is the content 2 of $(8, 12, 2)$, so this form itself composes $(16, 4, -1)$ and $(8, 4, -2)$. Division by 2 then implies that $(4, 6, 1)$ composes the given forms $(16, 4, -1)$ and $(4, 2, -1)$. The matrix of coefficients that describes the substitution that accomplishes the composition is

$$\begin{bmatrix} \frac{r\rho}{EF} & \frac{r(\sigma'-G)}{EF} & \frac{\rho(s'-G)}{EF} & \frac{s'\sigma'+A-G(s'+\sigma')}{EF} \\ 0 & \frac{r}{E} & \frac{\rho}{E} & \frac{s'+\sigma'}{E} \end{bmatrix} = \begin{bmatrix} 4 & -2 & -1 & 0 \\ 0 & 4 & 2 & 1 \end{bmatrix},$$

as can be found either by evaluating the matrix entries or by solving the equation

$$\left(16x + \left(2 + \sqrt{20}\right)y\right)\left(8u + \left(2 + \sqrt{20}\right)v\right) = 4\left(8X + \left(6 + \sqrt{20}\right)Y\right)$$

for X and Y in terms of x, y, u, and v. This is the third example of a composition given in Essay 3.6.

In the composition just given, both forms enter directly. The same method gives

$$\begin{bmatrix} 8 & 2 & -1 & 1 \\ 0 & 2 & 1 & 0 \end{bmatrix}$$

as a matrix that describes a substitution giving $(1, 0, -5)$ as a composition of $(16, -4, -1)$ and $(4, 2, -1)$ in which both forms enter directly. Reversing the sign of y then gives the substitution

$$\begin{bmatrix} 8 & 2 & 1 & -1 \\ 0 & 2 & -1 & 0 \end{bmatrix},$$

which gives $(1, 0, -5)$ as a composition of $(16, 4, -1)$ and $(4, 2, -1)$ in which the first form enters inversely and the second enters directly.

To find *all* compositions of $(16, 4, -1)$ and $(4, 2, -1)$, one would also find a composition in which the first entered directly and the second entered inversely, and a composition in which both entered inversely. Every composition of the two forms is then obtained from one of these four by a unimodular change of variables.

4

The Genus of an Algebraic Curve

Essay 4.1 Abel's Memoir

> *It appears to me that if one wants to make progress in the study of mathematics one should study the masters and not the pupils.*—Niels Henrik Abel, quoted from an unpublished source in [52, p. 138]

Niels Henrik Abel's submission of his *Mémoire sur une propriété générale d'une class très-étendue de fonctions transcendantes* to the Paris Academy in October 1826 should have been a high point in the history of mathematics. Instead, it was a low point in the history of the Paris Academy.

Abel, lonely and unknown, was temporarily in Paris thanks to a travel grant from the government of Norway, and he hoped to win recognition in the city that was then the mathematical capital of Europe. Unfortunately, he naively believed that recognition could be won by submitting a work of undeniable genius to Europe's leading mathematical institution. He did not understand that works of undeniable genius are inherently difficult to read, even for the most learned readers, and he did not understand that the members of Europe's leading mathematical institution would not devote the needed time and thought to the work of a 24-year-old mathematician who was unknown to them and who came from a country they had scarcely heard of.

Of course, one of the famous men of the Academy might by some lucky accident have taken notice of the memoir long enough to realize that it was worth pursuing, but none did. In 1837, eight years after Abel's untimely death, the Norwegian scholars charged with publishing Abel's collected works applied to the Academy *via the Norwegian government and its diplomatic representatives in Paris* for a copy of the memoir—Abel had apparently not kept a copy for himself—but the effort did not succeed, and the memoir is absent from the first publication of Abel's works in 1839. Finally, the Academy did publish the memoir in 1841, making it available to eager readers like C. G. J. Jacobi for the first time.

Fig. 4.1. Abel.

In the two and a half years Abel lived after submitting the memoir, he enjoyed a growing reputation based on his publications in Crelle's *Journal,* but he patiently awaited publication of the Paris memoir, believing it would ensure his fame. He even alluded to the memoir in one of his published works, piquing the curiosity and indignation of Jacobi, who read the allusion too late to write to Abel about it. That Abel's memoir remained unpublished in his lifetime* deprived him of the challenge and encouragement of readers' responses and therefore probably deprived mathematics of important further work.

(Incredibly, the tragedy was repeated only three years after Abel died when Galois went to an early grave ignored by the same Paris Academy.)

Abel's memoir deals with *integrals of algebraic differentials,* a topic that is not at all easy to understand from the point of view of naive geometry and integration along a curve. Because an algebraic differential like $dx/\sqrt{1-x^4}$ is "many-valued" and because, moreover, an integral of such a differential depends on choosing both a path and a constant of integration, modern readers

---

* The last work Abel published was a brief note that contained a theorem from the memoir. Abel's biographer Oystein Ore says that the theorem of that last brief note is the theorem of the memoir [52, p. 219], but it is far short of the theorem in the introduction of the memoir that I am discussing in this essay and that I take to be, in Ore's phrase, "the main theorem from the Paris memoir."

may well despair of understanding even what Abel *means* by the sum of a finite number of integrals of a given algebraic differential, much less why questions about such sums might be interesting or significant.

But there is another way to describe the main idea that makes better sense to modern readers and explains the main theorem of the memoir more clearly. Abel's "algebraic differentials" are differentials of the form $f(x, y)dx$, where f is a rational function of two variables and where y is an "algebraic function" of x. The notion of an "algebraic function" has become a source of unease for modern readers because an algebraic function is normally "many-valued" and the property of being single-valued is the essence of the set-theoretic notion of a "function." But of course there are modern ways to deal with algebraic functions. One is to give the functions their own special domain; this is the source of the theory of Riemann surfaces. The other is to regard an "algebraic function" not as a function at all, but simply as an element of an algebraic function field, which is to say an algebraic field whose transcendence degree is positive (see Essay 2.2). The subject of Abel's memoir is algebraic functions of *one variable,* which is to say, in the terminology of Essay 2.2, elements of an algebraic field of transcendence degree 1. In other words, Abel is dealing with *the field of rational functions on an algebraic curve defined over the rationals.*

The concept that I propose as an aid to understanding Abel's memoir is that of an **algebraic variation** of a set of points on an algebraic curve. Abel describes such a variation as the solutions of a pair of equations

$$\chi(x, y) = 0,$$
$$\theta(x, y, a, a', a'', \ldots) = 0,$$

where $\chi(x, y)$ is the irreducible polynomial with integer coefficients, monic in y, that defines the algebraic curve under discussion, and $\theta(x, y, a)$ is an auxiliary polynomial in x and y whose coefficients a, a', a'', ... are indeterminates. For each fixed value of the coefficients a, a', a'', ... the pair of equations determines a set of points $\{(x_k, y_k)\}$ on the curve $\chi = 0$, and as the coefficients vary, these points vary along the curve. A variation of points on the curve that can be generated in this way is an *algebraic variation.*

Somewhat more precisely, let C^N denote the set of all N-tuples of points on the curve C defined by $\chi(x, y) = 0$. An algebraic variation of a point of C^N is determined by choosing a $\theta(x, y, a)$ of the form $\theta(x, y, a) = \sum a_{ij} x^i y^j$, where the exponent pairs (i, j) are in some specified finite set. To say that $\theta(x, y, a) = 0$ at a particular point of the curve $\chi(x, y) = 0$ means that the parameters a_{ij} in θ satisfy a certain (linear) condition. Choose values for the a_{ij} that make $\theta = 0$ at all N of the given points. There will be *other* points of $\chi(x, y) = 0$ where $\theta = 0$ for these values of a_{ij}, say there are M of them. An allowable variation of the N given points is one that results when the a_{ij} are allowed to vary from their fixed values in such a way that the M additional zeros all remain at zero while the N original ones are allowed to move. For each point of C^N, the points of C^N that can be reached from it by a sequence of algebraic variations lie on an algebraic subvariety of C^N.

Abel probably had some geometric conception of such variations of sets of points on $\chi(x, y) = 0$, but exactly what it might have been can only be guessed. Today one would never discuss intersection points without first specifying an algebraically closed ground field, but Abel would probably not have thought of curves as ordered pairs of complex numbers in anything like the modern way. More likely, he would have just imagined sets of points of intersection of an ordinary plane curve with an auxiliary curve and considered constraints on variations of the intersection points produced by varying the auxiliary curve. In modern terms, the number of *constraints* on the variation of N points of a curve is the *codimension* of the subvariety of algebraic variations within the N-dimensional variety C^N of all variations. This codimension is very nearly the same as the **genus** of the curve, and whatever his geometric conception of the problem setting may have been, it is this number that Abel successfully investigated.

In terms somewhat closer to Abel's, if $f(x, y)dx$ is an algebraic differential (which is to say, a rational function f on the curve C times the symbol dx), and if an infinitesimal algebraic variation of the points $\{(x_k, y_k)\}$ is performed, Abel asserts that the resulting variation of $\sum f(x_k, y_k)dx_k$ is a differential that can be expressed *rationally* in terms of the parameters a_{ij} and their differentials.* Thus, if the point $(P_1, P_2, \ldots, P_N)$ can be moved to the point $(Q_1, Q_2, \ldots, Q_N)$ of C^N by an algebraic variation, then

$$\int_{P_1}^{Q_1} f(x, y)dx + \int_{P_2}^{Q_2} f(x, y)dx + \cdots + \int_{P_N}^{Q_N} f(x, y)dx$$

is equal to the integral of a *rational* differential in the a_{ij} and can therefore be expressed in terms of elementary functions—logarithms and trigonometric functions, as well as rational functions—of the a_{ij}.

Now let g be the codimension of the subvarieties of algebraic variations. Then $N - g$ of the points $(P_1, P_2, \ldots, P_N)$ can be moved in arbitrary ways by an algebraic variation, provided the remaining g points move in such a way as to keep the new $(P_1', P_2', \ldots, P_N')$ on the same subvariety. Thus, if O is a chosen base point on the curve C, there is an algebraic variation—or at least a succession of algebraic variations—of a point $(P_1, P_2, \ldots, P_N)$ of C^N that connects it to a point of C^N of the form $(O, O, \ldots, O, Q_1, Q_2, \ldots Q_g)$. Then

$$\int_{O}^{P_1} f(x, y)dx + \int_{O}^{P_2} f(x, y)dx + \cdots + \int_{O}^{P_{N-g}} f(x, y)dx$$
$$+ \int_{Q_1}^{P_{N-g+1}} f(x, y)dx + \cdots + \int_{Q_g}^{P_N} f(x, y)dx$$

can be expressed in terms of elementary functions of the parameters used in the variation, so that when the g integrals from O to Q_i are added, one obtains

---

* This, in essence, is the theorem of Abel's last published note that Ore mistook for the main theorem of the memoir. See the note above.

$$\int_O^{P_1} f(x,y)dx + \int_O^{P_2} f(x,y)dx + \cdots + \int_O^{P_N} f(x,y)dx$$

$$= \int_O^{Q_1} f(x,y)dx + \cdots + \int_O^{Q_g} f(x,y)dx + E,$$

where E can be expressed in terms of elementary functions of the parameters of the variation. (The paths of integration are, of course, the ones determined by the algebraic variation from $(O, O, \ldots, O, Q_1, Q_2, \ldots Q_g)$ to $(P_1, P_2, \ldots, P_N)$ that is assumed.) Thus, disregarding elementary functions, *a sum of any number N of integrals of $f(x,y)dx$ can be expressed as a sum of just g integrals,* where g depends only on the differential $f(x,y)dx$ being integrated—and in fact depends only on the algebraic curve $\chi(x,y) = 0$ on which the differential has its existence—not on N.

This is the main theorem of Abel's Paris memoir. In Abel's own words, "The number of these conditions [the number g above] does not depend at all on the number of summands, but only on the nature of the particular integrands that one considers. Thus, for example, for an elliptic integrand this number is 1; for an integrand that contains no irrationalities but a radical of the second degree, under which the variable has degree at most six, the number of necessary conditions is 2, and so forth."[*]

I have said above that the crucial number of conditions g is "roughly" the genus of the curve C. Abel's statement that g is 1 in the elliptic case, 2 in case y^2 is a polynomial of degree 5 or 6 in x, and so forth, of course suggests that g is connected to the genus and *is* the genus in many cases. It fails to be the genus only because Abel bases his variation of the points on the variation of parameters a_{ij} in functions of the form $\theta(x,y,a) = \sum a_{ij}x^iy^j$, which is not quite general enough and in some cases gives too large a value for g because it omits certain variations that deserve to be called algebraic variations. When θ is instead taken to have the form $\theta(x,y,a) = \sum a_i\theta_i(x,y)$ where the "functions" $\theta_i(x,y)$ are integral over x—which may reduce g because it may include more variations—g becomes the actual genus, as will be shown in Essay 4.6.

---

[*] In an effort to clarify Abel's statement, I have taken some liberties with the translation. His actual words were, "Le nombre de ces relations ne dépend nullement du nombre des fonctions, mais seulement de la nature des fonctions particulière qu'on considère. Ainsi, par exemple, pour une fonction elliptique ce nombre est 1; pour une fonction dont la dérivée ne contient d'autres irrationalités qu'un radical du second degré, sous lequel la variable ne passe pas le cinquième ou sixième degré, le nombre des relations nécessaires est 2, et ainsi de suite." His "fonctions" are the integrals above, and his "dérivées" are the integrands. What he is calling "une fonction elliptique" is what is today called an elliptic integral.

Essay 4.2 Euler's Addition Formula

Man sollte weniger danach streben, die Grenzen der mathematischen Wissenschaften zu erweitern, als vielmehr danach, den bereits vorhandenen Stoff aus umfassenderen Gesichtspunkten zu betrachten. (One should strive less to extend the boundaries of the mathematical sciences and much more to treat the already available material from more comprehensive viewpoints.)—E. Study [24, p. 140]

Euler stated his addition formula for elliptic integrals in a variety of ways, none of which shed enough light on the formula to suggest a generalization to other kinds of integrands. The great achievement of Abel's Paris memoir was to describe Euler's formula as the case $g = 1$ of a more general phenomenon.

It is customary today to describe an elliptic curve by a formula of the form $y^2 = x^3 + g_2 x + g_3$, in which g_2 and g_3 are rational numbers, called its "Weierstrass normal form." When the curve is written in this form, the "addition" or "group law" on the curve is described as follows: Let P and Q be given points on the curve, and let S be the third point in which the line through P and Q intersects the curve. (The curve, being a cubic, intersects a line in the xy-plane in three points when they are counted in the right way.) The sum $R = P + Q$ of P and Q is defined to be the third point in which the line through S and the point at infinity intersects the curve. (The lines through the point at infinity are the lines $x = $ constant—these are the lines that intersect the curve in only two finite points—so R is the point whose x-coordinate is the same as that of S and whose y-coordinate is the y-coordinate of S with the sign reversed.)

This construction is connected to the theorem of Abel's memoir in the following way: Let $\theta(x, y, a, b, c) = ax + by + c$. Algebraic variations of the given points P and Q are obtained by choosing initial values for a, b, and c that make θ zero at P and Q and allowing a, b, and c to vary in such a way that the third point of intersection of the line with the curve, call it S, remains fixed. In other words, the algebraic variations of the pair (P, Q) are the pairs of points (P', Q') on the curve for which P', Q', and S are colinear. In particular, if a point O of the curve is chosen as the origin—or the identity of the group law—then the algebraic variation of (P, Q) that carries P to O carries Q to the third point R in which the line through O and S intersects the curve. When O is chosen to be the point at infinity, R is the point $P + Q$ described above.

Abel's point of view explains why this "addition" is useful and shows that it is intrinsic to the curve. According to Abel, for any rational function $f(x, y)$ of x and $y = \sqrt{x^3 + g_2 x + g_3}$, the sum $\int_O^P f(x, y)dx + \int_R^Q f(x, y)dx$ can be expressed in terms of integrals of rational functions, or, what is the same, $\int_O^P f(x, y)dx + \int_O^Q f(x, y)dx = \int_O^R f(x, y)dx$ plus an integral of rational functions. In particular, in the special case $f(x, y) = \frac{1}{y}$, in which the integrand is *holomorphic* in the sense explained in Essay 4.6, the formula is

Fig. 4.2. Euler.

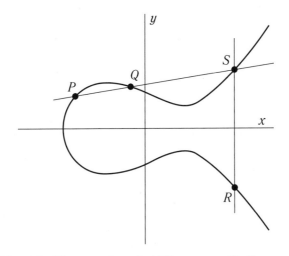

Fig. 4.3. The operation of addition on an elliptic curve.

$$\int_O^P \frac{dx}{y} + \int_O^Q \frac{dx}{y} = \int_O^R \frac{dx}{y},$$

which is one form of Euler's addition theorem. More precisely, these integrals depend on the paths of integration, and for the formula to hold, these paths must be chosen correctly. Thus, the sum of two integrals of dx/y can be expressed as just one integral of the same integrand, provided the limits of integration satisfy a certain algebraic relation and the paths of integration are chosen correctly.

Once it is known that two such integrals can be reduced to one, it follows that any number of such integrals can similarly be reduced to one. Abel's construction describes this all at once, rather than as a step-by-step reduction. An algebraic variation of a set of points $(P_1, P_2, \ldots, P_N)$ on the curve is described by a function $\theta(x, y, a)$ of the form $\sum a_{ij} x^i y^j$ for some selection of exponent pairs (i, j). Since y^2 is a polynomial in x, it is natural to assume that all of the chosen values of j are less than 2, so that θ takes the form $\phi_1(x) + \phi_2(x)y$, where ϕ_1 and ϕ_2 are polynomials in x containing terms of certain specified degrees whose coefficients are indeterminates a_{i1} and a_{i2}. The procedure is to give θ enough terms that values can be chosen for the parameters a_{ij} that make θ zero at the given points $P_1, P_2, \ldots, P_N$, and then to allow the parameters to vary from their chosen values in such a way that the value of θ remains at zero for the zeros of θ *other than* $P_1, P_2, \ldots, P_N$, while the N given zeros of θ are allowed to vary. The main question is, How many conditions are imposed on the variation of the N points along the curve by the requirement that the variation be describable in this way? That the answer is 1—that the genus of this curve is 1—can be seen in the following way.

The crucial step is to determine the number of zeros of $\theta(x, y) = \phi_1(x) + \phi_2(x)y$ on the curve. A simple way to do this is to make use of the idea that a rational function on an algebraic curve assumes each value the same number of times, when they are counted properly, and in particular that *the number of zeros is equal to the number of poles*. The function x assumes every value twice, and in particular, it has a double pole at the one point where $x = \infty$. The function y, on the other hand, assumes every value three times and has a triple pole at the one point where $x = \infty$. (These statements can be justified in various ways, but since they are used here only as heuristic devices, no formal justification will be given.) It follows that a polynomial $\phi(x)$ of degree ν has 2ν poles, all of them at the point where $x = \infty$, and that $\phi(x)y$ has $2\nu + 3$ poles, all at that same point. Consequently, if $\phi_1(x)$ has degree ν and $\phi_2(x)$ has degree $\nu - 2$, then $\theta(x, y) = \phi_1(x) + \phi_2(x)y$ has 2ν poles, so it also has that number of zeros. Since ϕ_1 has $\nu + 1$ variable coefficients and $\phi_2(x)$ has $\nu - 1$, θ has 2ν coefficients. If $2\nu > N$, the N conditions on the 2ν coefficients of θ imposed by the requirement that θ be zero at N given points can be satisfied by some choice of θ. Since θ has 2ν zeros, it has $2\nu - N$ zeros other than the N required ones, and the algebraic variations of the N given points are found by varying the 2ν coefficients of θ in such a way that these $2\nu - N$ extra zeros remain as zeros. The $2\nu - N$ conditions stating that θ must have these zeros are *independent*, so the coefficients of θ then vary with $2\nu - (2\nu - N) = N$ degrees of freedom. However, multiplication of θ by a constant does not change its zeros, so varying the coefficients of θ with N degrees of freedom varies its zeros with only $N - 1$ degrees of freedom. In short, an algebraic variation of N given points moves them in only $N - 1$ different directions, which is to say that algebraic variations satisfy *one* constraint in this case. Otherwise stated, algebraic variations describe subvarieties of codimension 1 in C^N.

This description of the phenomenon is in no way tied to the Weierstrass normal form. Gauss alludes indirectly to the elliptic curve $y^2 = 1 - x^4$ in the

introduction to Section 7 of the *Disquisitiones Arithmeticae* when he mentions
the transcendental functions related to integrals of $dx/\sqrt{1-x^4}$. Euler too
dealt with the curve $y^2 = 1 - x^4$ [26], for which explicit and beautiful formulas
can be developed for the addition law, and it is clear from Abel's published
papers that this particular curve is one that he studied intensely. To require
that it be put in Weierstrass normal form before the group law is described
loses certain symmetries that deserve to be kept. But the above heuristic
derivation of the fact that a curve in Weierstrass normal form has genus 1
also proves that $y^2 = 1 - x^4$ has genus 1, because in this case x is ∞ at two
points, both of them simple poles, whereas y has double poles at these points
$((\frac{y}{x^2})^2 = (\frac{1}{x})^4 - 1$ is finite when $x = \infty)$, so $\theta(x,y) = \phi_1(x) + \phi_2(x)y$ has a ν-fold
pole at each—and therefore 2ν zeros—when $\deg \phi_1 = \nu$ and $\deg \phi_2 = \nu - 2$.
Again the number of parameters in such a function $\phi_1(x) + \phi_2(x)y$ is 2ν,
and the same arguments then show that the algebraic variation of N points
on the curve moves them with only $N - 1$ degrees of freedom and therefore
determines subvarieties of C^N of codimension 1.

In the same way, Abel's construction generalizes the Euler addition formula
to any curve C for which the algebraic variations describe subvarieties of C^N
of codimension 1. If $(P_1, P_2, \ldots, P_N)$ is moved to $(O, O, \ldots, O, R)$ by means
of an algebraic variation, then, as before,

$$(1) \quad \int_O^{P_1} f(x,y)dx + \int_O^{P_2} f(x,y)dx + \cdots + \int_O^{P_N} f(x,y)dx = \int_O^R f(x,y)dx + E,$$

where E is a quantity that can be expressed in terms of integrals of rational
functions. (Moreover, E is zero when the integrand is *holomorphic* in the sense
defined in Essay 4.6. This is true, as will be shown, of the integrand dx/y for
curves in Weierstrass normal form or for the curve $y^2 = 1 - x^4$.)

Essay 4.3 An Algebraic Definition of the Genus

Modern treatments of the genus of a curve normally describe it in terms of the topology of the associated Riemann surface. Therefore, modern mathematicians are usually amazed to learn that the idea stems from Abel, who lived and worked at a time when even the notion of a complex function of a complex variable was in its early infancy and the notion of a Riemann surface was still in the future. (Riemann surfaces first appeared in Riemann's dissertation [57] of 1851.) But as the discussion in the preceding essay shows, Abel's point of view does not depend on complex numbers.

The geometric picture of N points on the curve varying with $N - g$ degrees of freedom that was presented in the preceding essays does depend on complex numbers, because the coordinates of the intersection points defined by $\chi = 0$, $\theta = 0$ exist only in some algebraically closed field, and the notion of continuous variation requires something like real numbers. But the actual determination of the genus depends on purely algebraic considerations, at least in the examples of the preceding essay. All that is needed is to construct, for a large number ν, a formula $\theta(x, y, a, a', a'', \ldots)$ for the most general "function" in the field that has poles only at points where $x = \infty$ and no longer has poles at those points when it is divided by x^ν (although this division will probably cause it to have poles at $x = 0$). The **genus** g is determined by the condition that the number of zeros of θ is $g - 1$ greater than the number of arbitrary constants in the formula for θ.

Of course elements of a function field are not really functions in the usual sense, so they do not really have zeros and poles, and the condition that an element have poles only where $x = \infty$ is far from rigorous. Therefore, this description of the genus needs more explanation. Starting with the field of rational functions on an algebraic curve $\chi(x, y) = 0$—which is simply the root field of a monic, irreducible polynomial $\chi(x, y)$ in y with coefficients in $\mathbf{Z}[x]$—one needs to define what it means to say that an element θ of the field has no poles where x is finite and that θ/x^ν has no poles where $1/x$ is finite, and then one needs to determine how many zeros such a θ has and how many arbitrary constants there are in the formula for the most general such θ.

The idea of an element θ having no poles where x is finite has a standard algebraic formulation: An element θ of the field of rational functions on a curve $\chi(x, y) = 0$ is **integral over** x if some power of θ is equal to a sum of multiples of lower powers in which the multipliers are elements of the ring $\mathbf{Q}[x]$ of polynomials in x with rational coefficients.* The justification of this

---

* One could also use the more restrictive, but perhaps more natural, definition in which the multipliers are required to be in $\mathbf{Z}[x]$. Then an element would be integral over x in the sense defined above if and only if some integer multiple of it was integral in the more restrictive sense. Since $\frac{1}{2}$ is certainly a "function" without poles, the definition given above is the one that describes "rational functions without poles" on an algebraic curve.

definition is, in the last analysis, pragmatic—it *works* in the sense that it suggests correct theorems and is useful in proofs.

(The analogous definition of an **algebraic integer** in an algebraic number field—see Essay 2.5—emerged in the work of Kronecker and Dedekind in the 1860s and 1870s. Bourbaki [6] claims it is in the work of Eisenstein as early as 1852, but I do not find it there. Kronecker [38, §1] used the above definition of integrality over x of an "algebraic function" in his study of function fields, but as far as I have found he does not explain or motivate it.)

It is easy to see that the elements of the field of rational functions on $\chi(x,y)$ that are integral over x form a ring in the field and that this ring contains $\mathbf{Q}[x]$.[†]

If θ is integral over x, then dividing an equation that demonstrates its integrality, say $\theta^n = a_1(x)\theta^{n-1} + \cdots + a_n(x)$, by $x^{n\mu}$ for μ larger than the maximum degree of the $a_i(x)$ shows that θ/x^μ is integral over $1/x$ for all such values of μ.[‡] The **order** of θ at $x = \infty$ is by definition the smallest ν for which θ/x^ν is integral over $1/x$.

Let $\Theta(x^\nu)$ denote the elements θ of the field of rational functions on $\chi(x,y) = 0$ that are integral over x and have order at most ν where $x = \infty$. The goal is to find a formula for the most general element θ of $\Theta(x^\nu)$, and to compare the "number of zeros" of θ to the "number of arbitrary constants it contains."

The "number of zeros" of such a θ has a very plausible meaning. By assumption, $\chi(x,y)$ is monic in y, say of degree n in y. Then the values of y for a given x are the roots of a monic polynomial of degree n, so there are n of them, counted with multiplicities. For this reason, there are n points on the curve for each x, so x assumes each value exactly n times on the curve, counted with multiplicities. For this reason, it is reasonable to take the view that x also assumes the value ∞ exactly n times—that x has n poles on the

---

[†] If $z_1^{n_1}$ can be expressed as a sum of multiples of lower powers of z_1 in which the multipliers are in $\mathbf{Q}[x]$, and $z_2^{n_2}$ can be expressed as a sum of multiples of lower powers of z_2 in which the multipliers are in $\mathbf{Q}[x]$, then every polynomial in z_1 and z_2 with coefficients in $\mathbf{Q}[x]$ can be expressed as a sum of multiples of $z_1^i z_2^j$ with coefficients in $\mathbf{Q}[x]$, where $i < n_1$ and $j < n_2$. Therefore, multiplication by any such polynomial in z_1 and z_2—in particular multiplication by $z_1 + z_2$ and $z_1 z_2$—can be represented by the $n_1 n_2 \times n_1 n_2$ matrix of elements of $\mathbf{Q}[x]$ that gives its effect on these $n_1 n_2$ monomials $z_1^i z_2^j$. Therefore, since the polynomial in z_1 and z_2 is a root of the (monic) characteristic polynomial of this $n_1 n_2 \times n_1 n_2$ matrix by the Cayley–Hamilton theorem, $z_1 + z_2$ and $z_1 z_2$, and, in the same way, all polynomials in z_1 and z_2 with coefficients in $\mathbf{Q}[x]$, are integral over x.

[‡] Note that $\chi(x,y)/x^\kappa$ has the form $\chi_1(1/x, y/x^\lambda)$, where χ_1 is irreducible with integer coefficients and monic in its second variable, when κ is large enough, and that x and y can be expressed rationally in terms of $u = 1/x$ and $v = y/x^\lambda$, so the field of rational functions on $\chi(x,y) = 0$ can also be regarded as the field of rational functions on the curve $\chi_1(u,v) = 0$. To say that an element of this field is integral over $1/x$ means, of course, that it is integral over u.

curve, counted with multiplicities. Therefore, x^ν has n poles of order ν for a total of $n\nu$ poles. Since θ has the same poles as x^ν (except when θ is in the subset of $\Theta(x^\nu)$ containing "functions" that have fewer than the maximum number of poles allowed, which is sparse in $\Theta(x^\nu)$), it follows that θ should be regarded as having $n\nu$ poles; reversing the above reasoning then leads to the conclusion that θ assumes each value $n\nu$ times, including the value zero. In short, it is plausible to take $n\nu$ to be the number of zeros of a typical element of $\Theta(x^\nu)$.

This analysis of the number of zeros of a typical element of $\Theta(x^\nu)$ for large ν overlooks, however, a phenomenon that is exhibited by the example $\chi(x,y) = (x^2+y^2)^2 - 2(x+y)^2$. The field of rational functions on this "curve"— the root field of this $\chi(x,y)$—contains the element $\frac{x^2+y^2}{x+y}$, which is a root of $X^2 - 2$ (by definition, the square of $x^2 + y^2$ is $2(x + y)^2$). Therefore, it is reasonable to let $\sqrt{2}$ denote this element of the field. Then $\chi(x,y) = (x^2 + y^2 - \sqrt{2}(x+y))(x^2+y^2+\sqrt{2}(x+y))$, which shows that the field is an extension of degree *two*, not four, of the field of rational functions in x with coefficients in the number field $\mathbf{Q}(\sqrt{2})$. Geometrically, the curve $(x^2+y^2)^2 - 2(x+y)^2 = 0$ is quite simple, because the reduction $(x^2+y^2-\sqrt{2}(x+y))(x^2+y^2+\sqrt{2}(x+y)) = 0$ shows that it is a *union of two circles,* namely, the circle whose diameter is the line from the origin to $(\sqrt{2}, \sqrt{2})$ and the one whose diameter is the line from the origin to $(-\sqrt{2}, -\sqrt{2})$. Geometers traditionally use algebraically closed ground fields in part to avoid situations like this in which a curve described by a simple irreducible polynomial becomes a union of two curves when the field of constants is extended.

The simple constructive solution to this difficulty is not to make the giant leap to an algebraically closed ground field—the usual choice being the field of complex numbers, which is not an algebraic but a transcendental extension— but to *adjoin new constants as needed.* In the example, the constant $\sqrt{2}$ is not just needed, it is already present as $\frac{x^2+y^2}{x+y}$, and when it is used the curve is reducible, and the geometric picture of the "curve" whose field of rational functions is the root field of $(x^2 + y^2)^2 - 2(x + y)^2$ is a single circle $x^2 + y^2 = \sqrt{2}(x+y)$. This revision of the picture makes it clear that the number of zeros of x on the curve is two, not four. Consequently, the number of poles of x^ν, which is the same as the number of zeros, is 2ν, not 4ν, and a typical element of $\Theta(x^\nu)$ has 2ν zeros, not 4ν.

More generally, one needs to take into consideration the possibility that the root field of $\chi(x,y)$ may contain constants other than the obvious constants in $\mathbf{Q}$. Here a "constant" is an element of the root field that is a root of a polynomial with integer coefficients, or, what is the same, an element of $\Theta(x^0)$. (A polynomial in x with rational coefficients is equal to a polynomial in $1/x$ with rational coefficients if and only if it is a rational number, and a root of a monic polynomial with rational coefficients is a root of a polynomial with integer coefficients.) For this reason, $\Theta(x^0)$ will be called the **field of**

constants of the root field of $\chi(x, y)$. (Note that $\Theta(x^\nu)$ is a vector space over $\mathbf{Q}$ for all ν, and that $\Theta(x^0)$ is in fact a *field*.)

The example then suggests the following definition: The **number of zeros** of a typical element of $\Theta(x^\nu)$ for large values of ν is $n_0\nu$, where n_0 is the degree of the root field of χ as an extension, not of the field $\mathbf{Q}(x)$ of rational functions in x with integer coefficients, but of the field of rational functions of x with coefficients in the field of constants $\Theta(x^0)$. When $\Theta(x^0) = \mathbf{Q}$, n_0 is simply the degree of χ in y, but in the general case it is this degree divided by $[\Theta(x^0) : \mathbf{Q}]$.

Similarly, when one counts the "number of constants" in a formula for a typical element of $\Theta(x^\nu)$, one thinks of the constants as being in $\Theta(x^0)$, not $\mathbf{Q}$. Because the product of an element of $\Theta(x^\nu)$ and an element of $\Theta(x^\mu)$ is an element of $\Theta(x^{\nu+\mu})$, $\Theta(x^\nu)$ is a vector space over $\Theta(x^0)$. The number of constants in a formula for a typical element of $\Theta(x^\nu)$ is quite simply the *dimension* of $\Theta(x^\nu)$ as a vector space over the field $\Theta(x^0)$.

In this way, Abel's conception of the number of integrals to which a sum of integrals of an algebraic integrand can be reduced leads to the definition of the **genus** of the root field of $\chi(x, y)$ as the number g for which $\dim \Theta(x^\nu) = n_0\nu - g + 1$, where ν is a large integer, where n_0 is the degree n of χ in y divided by the degree c of the field of constants $\Theta(x^0)$ as an extension of $\mathbf{Q}$, where $\Theta(x^\nu)$ denotes the subset of the root field containing elements θ that are integral over x and have order at most ν where $x = \infty$, and where the dimension is the dimension of $\Theta(x^\nu)$ as a vector space over the field of constants $\Theta(x^0)$. The underlying idea is that an element of $\Theta(x^\nu)$ has $n_0\nu$ zeros and contains $\dim \Theta(x^\nu)$ parameters; variation of all $\dim \Theta(x^\nu)$ parameters in such a θ varies its zeros with only $\dim \Theta(x^\nu) - 1$ degrees of freedom—one degree of freedom is lost because multiplication of a function by a constant does not change its zeros—so the number of constraints g on the motion of the $n_0\nu$ zeros under an algebraic variation is determined by the equation $\dim \Theta(x^\nu) - 1 = n_0\nu - g$.

The main theorem will be to show that this genus is **intrinsic** to the curve $\chi(x, y) = 0$ in the sense that if the root fields of two polynomials $\chi(x, y)$ are isomorphic—if the two corresponding curves are **birationally equivalent**—then the fields have the same genus. Although the proof will be somewhat long, the underlying reason that the genus is intrinsic stems from the above discussions: It is the codimension of the subvarieties of C^N determined by the algebraic variations of N points on the curve.

Essay 4.4 Newton's Polygon

...ses [Newton's] *principaux Guides dans ces Recherches* [on cubic curves] *ont été la Doctrine des* Séries infinies, *qui lui doit presque tout, & l'usage du* Parallélogramme analytique, *dont il est l'Inventeur. ...Il est facheux que Mr. Newton se soit contenté d'étaler ses découvertes sans y joindre les Démonstrations, et qu'il ait préféré le plaisir de se faire admirer à celui d'instruire.* (Newton's main guides in his researches on cubic curves were the doctrine of *infinite series,* which owes him practically everything, and the use of the *analytic parallelogram,* of which he is the inventor. ...It is annoying that Mr. Newton contented himself with laying out his discoveries without accompanying them with proofs, and that he preferred the pleasure of making himself admired to that of instructing.)—Gabriel Cramer, [11, Preface]

The program outlined at the end of the last essay for constructing the genus of an algebraic curve—or, more precisely, the genus of the root field of a given $\chi(x, y)$—will be carried out in the following essays using an algorithm of Isaac Newton* for expanding an algebraic function of x as a power series in *fractional* powers of x. Known as Newton's *polygon,* or sometimes Newton's *parallelogram,* it constructs, for a given polynomial equation $\chi(x, y) = 0$, infinite series expansions of y in fractional powers of x. It involves *choices* and results in n *different* expansions, where $n = \deg_y \chi$.

It will be useful to expand y not only in powers of x but also in powers of $x - \alpha$ for various algebraic numbers α, something that can be accomplished by the same method, since setting $x_1 = x - \alpha$ and $\chi_1(x_1, y) = \chi(x - \alpha, y)$ gives an algebraic relation between x_1 and y that can be used to expand y in (fractional) powers of x_1 using Newton's polygon.

Let $\chi(x, y)$ be an irreducible polynomial with integer coefficients that has positive degree in both x and y and is monic in y, and let α be a given value for x. The objective is to find infinite series "solutions" $y = \theta_0(x - \alpha)^{\varepsilon_0} + \theta_1(x - \alpha)^{\varepsilon_1} + \theta_2(x - \alpha)^{\varepsilon_2} + \cdots$ of $\chi(x, y) = 0$ in which the coefficients θ_i are algebraic numbers and the exponents $\varepsilon_0 < \varepsilon_1 < \cdots < \varepsilon_k < \cdots$ are an increasing sequence of rational numbers. It will also be assumed that the exponents increase without bound in the sense that for any given N one can find a value of k for which $\varepsilon_k > N$. The meaning of the statement that such a

---

* Newton's presentation is quite sketchy. My main source was Walker [61]. See also Newton [51, vol. 3, p. 50 and p. 360, vol. 4, p. 629], Hensel–Landsberg [31], and Chebotarev [9]. Chebotarev cites (end of §2) the Hensel–Landsberg book as his basic source, but he examines the Newton polygon much more fully than that book does, dealing thoroughly, for instance, with the history of the method. Unfortunately, his article is available only in Russian, and is difficult to find. Chebotarev advocates calling it Newton's "diagram" as Hensel and Landsberg do, saying that the "polygon" was not present in Newton's formulation, but the name "Newton's polygon" now seems firmly established.

Fig. 4.4. Newton.

series "solves" $\chi(x, y) = 0$ is clear, if somewhat nonconstructive: Such series can be added and multiplied term by term, and $x = \alpha + (x - \alpha)$ is such a series (a terminating one), so $\chi(x, y)$ represents such a series, the coefficients of which can be computed in an open-ended way by finding, for any given upper bound, all terms of the series $\chi(x, y)$ in which the exponent is less than that bound. To say that $\chi(x, y) = 0$ means simply that the result is always zero.

Since y is integral over x, the exponents ε_i are to be expected to be nonnegative. Therefore, a knowledge of the terms of the series for y through the term $\theta_k(x - \alpha)^{\varepsilon_k}$ is all that is needed to compute all terms of $\chi(x, y)$ in which the exponents are less than or equal to ε_k, because all omitted terms contain $(x - \alpha)^{\varepsilon_{k+i}}$ for some $i > 0$ and $\varepsilon_{k+i} > \varepsilon_k$. What is sought, then, are infinite sequences $\theta_0, \theta_1, \theta_2, \ldots$ and $0 \leq \varepsilon_0 < \varepsilon_1 < \cdots$ for which all terms of the *terminating* sequence $\chi(x, \theta_0(x - \alpha)^{\varepsilon_0} + \theta_1(x - \alpha)^{\varepsilon_1} + \cdots + \theta_k(x - \alpha)^{\varepsilon_k})$ have exponents greater than ε_k. A constructive solution of this problem must of course be an *algorithm* for *generating* such sequences. "Newton's polygon" is such an algorithm.

More specifically, given the initial terms $\theta_0(x - \alpha)^{\varepsilon_0} + \theta_1(x - \alpha)^{\varepsilon_1} + \cdots + \theta_k(x - \alpha)^{\varepsilon_k}$ of an infinite series solution y of $\chi(x, y) = 0$ in the sense just described, the algorithm should give *all possible* values $\theta_{k+1}(x - \alpha)^{\varepsilon_{k+1}}$ for the next term of the sequence. They can be completely described in the following way: To avoid fractional exponents, let m be the least common denominator of $\varepsilon_0, \varepsilon_1, \ldots, \varepsilon_k$ and let $s = (x - \alpha)^{1/m}$, so that the initial terms that are assumed to be known take the form $\beta_0 + \beta_1 s + \cdots + \beta_h s^h$, where h is the integer $m\varepsilon_k$ and where β_i is zero unless i is of the form $m\varepsilon_j$ for some j, in which case $\beta_i = \theta_j$. Let the term following $\beta_h s^h$ be $\gamma s^{\rho+h}$, so that the required equation is $\chi(\alpha + s^m, \beta_0 + \beta_1 h + \cdots + \beta_h s^h + \gamma s^{h+\rho} + \cdots) = 0$, where ρ is a positive rational number. To determine the possible values of γ and ρ expand $\chi(\alpha + s^m, \beta_0 + \beta_1 s + \cdots + \beta_h s^h + ts^h)$, a polynomial in s and t whose coefficients are algebraic numbers (because they are polynomials in $\beta_0, \beta_1, \ldots, \beta_h, \alpha$, and the coefficients of χ), as a polynomial in t, $\Phi_0(s) + \Phi_1(s)t + \Phi_2(s)t^2 + \cdots + \Phi_n(s)t^n$, whose coefficients $\Phi_i(s)$ are polynomials in s. Again, to avoid fractional exponents, let ρ be written $\rho = \frac{\sigma}{\tau}$, where σ and τ are positive integers, and let $s_1 = s^{1/\tau}$, so that the required identity becomes $\chi(\alpha + s_1^{m\tau}, \beta_0 + \beta_1 s_1^\tau + \cdots + \beta_h s_1^{h\tau} + \gamma s_1^{h\tau+\sigma} + \cdots) = 0$, which is to say

$$\Phi_0(s_1^\tau) + \Phi_1(s_1^\tau)(\gamma s_1^\sigma + \cdots) + \Phi_2(s_1^\tau)(\gamma s_1^\sigma + \cdots)^2 + \cdots + \Phi_n(s_1^\tau)(\gamma s_1^\sigma + \cdots)^n = 0.$$

The simple idea that underlies Newton's polygon is the observation that this infinite series in s_1 with algebraic number coefficients, which is a sum of $n+1$ such series, can be identically zero only if all terms in the sum cancel, and, in particular, only if the *lowest-order terms* of these series cancel. If the polynomial $\Phi_i(s)$ is nonzero, it has the form $\zeta_i s^{j_i} + \cdots$, where $\zeta_i \neq 0$ and the omitted terms all have degree greater than j_i. With this notation, the term of $\Phi_i(s_1^\tau)(\gamma s_1^\sigma + \cdots)^i$ of lowest degree, when $\Phi_i(s) \neq 0$, is $\zeta_i \gamma^i s_1^{\sigma i + \tau j_i}$. The required cancellation dictates that the positive integers σ and τ must have the property that $\sigma i + \tau j_i$ assumes its minimum value for at least two different pairs (i, j_i) (note that these pairs are determined by χ, m, and $\beta_0 + \beta_1 s + \cdots + \beta_h s^h$). These conditions limit the pairs (σ, τ) to a finite number of possibilities— the geometrical picture is the one described below—and even gives strong information about the coefficient γ of the next term, namely, that it is a nonzero root of the polynomial $\sum \zeta_i \gamma^i$, where the sum is extended over just those values of i for which $\Phi_i(s) \neq 0$ and $\sigma i + \tau j_i$ assumes its minimum value.

Some term of some series $\Phi_i(s_1^\tau)(\gamma s_1^\tau + \cdots)^i$ for $i > 0$ must cancel the first term $\zeta_0 s_1^{\tau j_0}$ of $\Phi_0(s_1^\tau)$, so $\tau j_0 \geq \sigma i + \tau j_i$ for some $i > 0$. Since σ and τ are both positive, j_0 must be greater than j_i for at least one $i > 0$. Therefore, the above discussion shows that the series $\beta_0 + \beta_1 s + \cdots + \beta_h s^h$ can be extended to be an infinite series solution y of $\chi(x,y) = 0$ when $x = \alpha + s^m$ only if the polynomial in two variables $\chi(\alpha + s^m, \beta_0 + \beta_1 s + \cdots + \beta_h s^h + ts^h) = \Phi_0(s) + \Phi_1(s)t + \Phi_2(s)t^2 + \cdots + \Phi_n(s)t^n$ has the property that s divides $\Phi_0(s)$ more times than it divides $\Phi_i(s)$ for at least one $i > 0$. Otherwise stated, the term or terms of this polynomial of lowest degree in s must all contain t.

As will be shown, these *necessary* conditions on the constants that describe the next term when a certain number of terms are known permit one to construct *all possible* solutions y of $\chi(x,y) = 0$ as infinite series of fractional powers of $x - \alpha$.

A **truncated solution** y of $\chi(x,y) = 0$ at $x = \alpha$ will by definition consist of (1) an algebraic number field $\mathbf{A}$ containing α, (2) a positive integer m, and (3) a finite sequence $\beta_0, \beta_1, \ldots, \beta_h$ in $\mathbf{A}$ with the property that the term or terms of $\chi(\alpha + s^m, \beta_0 + \beta_1 s + \cdots + \beta_h s^h + t s^h)$ of lowest degree in s all contain t. In addition, it will be assumed that the result $\Phi_0(s)$ of setting $t = 0$ in this polynomial is not zero; otherwise, $y = \beta_0 + \beta_1(x - \alpha)^{1/m} + \cdots + \beta_h(x - \alpha)^{h/m}$ is an **actual solution** of $\chi(x, y) = 0$ and there is no need to use higher powers of $x - \alpha$.

Newton's Polygon

Input: A truncated solution y of $\chi(x, y) = 0$ at $x = \alpha$, as that term was just defined.

Algorithm: As above, let $\chi(\alpha + s^m, \beta_0 + \beta_1 s + \cdots + \beta_h s^h + t s^h)$ be written in the form $\Phi_0(s) + \Phi_1(s)t + \Phi_2(s)t^2 + \cdots + \Phi_n(s)t^n$ of a polynomial in t whose coefficients $\Phi_i(s)$ are polynomials in s with coefficients in the field $\mathbf{A}$ specified by the input. Consider the set of pairs (i, j_i) of integers, where i is in the range $0 \le i \le n$, where $\Phi_i(s) \ne 0$, and where j_i is the number of times that s divides $\Phi_i(s)$. By assumption, j_0 is defined and greater than at least one other j_i. The **segments of the Newton polygon** *corresponding to this input are the line segments that join two points (i, j_i), say those corresponding to the indices i_1 and $i_2 > i_1$, in such a way that (1) the segment has negative slope, so it is described by the equation $\sigma i + \tau j = k$ where $\sigma = j_{i_1} - j_{i_2}$ and $\tau = i_2 - i_1$ are both positive and where k is the common value of $\sigma i + \tau j$ for these two indices, (2) $\sigma i + \tau j_i \ge k$ for all indices i for which j_i is defined, and (3) $\sigma i + \tau j_i > k$ whenever j_i is defined and $i < i_1$ or $i > i_2$. With each such segment, associate the polynomial*

$$\eta(c) = \sum_{\sigma i + \tau j_i = k} \zeta_i c^i$$

with coefficients in $\mathbf{A}$, where the sum is over just those values of i for which (i, j_i) lies on the segment, of which there are at least two, and where ζ_i is the coefficient of s^{j_i} in $\Phi_i(s)$. Extend the input field $\mathbf{A}$, if necessary, to split all polynomials $\eta(c)$ that result in this way from segments of the polygon.

(Geometrically, the segments join to form a polygonal path that joins the point $(0, j_0)$ to the first point, call it (I, J), of the form (i, j_i) at which j_i assumes its minimum value. This path is determined by the fact that it joins points of the form (i, j_i) in such a way that none of these points are in the interior of the closed polygon formed by it and the segments from $(0, j_0)$ to $(0, J)$ and from $(0, J)$ to (I, J).)

Output: A truncated solution of $\chi(x, y) = 0$ for $x = \alpha$ for each *nonzero* root γ of each polynomial $\eta(c)$ in the extended field constructed by the algorithm, namely, the truncated solution

(1)
$$y = \beta_0 + \beta_1(x - \alpha)^{1/m} + \beta_2(x - \alpha)^{2/m}$$
$$+ \cdots + \beta_h(x - \alpha)^{h/m} + \gamma(x - \alpha)^{(h\tau + \sigma)/\tau m}$$

in which one term with coefficient γ and exponent $\frac{h\tau + \sigma}{\tau m} = \frac{h + \rho}{m}$ is added to the input truncated solution where $\rho = \frac{\sigma}{\tau}$. In other words, the output

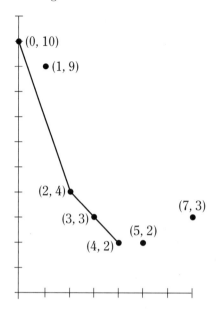

Fig. 4.5. When there are seven points $(i, j_i) = (0, 10)$, $(1, 9)$, $(2, 4)$, $(3, 3)$, $(4, 2)$, $(5, 2)$, $(7, 3)$, Newton's polygon has two segments. They join $(0, 10)$ and $(4, 2)$ via $(2, 4)$.

truncated solution corresponding to γ consists of the (possibly) extended field **A** constructed by the algorithm, the positive integer τm, and the sequence β_0', β_1', β_2', ..., $\beta_{h\tau+\sigma}'$ in which $\beta_{i\tau}' = \beta_i$ for $i = 0, 1, \ldots, h$ and $\beta_{h\tau+\sigma}' = \gamma$ but all other coefficients β_i' are zero.

That each output (1) is a truncated solution—unless, of course, it is an *actual* solution—can be proved as follows: Set $s = s_1^\tau$ and $t = s_1^\sigma(\gamma + t_1)$ in the definition of $\Phi_0(s)$, $\Phi_1(s)$, ..., $\Phi_n(s)$ to put the new equation in the form

$$(2) \quad \chi(\alpha + s_1^{\tau m}, \beta_0 + \beta_1 s_1^\tau + \beta_2 s_1^{2\tau} + \cdots + \beta_h s_1^{h\tau} + \gamma s_1^{h\tau+\sigma} + t_1 s_1^{h\tau+\sigma})$$

$$= \sum_{i=0}^{n} \Phi_i(s_1^\tau)(s_1^\sigma)^i(\gamma + t_1)^i.$$

By the choice of σ and τ, no term on the right contains s_1 to a power less than the minimum value of $\sigma i + \tau j_i$, call it k, and the terms that contain s_1 to the power k exactly are $s_1^k \eta(\gamma + t_1)$ by the definition of η. Since $\eta(\gamma + t_1)$ is a nonzero polynomial (its degree in t_1 is the largest value of i for which the point (i, j_i) lies on the corresponding segment of the Newton polygon) with constant term zero (by the choice of γ), (2) is a polynomial in which the terms of lowest degree k in s_1 all contain t_1, as was to be shown.

The **ambiguity** of a truncated solution is, in the notation used above, the least index i for which j_i attains its minimum. Otherwise stated, it is the i-coordinate I of the endpoint (I, J) of the Newton polygon other than $(0, j_0)$.

A truncated solution will be called **unambiguous** if its ambiguity is 1. In this case, the polygon consists of a single line segment, and $\eta(c)$ is a polynomial of degree 1 whose single root is nonzero, so the algorithm produces a single output; moreover, the algorithm does not increase m, and it results in no extension of **A** because the root of $\eta(c)$ is already in **A**.

The ambiguity of an output solution is the multiplicity of its γ as a root of its $\eta(c)$, as follows from the above observation that the terms of lowest degree in s_1 are $s_1^k \eta(\gamma + t_1)$, because the multiplicity of γ as a root of $\eta(c)$ is the number of times t_1 divides $\eta(\gamma + t_1)$. Thus, among the nonzero terms $\epsilon s_1^p t_1^q$ of (2) in which p assumes its minimum value k, the one in which q has its least value is the one in which q is the multiplicity of γ as a root of its $\eta(c)$.

In particular, *if the input truncated solution is unambiguous, so is the output truncated solution.* Thus, if it begins with an unambiguous truncated solution, the algorithm constructs an infinite series solution (which may in rare cases be an actual terminating solution) with coefficients in the same **A**. In short, the construction of infinite series solutions is reduced by the Newton's polygon algorithm to the construction of unambiguous truncated solutions.

Theorem 1.   *Construct n distinct infinite series solutions y of $\chi(x, y) = 0$ at $x = \alpha$.*

As above, α is a given algebraic number and $\chi(x, y)$ is a given polynomial with integer coefficients that is irreducible, contains both x and y, and is monic of degree n in y. For the reason just stated, an infinite series solution can be regarded as having been constructed when an unambiguous truncated solution has been constructed. The proof of the theorem will follow an example:

Let $\chi(x, y) = y^3 - xy + x^3$ (the curve $\chi = 0$ is the folium of Descartes—see Fig 4.6, p. 151) and let $\alpha = 0$. If one begins with the truncated solution $m = 1$, $y = 0$, one begins with $\chi(0 + s, 0 + t) = s^3 - st + t^3$, and Newton's polygon joins the points $(0, 3)$ and $(3, 0)$ via the point $(1, 1)$. The two segments of the polygon are described by the equations $2i + j = 3$ and $i + 2j = 3$.

The first segment gives just one output truncated solution $y = x^2$, because $m = 1$, $h = 0$, $\sigma = 2$, $\tau = 1$ and because the polynomial $\eta(c) = 1 - c$ in this case has just one nonzero root 1. It is a simple root, so this output solution is unambiguous.

The second segment gives two output truncated solutions $y = \gamma \sqrt{x}$ (because $m = 1$, $h = 0$, $\sigma = 1$, $\tau = 2$), where γ is a nonzero root of $\eta(c) = -c + c^3$. Thus $\gamma = \pm 1$, and the output consists of two truncated solutions $y = \pm \sqrt{x}$. Both are unambiguous.

Thus, Newton's polygon constructs three unambiguous truncated solutions and therefore constructs the three required infinite series solutions of $y^3 - xy + x^3 = 0$ for $x = 0$. These infinite series solutions can be found by repeated application of the Newton polygon algorithm, but the first few terms can be found more easily by the following method.

The truncated solution $y = \pm \sqrt{x}$ calls for the computation of $\chi(s^2, \pm s + st) = (s^2)^3 - s^2(\pm s + st) + (\pm s + st)^3 = s^3 \left(s^3 - (\pm 1 + t) + (\pm 1 + t)^3 \right) = $

$s^3(s^3 + 2t \pm 3t^2 + t^3)$. The term $2t$ shows that this truncated solution is unambiguous. The continuation of the truncated solution $y = \pm\sqrt{x} + \cdots$ can be found using the equation $s^3 + 2t \pm 3t^2 + t^3 = 0$ to express t as a power series in $s = \sqrt{x}$ and substituting the result in $y = \pm s + st$. Consider first the case in which the sign is plus. The relation $s^3 + 2t + 3t^2 + t^3 = 0$ can be written $t = -\frac{1}{2}s^3 + t^2(-\frac{3}{2} - \frac{t}{2})$ to find that $t = -\frac{1}{2}s^3 + (-\frac{1}{2}s^3 + t^2(-\frac{3}{2} - \frac{t}{2}))^2(-\frac{3}{2} - \frac{t}{2}) = -\frac{1}{2}s^3 + (\frac{1}{4}s^6 + \frac{3}{2}s^3t^2 + \cdots)(-\frac{3}{2} - \frac{t}{2}) = -\frac{1}{2}s^3 - \frac{3}{8}s^6 - \frac{9}{4}s^3t^2 - \frac{1}{8}s^6t + \cdots = -\frac{1}{2}s^3 - \frac{3}{8}s^6 - \frac{9}{4}\cdot\frac{1}{4}\cdot s^9 + \frac{1}{8}\cdot\frac{1}{2}s^9 + \cdots = -\frac{1}{2}s^3 - \frac{3}{8}s^6 - \frac{1}{2}s^9 + \cdots$, where the omitted terms all contain s^{12}, from which $y = s + st = s - \frac{1}{2}s^4 - \frac{3}{8}s^7 - \frac{1}{2}s^{10} + \cdots$. When $s^3 + 2t + 3t^2 + t^3 = 0$ is changed to $s^3 + 2t - 3t^2 + t^3 = 0$, the corresponding solution is found by changing s to $-s$ and t to $-t$. In summary, the second segment $i + 2j = 3$ corresponds to two infinite series solutions of $y^3 - xy + x^3 = 0$; they begin

$$y = \pm\sqrt{x} - \frac{1}{2}x^2 \mp \frac{3}{8}x^3\sqrt{x} - \frac{1}{2}x^5 + \cdots.$$

The infinite series solution $y = x^2 + \cdots$ corresponding to the first segment $2i + j = 3$ calls for computing the polynomial $\chi(s, s^2 + s^2t) = s^3 - s^3(1 + t) + s^6(1 + t)^3 = s^3(-t + s^3(1 + t)^3)$. The term $-t$ shows that the truncated solution $y = x^2$ is unambiguous. The expansion of y in powers of x can be found by using the relation $-t + s^3(1 + t)^3 = 0$ to expand t in powers of $s = x$ and substituting the result in $y = s^2 + s^2t$. Now, $t = s^3(1 + t)^3$ implies

$$
\begin{aligned}
t &= s^3(1 + s^3(1 + t)^3)^3 = s^3(1 + 3s^3(1 + t)^3 + 3s^6(1 + t)^6 + s^9(1 + t)^9) \\
&= s^3 + 3s^6(1 + t)^3 + 3s^9(1 + t)^6 + s^{12}(1 + t)^9 \\
&= s^3 + 3s^6 + 9s^6t + 9s^6t^2 + \cdots + 3s^9 + 18s^9t + \cdots + s^{12} + \cdots \\
&= s^3 + 3s^6 + 12s^9 + 28s^{12} + \cdots,
\end{aligned}
$$

so

$$y = x^2 + x^5 + 3x^8 + 12x^{11} + 28x^{14} + \cdots$$

is the beginning of this infinite series solution of $y^3 - xy + x^3 = 0$.

(Note that the sum of the three series is zero, at least up to the terms in x^5, in accord with the fact that the coefficient of y^2 in $y^3 - xy + x^3$ is zero.)

Proof of Theorem 1. A truncated solution of $\chi(x, y)$ at $x = \alpha$ in which $m = 1$ and $h = 0$ is an algebraic number β_0 for which the terms of $\chi(\alpha + s, \beta_0 + t)$ of lowest degree in s all contain t; since $\chi(\alpha + s, \beta_0 + t)$ contains the term t^n with no s at all, $y = \beta_0$ is a truncated solution if and only if $\chi(\alpha + s, \beta_0)$ does not contain a term without s or, to put it more simply, if and only if $\chi(\alpha, \beta_0) = 0$. In short, these truncated solutions $y = \beta_0$ are the roots of $\chi(\alpha, y)$.

The *ambiguity* of such a truncated solution $y = \beta_0$ of $\chi(x, y)$ at $x = \alpha$ is equal to the *multiplicity* of β_0 as a root of $\chi(\alpha, y)$, because the ambiguity of the truncated solution is by definition the least index i for which $\Phi_i(0) \neq 0$, where $\chi(\alpha + s, \beta_0 + t) = \Phi_0(s) + \Phi_1(s)t + \Phi_2(s)t^2 + \cdots + t^n$, which is the multiplicity of β_0 as a root of $\chi(\alpha, y)$. In particular, if all roots of $\chi(\alpha, y)$ are simple, the

Newton polygon algorithm applied to any one of the n unambiguous truncated solutions $y = \beta_0$ generates an infinite series solution y of $\chi(x, y) = 0$ at $x = \alpha$, which proves the theorem in this case. In the general case, one can apply the following algorithm:

Input: A set of truncated solutions of $\chi(x, y) = 0$ at $x = \alpha$.

Algorithm: While the set contains a truncated solution whose ambiguity is greater than 1, let the Newton polygon algorithm be used to replace one such truncated solution with one or more longer truncated solutions.

The theorem will be proved by proving that this algorithm *terminates*—that is, it reaches a stage at which all truncated solutions in the set that has been found are unambiguous—and by proving that each step leaves the sum of the ambiguities unchanged, so that if the algorithm starts with the truncated solutions $y = \beta_0$, the sum of whose ambiguities is $\deg_y \chi(\alpha, y) = n$ (because this sum is the sum of the multiplicities of the roots β_0 of $\chi(\alpha, y)$), it terminates with a set of n unambiguous truncated solutions, which then imply n infinite series solutions.

That the sum of the ambiguities does not change can be seen as follows: Let the notation be as in the description of Newton's polygon. The ambiguity of the input truncated solution is the least index I for which j_i attains its minimum value J. Since the segments of the Newton polygon join $(0, j_0)$ to (I, J) and since the number of nonzero roots—counted with multiplicities—of any $\eta(c)$ is its degree minus the number of times c divides it, which is the difference $i_2 - i_1$ of the i-coordinates of the endpoints of the corresponding segment, the ambiguity of the input truncated solution is the total number of nonzero roots, counted with multiplicities, of the polynomials $\eta(c)$ corresponding to segments of the polygon. Since, as was noted above, the multiplicity of γ as a root of $\eta(c)$ is the ambiguity of the output truncated solution corresponding to γ, the sum of the ambiguities of the output solutions is the ambiguity of the input solution, as was to be shown.

Each step of the above algorithm increases the number of truncated solutions in the list unless the input truncated solution, which has ambiguity greater than 1 by assumption, yields a single output truncated solution, which means that $\eta(c)$ is a constant times $(c - \gamma)^\mu$ for some nonzero algebraic number γ, where μ is the ambiguity of the input solution. It will be shown that the number of steps of this type is bounded above, so that repeated application of the Newton polygon algorithm eventually must increase the number of truncated solutions in the set. Since the total of their ambiguities is n at each step, it will follow that the process must terminate with n unambiguous truncated solutions, and the theorem will be proved.

Suppose, therefore, that the (ambiguous) input truncated solution is one that produces a single output truncated solution. It is to be shown that iteration of the algorithm eventually produces more than one output truncated solution. Let μ be the ambiguity of the input solution, which is therefore the ambiguity of each subsequent output solution as long as there is only one of them. As was just noted, when there is only one output truncated solution, $\eta(c)$ is a constant times $(c - \gamma)^\mu$ for some algebraic number γ, which implies that Newton's polygon consists of a single segment that passes through pairs (i, j_i) in which i has all values from 0 to μ because $\eta(c)$ contains terms in which c has all of these exponents. The single segment of the polygon is $(j_0 - j_\mu)i + \mu j = k$, where $k = \mu j_0$. Then $(j_0 - j_\mu) \cdot 1 + \mu j_1 = \mu j_0$, which shows that $j_0 - j_1$ is divisible by μ. Therefore, the segment can also be written $\sigma i + j = j_0$, where $\sigma = \frac{j_0 - j_\mu}{\mu}$; that is, τ can be taken to be 1, so that $s_1 = s$. Then (2) is divisible at least j_0 times by s (because $k = j_0$), whereas $\chi(\alpha + s^m, \beta_0 + \beta_1 s + \cdots + \gamma s^h + t s^h)$ is, by the definition of j_μ, divisible exactly j_μ times by s. In other words, adding the next term $\gamma s^{h+\sigma}$ to the truncated solution increases the number of times s divides $\chi(\alpha + s^m, \beta_0 + \beta_1 s + \cdots + \beta_h s^h + t s^h)$ from j_μ to at least $j_0 = j_\mu + \sigma\mu$.

Thus, if ν successive steps repeat the phenomenon of producing a single output truncated solution, it produces a truncated solution $y = \beta_0 + \cdots + \beta_h s^h + \gamma_1 s^{h+\sigma_1} + \gamma_2 s^{h+\sigma_1+\sigma_2} + \cdots + \gamma_\nu s^{h+\Sigma}$, where $\Sigma = \sigma_1 + \sigma_2 + \cdots + \sigma_\nu$, for which $\chi(\alpha + s^m, y + t s^{h+\Sigma})$ is divisible $j_\mu + \mu\Sigma$ times by s, say

$$\chi\left(\alpha + s^m, \beta_0 + \beta_1 s + \cdots + \beta_h s^h + \cdots + \gamma_\nu s^{h+\Sigma} + t s^{h+\Sigma}\right) = s^{j_\mu + \mu\Sigma} q(s, t).$$

Differentiation with respect to t gives

$$s^{h+\Sigma} \frac{\partial\chi}{\partial y}\left(\alpha + s^m, \beta_0 + \beta_1 s + \cdots + \beta_h s^h + \cdots + \gamma_\nu s^{h+\Sigma} + t s^{h+\Sigma}\right)$$
$$= s^{j_\mu + \mu\Sigma} \frac{\partial q}{\partial t}(s, t).$$

On the other hand, elimination of y between $\chi(x, y)$ and $\frac{\partial\chi}{\partial y}(x, y)$ (see Essay 1.3) gives, because the irreducibility of χ implies that these polynomials are relatively prime, an equation of the form

$$A(x, y)\chi(x, y) + B(x, y)\frac{\partial\chi}{\partial y}(x, y) = D(x),$$

in which $A(x, y)$, $B(x, y)$, and $D(x)$ are polynomials with integer coefficients. Substitution of $x = \alpha + s^m$ and $y = \beta_0 + \beta_1 s + \cdots + \beta_h s^h + \cdots + \gamma_\nu s^{h+\Sigma} + t s^{h+\Sigma}$ in $A(x, y)\chi(x, y) + B(x, y)\frac{\partial\chi}{\partial y}(x, y) = D(x)$ gives $D(\alpha + s^m)$ on the right and on the left gives a polynomial in s and t that is divisible at least $(j_\mu + \mu\Sigma) - (h + \Sigma)$ times by s. Thus $j_\mu + (\mu - 1)\Sigma - h$ is bounded above by the number of times s divides $D(\alpha + s^m)$. Since $\mu > 1$, this implies an upper bound on Σ; but $\Sigma \geq \nu$, because Σ is a sum of ν terms, each of which is at least 1, so ν is bounded above, and the proof of Theorem 1 is complete.[*]

---

[*] Walker's proof of this point [61, p. 102] is not constructive, because he jumps from the observation that the ambiguity can never increase and can never go below 1 to the conclusion that he can find a step beyond which the ambiguity never decreases.

Theorem 2. *Every truncated solution of $\chi(x, y) = 0$ for $x = \alpha$ is a truncation of one of the infinite series solutions constructed by Theorem 1.*

Proof. As was shown prior to the statement of the Newton polygon algorithm, if an infinite series solution is truncated, and the algorithm is applied to the result, one of the outputs is the truncated series with the next nonzero term after the truncation added. Therefore, any truncated solution is among the outputs if one starts with the truncated solutions $y = \beta_0$ in which $\chi(\alpha, \beta_0) = 0$ and repeatedly applies the algorithm. Since these are the truncated solutions constructed by Theorem 1, Theorem 2 follows.

Essay 4.5 Determination of the Genus

> *On doit donner au problème une forme telle qu'il soit toujours possible de*
> *le résoudre, ce qu'on peut toujours faire d'un problème quelquonque.* (One
> should give the problem a form in which it will always be possible to solve
> it, which can always be done for any problem whatever.)—Niels Henrik
> Abel [2, p. 217]

I confess that the meaning of this dictum of Abel's is not altogether clear
to me. Certainly it sounds like good advice, if one can understand what it
means. My best guess is that he means something like what Kronecker meant
when he said that one should require of one's definitions that one be able
to determine by a finite calculation whether the definition is fulfilled in any
given case. In the case of the determination of the genus of an algebraic field
of transcendence degree one—the genus of a given algebraic curve—I believe
both men would focus on constructive techniques like the ones given in this
essay.

The genus was described in Essay 4.3 in terms of the dimensions of the
spaces $\Theta(x^\nu)$ of elements of the root field of $\chi(x, y)$ (as always, an irreducible
polynomial with integer coefficients that contains both x and y and is monic
in y) that are integral over x and become integral over $\frac{1}{x}$ when they are divided
by x^ν. These dimensions can be determined easily once one constructs what
Dedekind and Weber [14] called a **normal basis** of the root field.

Theorem. *For a given $\chi(x, y)$, construct a subset y_1, y_2, ..., y_n of its root
field with the property that y_1, y_2, ..., y_n is a **basis** of the field over the field
of rational functions in x in the sense that each element w of the root field
has a unique representation in the form*

$$(1) \qquad w = \phi_1(x)y_1 + \phi_2(x)y_2 + \cdots + \phi_n(x)y_n,$$

*where the coefficients $\phi_i(x)$ are rational functions of x, and is an **integral
basis** in the sense that w is integral over x if and only if each coefficient
$\phi_i(x)$ in its representation (1) is a polynomial with rational coefficients, and
further is a **normal basis** in the sense that w is in $\Theta(x^\nu)$ if and only if each
coefficient $\phi_i(x)$ in its representation (1) is not only a polynomial but also
satisfies $\deg \phi_i + \lambda_i \leq \nu$, where λ_i is the order of y_i at $x = \infty$ for each i, that
is, the least integer for which y_i is in $\Theta(x^{\lambda_i})$.*

Proof. Dedekind and Weber gave what appears to be an algorithm for constructing an integral basis (their §3), but their construction relies on the assumption that for a given constant α one can either find an element y that
is integral over x and remains integral over x when it is divided by $x - \alpha$ or
prove that there is no such y. The proof below uses, in essence, the method
of Newton's polygon to justify this assumption and then constructs an integral basis using a method similar to theirs. However, they also assume that

a polynomial with rational coefficients can be written as a product of linear factors—they assume complex number coefficients—and the proof below is a modified version of theirs that adjoins only the constants that are needed. The first step will be to find a *common denominator* of the elements integral over x.

The operation of multiplication by an element of the field can be described by the $n \times n$ matrix of rational functions of x that describes it with respect to the basis $1, y, \ldots, y^{n-1}$ of the root field as a vector space over the field of rational functions in x. In other words, an element z of the root field of $\chi(x, y)$ can be described by the matrix whose entry m_{ij} in the ith row of the jth column is the rational function of x that is the coefficient of y^{j-1} in the representation of zy^{i-1} with respect to the basis $1, y, \ldots, y^{n-1}$. The trace $\sum_{i=1}^{n} m_{ii}$ of the matrix obtained in this way is the **trace** of z with respect to x.

Lemma. *If an element of the root field of $\chi(x, y)$ is integral over x, its trace is a polynomial in x with rational coefficients.*

Let $\psi = p(x, y)/q(x)$ be integral over x. Then, by the definition of integrality, there is a relation of the form $F(\psi) = 0$, in which F is a monic polynomial with coefficients in $\mathbf{Q}[x]$. Since F can be written as a product of irreducible, monic polynomials with coefficients in $\mathbf{Q}[x]$, ψ must be a root of an *irreducible* monic polynomial with coefficients in $\mathbf{Q}[x]$; call it F_1.

By the proposition of Essay 2.3, the root field of $\chi(x, y)$, because it contains ψ and is generated over $\mathbf{Q}(x)$ by y, can be described by two adjunction relations $f_1(\psi) = 0$ and $f_2(y, \psi) = 0$, where f_1 and f_2 have coefficients that are rational functions of x, f_1 is monic of degree ν_1, say, and is irreducible, while f_2 is monic of degree ν_2, say, in y and is irreducible as a polynomial in y with coefficients in the field of rational functions in x with ψ adjoined. Because f_1 and F_1 both have ψ as a root, because both are monic with coefficients that are rational functions of x, and because both are irreducible over the field of rational functions in x (F_1 is irreducible in this sense by virtue of Gauss's lemma), $f_1 = F_1$. In particular, the coefficients of f_1 are not just rational functions of x, they are polynomials in x with rational coefficients.

The trace of ψ is by definition the trace of the matrix that represents multiplication by ψ relative to the basis $1, y, \ldots, y^{n-1}$ of the root field over the field of rational functions in x. Therefore (because $\mathrm{tr}\,(AB) = \mathrm{tr}\,(BA)$, so $\mathrm{tr}\,(M^{-1}AM) = \mathrm{tr}\,(AMM^{-1}) = \mathrm{tr}\,(A)$), it is the trace of the matrix that represents multiplication by ψ relative to *any* basis. In particular, it is the trace of the matrix that represents multiplication by ψ relative to the basis $\psi^i y^j$, $0 \le i < \nu_1$, $0 \le j < \nu_2$. When the elements of this basis are suitably ordered, the matrix that represents multiplication by ψ becomes a $\nu_2 \times \nu_2$ matrix of $\nu_1 \times \nu_1$ blocks (note that $\nu_1 \nu_2 = n$) in which the blocks off the diagonal are all 0 and the blocks on the diagonal are all the same matrix: Its first $\nu_1 - 1$ rows are the last $\nu_1 - 1$ rows of I_{ν_1}, and its last row contains the negatives of the coefficients (after the first) of the polynomial $f_1 = F_1$, listed in reverse order. In particular, its entries are all in $\mathbf{Q}[x]$, so its trace is in $\mathbf{Q}[x]$. (In fact, its trace—and therefore the trace of ψ—is simply $-\nu_2$ times the second coefficient of F_1.)

The matrix, call it S, whose entry in the ith row of the jth column is the trace of y^{i+j-2}, is a matrix of polynomials in x with integer coefficients. Therefore, its determinant, call it $D(x)$, is a polynomial in x with integer coefficients. The lemma implies that $D(x)$ *is a common denominator of the elements of the root field integral over x.* In fact, if $p(x,y)/q(x)$ is integral over x, where $p(x,y)$ and $q(x)$ are polynomials with rational coefficients and $q(x) \neq 0$, and if it is in lowest terms, then not only does $q(x)$ divide $D(x)$, but so does $q(x)^2$. This can be proved as follows:

The matrix S of which $D(x)$ is the determinant represents the bilinear form "the trace of the product" on the root field of $\chi(x,y)$ relative to the basis 1, y, ..., y^{n-1}. This observation implies that $D(x) \neq 0$, because if $D(x)$ were zero, there would be a solution $v(x)$ of $S \cdot v(x) = 0$ that was a nonzero column matrix whose entries $v_i(x)$ were rational functions of x, and this would imply $\operatorname{tr}_x(\hat{w}\hat{v}) = 0$ for all elements $\hat{w}$ of the root field, where $\hat{v} = \sum_{i=1}^{n} v_i(x)y^{i-1}$, contrary to the fact that $\operatorname{tr}_x(\hat{w}\hat{v}) = n$ when $\hat{w}$ is the reciprocal of $\hat{v}$.

If $p(x,y)/q(x)$ is integral over x and in lowest terms in the sense that $q(x)$ and the coefficients $p_i(x)$ of $p(x,y) = p_0(x) + p_1(x)y + \cdots + p_{n-1}(x)y^{n-1}$ have no common divisor of positive degree, and if one of the coefficients $p_i(x)$ is nonzero, then a new basis of integers is obtained by replacing y^i with $p(x,y)/q(x)$ in the basis 1, y, ..., y^{n-1}. The entries of the matrix that represents the bilinear form "the trace of the product" relative to this new basis are polynomials in x with rational coefficients, because they are traces of elements integral over x. Therefore, its determinant is a polynomial in x. On the other hand, its determinant is $\left(\frac{p_i(x)}{q(x)}\right)^2 D(x)$, because the matrix that makes the transition from one basis to the other is the identity matrix with the $(i+1)$st row replaced by a new row consisting of the coefficients of $p(x,y)/q(x)$ and which therefore has $\frac{p_i(x)}{q(x)}$ in its $(i+1)$st column, so both the transition matrix and its transpose have determinant $\frac{p_i(x)}{q(x)}$. Therefore, $q(x)^2$ divides $p_i(x)^2 D(x)$ for each i (trivially so when $p_i(x) = 0$). Thus, $q(x)^2$ divides the greatest common divisor of these polynomials $p_i(x)^2 D(x)$, which is the greatest common divisor of the $p_i(x)^2$ times $D(x)$. Since $p(x,y)/q(x)$ is in lowest terms by assumption, $q(x)^2$ is relatively prime to the greatest common divisor of the $p_i(x)^2$, so $q(x)^2$ divides $D(x)$, as was to be shown.

Since every element $p(x,y)/q(x)$ of the root field can be written in the form $P(x,y) + \frac{r(x,y)}{q(x)}$, where $P(x,y)$ is a polynomial in x and y and where $\frac{r(x,y)}{q(x)}$ is a proper fraction in the sense that $\deg_x r < \deg q$ (and, as it is natural to assume, $\deg_y r < n = \deg_y \chi$), in order to determine which elements $p(x,y)/q(x)$ are integral over x it will suffice to determine which elements $r(x,y)/q(x)$ are integral over x, because polynomials $P(x,y)$ are always integral over x. But since $r(x,y)/q(x)$ for a given $q(x)$ contains just $n \cdot \deg q$ unknown rational co-

efficients, the following proposition reduces this determination to the solution
of a system of homogeneous linear equations.

Proposition. *A rational function $p(x,y)/q(x)$ with rational coefficients is
integral over x if and only if for each algebraic number α that is a root of $q(x)$
and for each infinite series solution y of $\chi(x,y) = 0$ in fractional powers of
$x - \alpha$ given by Newton's polygon, all terms of the power series in s that results
from substituting the series for y in $p(x,y)$ and then substituting $\alpha + s^m$ for
x, where m clears the denominators in the fractional exponents, are divisible
by s at least as many times as the polynomial $q(\alpha + s^m)$ is.*

Loosely speaking, the condition is that each expression of $p(x,y)/q(x)$
obtained by using an expansion of y as a power series in fractional powers
of $x - \alpha$, where α is a root of $q(x)$, and writing the reciprocal of $q(x)$ as a
negative power of $x - \alpha$ times a power series in $x - \alpha$ with nonzero constant
term, contains nonnegative exponents exclusively; in short, $p(x,y)/q(x)$ *has
no poles where x is finite.*

Proof. Let $\Psi(x,p) = p^\nu + c_1(x)p^{\nu-1} + \cdots + c_\nu(x)$ be the irreducible, monic polynomial* whose coefficients $c_i(x)$ are rational functions of x of which $p(x,y)$—
regarded as an element of the root field—is a root. Because $p(x,y)$ is integral
over x (it is a polynomial in y with coefficients in $\mathbf{Q}[x]$), the coefficients $c_i(x)$
are polynomials in x with rational coefficients. To say that $p(x,y)/q(x)$ is integral over x means that $c_i(x)$ is divisible by $q(x)^i$ for each i. It is to be shown
that this is true if and only if $p(\alpha + s^m, \beta_0 + \beta_1 s + \beta_2 s^2 + \cdots) \equiv 0 \bmod s^e$ for
all infinite series solutions $y = \beta_0 + \beta_1 s + \beta_2 s^2 + \cdots$ of $\chi(x,y) = 0$, where α
is a root of $q(x)$, where $s = (x - \alpha)^{1/m}$, and where e is the number of times
that s divides $q(\alpha + s^m)$.

By definition, to say that $\Psi(x,p) = 0$, where x and p are regarded as
elements of the root field, means that $\Psi(x,p(x,y)) \equiv 0 \bmod \chi(x,y)$. In other
words, it means that $\Psi(x,p(x,y)) = q(x,y)\chi(x,y)$ for some polynomial $q(x,y)$
with rational coefficients. Since $\chi(\alpha + s^m, \beta_0 + \beta_1 s + \cdots + \beta_h s^h) \equiv 0 \bmod s^{h+1}$
for each h, it follows that $\Psi(\alpha + s^m, p(\alpha + s^m, \beta_0 + \beta_1 s + \cdots + \beta_h s^h)) \equiv$
$0 \bmod s^{h+1}$ for each h. Therefore, $p(\alpha + (x - \alpha), \beta_0 + \beta_1(x - \alpha)^{1/m} + \cdots +$
$\beta_h(x-\alpha)^{h/m})$, when it is regarded as a polynomial in $(x-\alpha)^{1/m}$ and truncated
by omitting all terms in which the exponent is larger than h/m, is a truncated
solution of $\Psi(x,p) = 0$ for $x = \alpha$. By Theorem 2 of the last essay, it is therefore
a truncation of one of the infinite series solutions p of $\Psi(x,p) = 0$ for $x = \alpha$
found by the construction[†] of Theorem 1 of the last essay. It is to be shown,

---

* This polynomial can be found because $\Psi(x,p)$ is a factor of the characteristic
polynomial of the matrix that represents multiplication by $p(x,y)$ relative to the
basis $1, y, \ldots, y^{n-1}$.

† Strictly speaking, this construction does not apply to $\Psi(x,p)$, because its coefficients are rational and the description of the Newton polygon algorithm in Essay
4.4 assumes that the coefficients of the given equation $\chi(x,y) = 0$ are integers,
but the algorithm applies without modification to the case of rational coefficients.

therefore, that $q(x)^i$ divides $c_i(x)$ for each i if and only if for each root α of $q(x)$ in an algebraic number field, every infinite series solution p of $\Psi(x,p) = 0$ in fractional powers of $x - \alpha$ is divisible by the highest power of $x - \alpha$ that divides $q(x) = q(\alpha + (x - \alpha))$.

Suppose first that $q(x)^i$ divides $c_i(x)$ for each i. For a given root α of $q(x)$ whose multiplicity is e, $(x - \alpha)^{ei}$ then divides $c_i(x)$. The initial term of any infinite series solution p of $\Psi(x,p) = 0$ in fractional powers of $x - \alpha$ can be found using the method by which the Newton's polygon algorithm finds the next term of a truncated series solution. Specifically, the equation $\Psi(\alpha + s, p) = c_\nu(\alpha + s) + c_{\nu-1}(\alpha + s)p + \cdots + c_1(\alpha + s)p^{\nu-1} + p^\nu = 0$ shows, because the terms of lowest degree cancel, that the lowest order term of a series expansion $p = \gamma s^{\sigma/\tau} + \cdots$ corresponds to a segment of the "Newton polygon" dictated by the points (i, j_i), where j_i for $i = 0, 1, \ldots, \nu$ is the number of times $s = x - \alpha$ divides $c_{\nu-i}(\alpha + s)$, except that j_i is undefined when $c_{\nu-i}(x) = 0$. Since $(x-\alpha)^{e(\nu-i)}$ divides $c_{\nu-i}(x)$, j_i is at least $e(\nu - i)$ whenever it is defined. In particular, the minimum value 0 of j_i occurs only for $i = \nu$. The rightmost segment of the polygon, call it $\sigma i + \tau j = k$, therefore has $(\nu, 0)$ as its right end; its other end is at a point (i, j_i) for which $\sigma i + \tau j_i = k = \sigma \nu + \tau \cdot 0$. For this index i, both $j_i = \frac{\sigma}{\tau}(\nu - i)$ and $j_i \geq e(\nu - i)$ hold. Therefore, for this segment of the polygon, $\frac{\sigma}{\tau} \geq e$. All infinite series solutions $p = \gamma(x - \alpha)^{\sigma/\tau} + \cdots$ that correspond to this segment of the polygon are therefore divisible by $(x - \alpha)^e$. As is easily shown, the ratio $\frac{\sigma}{\tau}$ is *smallest* for this rightmost segment,[*] so all solutions $p = \gamma(x - \alpha)^{\sigma/\tau} + \cdots$ are divisible by $(x - \alpha)^e$, as was to be shown.

Conversely, if $q(x)^i$ fails to divide $c_i(x)$ for some i, then $(x - \alpha)^{ei}$ fails to divide $c_i(x)$ for some root α of multiplicity e of $q(x)$ and some index i.

Moreover, $\chi(x,y)$ was assumed in Essay 4.4 to be irreducible. The series expansions of a reducible polynomial can be found by finding the expansions of its irreducible factors.

[*] What is to be shown is that the ratio σ/τ for any segment of the polygon is *larger* than the ratio σ/τ for the segment to its right. Since σ/τ is minus the slope of the segment, this is the statement that the slopes of the segments *increase* as one moves from left to right, which is evident. In actual inequalities, the three endpoints of two successive segments of Newton's polygon, call them (r, j_r), (s, j_s), (t, j_t), satisfy

$$\sigma r + \tau j_r = \sigma s + \tau j_s < \sigma t + \tau j_t,$$
$$\sigma' r + \tau' j_r > \sigma' s + \tau' j_s = \sigma' t + \tau' j_t,$$

where σ' and τ' pertain to the segment from (s, j_s) to (t, j_t), from which

$$\tau(j_r - j_s) = \sigma(s - r) \quad \text{and} \quad \tau'(j_r - j_s) > \sigma'(s - r)$$

follow. Therefore

$$\frac{\sigma}{\tau} = \frac{j_r - j_s}{s - r} > \frac{\sigma'}{\tau'},$$

as was to be shown.

For such an α the points (i, j_i) of the polygon arising from $\Psi(\alpha + s, p) = c_\nu(\alpha+s) + c_{\nu-1}(\alpha+s)p + \ldots + p^\nu$ include at least one for which $e(\nu - i) > j_i$. If $j_i = 0$ for some $i < \nu$, then $\Psi(\alpha, p)$ contains a term of degree less than ν in p, so this polynomial in p has a nonzero root, call it β_0, and there is a solution $p = \beta_0 + \cdots$ of $\Psi(\alpha, p) = 0$ that is not divisible by $s = x - \alpha$, and therefore not divisible by $(x - \alpha)^e$. Otherwise, as before, the rightmost segment of the polygon, call it $\sigma i + \tau j = k$, passes through $(\nu, 0)$ and at least one other point of the form (i, j_i). At least one point (i, j_i) lies below the line $j = e(\nu - i)$ of slope $-e$ passing through $(\nu, 0)$; since all points (i, j_i) lie on or above any segment of the polygon, the rightmost segment $j = \frac{\sigma}{\tau}(\nu - i)$ must lie under the line $j = e(\nu - i)$ for $i < \nu$. Thus, $\frac{\sigma}{\tau} < e$, so no solution $p = \gamma(x - \alpha)^{\sigma/\tau} + \cdots$ arising from this segment of the polygon is divisible by $(x - \alpha)^e$, and the proof is complete.

Thus, in a proper fraction $r(x, y)/q(x)$ that is integral over x, the coefficients of $r(x, y)$ satisfy a homogeneous system of linear equations, so the most general such fraction can be written as a linear combination of a finite number of them, say of $\xi_1, \xi_2, \ldots, \xi_k$, with rational coefficients. When these elements $\xi_1, \xi_2, \ldots, \xi_k$ together with $1, y, y^2, \ldots, y^{n-1}$ are taken as input to the following algorithm of Kronecker ([39, §7]), the algorithm produces an *integral basis* of the root field of $\chi(x, y)$ as described in the statement of the theorem.

Construction of an Integral Basis

Input: Elements $y_1, y_2, \ldots, y_l$ of the root field of $\chi(x, y)$ integral over x that span the elements integral over x in the sense that each element integral over x can be expressed in the form $\sum_{i=1}^l \phi_i(x) y_i$ where the coefficients $\phi_i(x)$ are polynomials in x with rational coefficients. (At the outset, $l = n + k$, and the coefficients of the ξ_i can be taken to be rational numbers.)

Algorithm: As long as the number l of elements in the spanning set is greater than n, carry out the following operations. Consider the $l \times l$ symmetric matrix $[\mathrm{tr}_x(y_i y_j)]$ and consider its symmetric $n \times n$ minor determinants—those $n \times n$ minor determinants in which the indices of the n columns selected coincide with those of n the rows selected. Each such minor determinant is a polynomial in x with rational coefficients because all of its entries are. Rearrange $y_1, y_2, \ldots, y_l$, if necessary, to make the first such minor—the one formed by selecting the first n rows and columns—nonzero and of degree no greater than that of any other nonzero symmetric $n \times n$ minor. Then the first n entries of $y_1, y_2, \ldots, y_l$ are linearly independent over $\mathbf{Q}(x)$, which means that each remaining entry $y_{n+1}, y_{n+2}, \ldots, y_l$ can be expressed as a sum of multiples of the first n in which the multipliers are rational functions of x. Each multiplier in each of these expressions can be written as a polynomial in x plus a proper rational function of x, one in which the degree of the numerator is less than the degree of the denominator. Let polynomial multiples of the first n of the y's be subtracted from the later y's in order to make the multipliers in the *representations of the later y's in terms of the first n all proper rational functions. Delete any y's that have become zero as a result of these subtractions, rearrange the list again, and repeat.*

Output: A list y_1, y_2, ..., y_n of just n elements integral over x that span, over $\mathbf{Q}[x]$, the set of all elements integral over x.

The operations of the algorithm—rearrange the y's, delete zeros, and subtract one y times a polynomial in x with rational coefficients from another y—do not change the conditions satisfied by the original set of y's that they span the elements integral over x when coefficients that are polynomials in x with rational coefficients are used.

An argument like the one above that proves that $D(x)$ is a common denominator of the elements integral over x proves that *each iteration of the algorithm reduces the degree of the determinant of the first $n \times n$ symmetric minor.* Specifically, if, after the multipliers in the representations of y_{n+1}, y_{n+2}, ..., y_l as sums of multiples of y_1, y_2, ..., y_n have been reduced so that they are proper rational functions, and after zeros have been deleted, there are more than n items in the list, then one of the coefficients—say the coefficient of y_1—in the representation of y_{n+1} is a nonzero proper fraction, call it $\frac{p(x)}{q(x)}$, where $\deg p < \deg q$. The symmetric $n \times n$ minor for any selection of n indices is a polynomial. As before, $M_1 = \left(\frac{p(x)}{q(x)}\right)^2 M_0$ when M_1 is the minor in which the selected indices are 2, 3, ..., $n+1$ and M_0 is the one in which they are 1, 2, ..., n. Thus, $q(x)^2 M_1 = p(x)^2 M_0$, which shows that $\deg M_1 < \deg M_0$. Thus, the minor of *least* degree has degree less than $\deg M_0$, and $\deg M_0$ decreases with each step, as was to be shown.

In this way, the algorithm continues to reduce the degree of the first $n \times n$ minor. By the principle of infinite descent, the algorithm must terminate. In other words, a stage must be reached at which the list contains only n elements. Clearly, they are an integral basis of the root field.

The proof of the theorem will be completed by a second algorithm, which starts with an integral basis and produces a normal basis. It requires that one also construct an integral basis relative to the parameter $u = \frac{1}{x}$; in other words, it uses a set z_1, z_2, ..., z_n of elements of the root field of $\chi(x,y)$ with the property that every element of the root field has a unique representation in the form $\sum \psi_i(x) z_i$, where the coefficients $\psi_i(x)$ are rational functions of x, and that the element is integral over $u = \frac{1}{x}$ if and only if each $\psi_i(x)$ is a polynomial in $\frac{1}{x}$. The algorithm just given can be used to construct such a set z_1, z_2, ..., z_n; simply describe the root field as the root field of $\chi_1(u,v) = \chi(x,y)/x^{n\lambda}$, where $u = \frac{1}{x}$, $v = \frac{y}{x^\lambda}$, and λ is large enough to make χ_1 a polynomial in u and v.

Such an integral basis z_1, z_2, ..., z_n relative to $\frac{1}{x}$ will be used to determine, given an integral basis y_1, y_2, ..., y_n, whether the basis

$$\frac{y_1}{x^{\lambda_1}}, \quad \frac{y_2}{x^{\lambda_2}}, \quad \cdots, \quad \frac{y_n}{x^{\lambda_n}},$$

is an integral basis relative to $\frac{1}{x}$, where λ_i, for each i, is the order of y_i at $x = \infty$; that is, λ_i is the least integer for which y_i/x^{λ_i} is integral over $\frac{1}{x}$.

Construction of a Normal Basis

Input: An integral basis y_1, y_2, $\ldots$, y_n of the root field of $\chi(x, y)$ relative to x.

Algorithm: Find the orders λ_1, λ_2, $\ldots$, λ_n *of* y_1, y_2, $\ldots$, y_n *at* $x = \infty$. *As long as* $\frac{y_1}{x^{\lambda_1}}$, $\frac{y_2}{x^{\lambda_2}}$, $\ldots$, $\frac{y_n}{x^{\lambda_n}}$ *(which is a basis consisting of elements integral over* $\frac{1}{x}$*) is not an integral basis relative to* $\frac{1}{x}$, *construct a new integral basis in which one* y_k *is replaced by a new* y'_k *whose order* λ'_k *at* $x = \infty$ *is less than* λ_k *in the following way. Write each* z_i *of an integral basis relative to* $\frac{1}{x}$ *in the form* $\sum_j \psi_{ij}(x)\frac{y_j}{x^{\lambda_j}}$, *where the* $\psi_{ij}(x)$ *are rational functions of* x. *By assumption, at least one* $\psi_{ij}(x)$ *is not a polynomial in* $\frac{1}{x}$. *(If all were polynomials in* $\frac{1}{x}$, *then each* z_i *and therefore each element integral over* $\frac{1}{x}$ *would be a sum of multiples of the* $\frac{y_i}{x^{\lambda_i}}$ *with coefficients that were polynomials in* $\frac{1}{x}$.) *Choose a value of* i *for which at least one* $\psi_{ij}(x)$ *is not a polynomial in* $\frac{1}{x}$. *Since* $x^\nu z_i = \sum \psi_{ij}(x)x^{\nu-\lambda_j}y_j$ *is integral over* x *for sufficiently large* ν, *and* y_1, y_2, $\ldots$, y_n *is an integral basis, the denominator of* $\psi_{ij}(x)$ *is a power of* x *for each* $j = 1, 2, \ldots, n$, *say* $\psi_{ij}(x) = x\xi_j(x) + \theta_j(\frac{1}{x})$, *where* $\xi_j(x)$ *is a polynomial in* x, *and* $\theta_j(\frac{1}{x})$ *is a polynomial in* $\frac{1}{x}$. *By the choice of* i, $\xi_j(x) \neq 0$ *for at least one* j. *Let* $\sigma > 0$ *be the maximum of the degrees of* $\xi_1(x)$, $\xi_2(x)$, $\ldots$, $\xi_n(x)$. *Among those indices* j *for which* $\deg \xi_j = \sigma$, *let* k *be one for which* λ_k *is as large as possible and set* $y'_k = \sum c_j x^{\lambda_k - \lambda_j} y_j$, *where* c_j *is the coefficient of* x^σ *in* $\xi_j(x)$ *(which is zero if* $\deg \xi_j \neq \sigma$).

Output: An integral basis y_1, y_2, $\ldots$, y_n with the property that

$$\frac{y_1}{x^{\lambda_1}}, \quad \frac{y_2}{x^{\lambda_2}}, \quad \ldots \quad \frac{y_n}{x^{\lambda_n}},$$

is an integral basis relative to $\frac{1}{x}$.

Justification. Replacement of y_k with y'_k gives an integral basis, as is shown by the two formulas $y'_k = \sum_j c_j x^{\lambda_k - \lambda_j} y_j$ (note that $\lambda_k \geq \lambda_j$ for all j by the choice of k) and $y_k = \frac{1}{c_k}y'_k - \sum_{j\neq k}\frac{c_j}{c_k}y_j$ (note that $c_k \neq 0$ by the choice of k). All that is to be shown, then, is that $\lambda'_k < \lambda_k$. To this end, note that $\frac{z_i}{x^{\sigma+1}} = \sum(c_j + \cdots)\cdot\frac{y_j}{x^{\lambda_j}}$, where the omitted terms contain $\frac{1}{x}$, $\frac{1}{x^2}$, $\frac{1}{x^3}$, $\ldots$. Multiply by x and use the definition of y'_k to obtain $\frac{z_i}{x^\sigma} = x\cdot\frac{y'_k}{x^{\lambda_k}} + \sum \eta_j(\frac{1}{x})\cdot\frac{y_j}{x^{\lambda_j}}$, where $\eta_j(\frac{1}{x})$ for each j is $x \cdot \frac{\psi_{ij}(x) - c_j x^{\sigma+1}}{x^{\sigma+1}}$, which is a polynomial in $\frac{1}{x}$. Thus, $x\cdot\frac{y'_k}{x^{\lambda_k}}$ is a difference of elements integral over $\frac{1}{x}$, which implies that the order of y'_k at $x = \infty$ is at most $\lambda_k - 1$, as was to be shown.

Since the algorithm reduces the sum of the λ_i at each step, it must terminate by the principle of infinite descent. When it terminates, the integral basis y_1, y_2, $\ldots$, y_n is a normal basis, because $w = \sum \phi_i(x)y_i$ has order at most ν if and only if all coefficients of $\frac{w}{x^\nu} = \sum \frac{\phi_i(x)}{x^{\nu-\lambda_i}}\cdot\frac{y_i}{x^{\lambda_i}}$ are polynomials in

$\frac{1}{x}$, which is true if and only if $\deg \phi_i \leq \nu - \lambda_i$, and the proof of the theorem is complete.

If $y_1, y_2, \ldots, y_n$ is a normal basis of the root field of $\chi(x,y)$, the elements of $\Theta(x^\nu)$ are those whose representations in the form $\sum_i \phi_i(x)y_i$ have coefficients $\phi_i(x)$ that are polynomials in x, with rational coefficients, of degree at most $\nu - \lambda_i$ for each i. When $\nu < \lambda_i$ this condition of course means that $\phi_i(x) = 0$. Therefore, the dimension of $\Theta(x^\nu)$ as a vector space over $\mathbf{Q}$ is the sum of the numbers $\nu - \lambda_i + 1$ over all indices i for which $\lambda_i \leq \nu$. For large ν, then, the dimension of $\Theta(x^\nu)$ as a vector space over $\mathbf{Q}$ is exactly $(\nu+1)n - \sum \lambda_i$. At the other extreme, when $\nu = 0$ this dimension—which is the degree of the field of constants $\Theta(x^0)$ as an extension of $\mathbf{Q}$, denoted by c in Essay 4.3—is simply the number of indices i for which $\lambda_i = 0$.

In the notation of Essay 4.3, the genus of the root field of $\chi(x,y)$ is $g = n_0\nu - \dim \Theta(x^\nu) + 1$ for all sufficiently large ν, where $n_0 = n/c$ and the dimension is the dimension as a vector space over the field of constants, which is the dimension as a vector space over $\mathbf{Q}$ divided by c; thus,

$$g = n_0\nu - \frac{1}{c}\left((\nu+1)n - \sum \lambda_i\right) + 1 = \frac{\sum \lambda_i}{c} - (n_0 - 1).$$

In particular, when $\mathbf{Q}$ is the field of constants of the root field of $\chi(x,y)$, the genus of the root field is simply

$$\left(\sum \lambda_i\right) - (n - 1),$$

where $n = \deg_y \chi$ and $\lambda_1, \lambda_2, \ldots, \lambda_n$ are the orders of the elements $y_1, y_2, \ldots, y_n$ of a normal basis of the field.

As the discussion of Essay 4.3 already shows, the natural description of the genus uses the field of constants of the root field under consideration instead of the field of rational numbers:

Determination of the Genus. *As was just explained, the construction of the theorem gives a basis over* $\mathbf{Q}$ *of the field of constants of the root field of* $\chi(x,y)$, *namely, the elements* y_i *of order zero in a normal basis. When the field* $\mathbf{Q}$ *is replaced by the (possibly) larger field of constants in the theorem, the construction gives a subset* $y_1, y_2, \ldots, y_{n_0}$ *of the root field of* $\chi(x,y)$ *and nonnegative integers* $\mu_1, \mu_2, \ldots, \mu_{n_0}$ *with the property that the elements of* $\Theta(x^\nu)$ *for any given* ν *are precisely those of the form*

$$\phi_1(x)y_1 + \phi_2(x)y_2 + \cdots + \phi_{n_0}(x)y_{n_0}$$

where $\phi_i(x)$ *is a polynomial of degree at most* $\nu - \mu_i$ *in* x *whose coefficients are in the field of constants of the root field of* $\chi(x,y)$. *Thus, for large* ν, *the dimension of* $\Theta(x^\nu)$ *as a vector space over the field of constants is* $\sum_{i=1}^{n_0}(\nu - \mu_i + 1) = n_0\nu - \sum \mu_i + n_0$. *By the definition of the genus, this dimension is* $n_0\nu - g + 1$, *from which it follows that*

$$g = \left(\sum_{i=1}^{n_0} \mu_i\right) - (n_0 - 1).$$

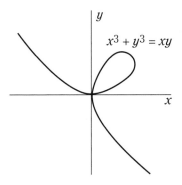

Fig. 4.6. The folium of Descartes.

In particular, $\sum \mu_i \geq n_0 - 1$.

Example 1: $\chi(x, y) = y^3 - xy + x^3$ (the folium of Descartes).
Multiplication by y is represented by

$$\begin{bmatrix} 0 & 1 & 0 \\ 0 & 0 & 1 \\ -x^3 & x & 0 \end{bmatrix}$$

relative to the basis 1, y, y^2 of the root field over $\mathbf{Q}(x)$. Therefore, the trace of y is 0. The trace of y^2 is the trace of

$$\begin{bmatrix} 0 & 1 & 0 \\ 0 & 0 & 1 \\ -x^3 & x & 0 \end{bmatrix}^2 = \begin{bmatrix} 0 & 0 & 1 \\ -x^3 & x & 0 \\ 0 & -x^3 & x \end{bmatrix},$$

which is $2x$. The trace of $y^3 = xy - x^3$ is x times the trace of y plus $-x^3$ times the trace of 1, which is $x \cdot 0 - x^3 \cdot 3$. Similarly, the trace of $y^4 = xy^2 - x^3 y$ is $2x^2$, from which it follows that

$$S = \begin{bmatrix} 3 & 0 & 2x \\ 0 & 2x & -3x^3 \\ 2x & -3x^3 & 2x^2 \end{bmatrix} \quad \text{and} \quad D(x) = 12x^3 - 8x^3 - 27x^6 = x^3(4 - 27x^3).$$

The square of the denominator $q(x)$ of an element of the root field integral over x must divide $x^3(4 - 27x^3)$, so x is a common denominator of these integral elements. A proper fraction integral over x must therefore be of the form $\frac{a + by + cy^2}{x}$, where a, b, and c are rational numbers.

By the proposition, and by the fact that $y = \pm\sqrt{x} - \cdots$ and $y = x^2 + \cdots$ are the series expansions of y in fractional powers of x, such an expression is integral over x if and only if $a + b(\pm s) + c(\pm s)^2 \equiv 0 \bmod s^2$ and $a + bs^2 + cs^4 \equiv 0 \bmod s$. These conditions hold if and only if $a = b = 0$, so the proper fractions integral over x are the rational multiples of $\frac{y^2}{x}$. Thus, 1, y, $\frac{y^2}{x}$ are an integral basis. For this basis, $\lambda_1 = 0$ and $\lambda_2 = 1$. To find the order λ_3 of $y_3 = \frac{y^2}{x}$ at $x = \infty$, one needs to find the equation of which it is a root, which is the characteristic polynomial of $\frac{1}{x} \begin{bmatrix} 0 & 0 & 1 \\ -x^3 & x & 0 \\ 0 & -x^3 & x \end{bmatrix}$. This characteristic polynomial is $X^3 - 2X^2 + X - x^3$, so $y_3^3 - 2y_3^2 + y_3 - x^3 = 0$, and $\left(\frac{y_3}{x}\right)^3 - 2 \cdot \frac{1}{x} \cdot \left(\frac{y_3}{x}\right)^2 + \frac{1}{x^2} \cdot \left(\frac{y_3}{x}\right) - 1 = 0$, which makes it clear that $\lambda_3 = 1$.

With $u = \frac{1}{x}$ and $v = \frac{y}{x}$ the equation $v^3 - uv + 1 = 0$ holds. That $1, v, v^2$ is an integral basis of the root field of $v^3 - uv + 1$ follows from the fact that in this case

$$S = \begin{bmatrix} 3 & 0 & 2u \\ 0 & 2u & -3 \\ 2u & -3 & 2u^2 \end{bmatrix}, \quad \text{from which} \quad D(u) = 4u^3 - 27.$$

Since $D(u)$ is square-free, $1, v, v^2$ is an integral basis over u. Thus, $1, y, y^2/x$ is a normal basis, because $1, \frac{y}{x}, \frac{y^2/x}{x}$ is the integral basis $1, v, v^2$ over u.

In this case, then, $\mathbf{Q}$ is the field of constants, and the genus is $(0 + 1 + 1) - (3 - 1) = 0$.

Example 2: $\chi(x, y) = y^3 + x^3 y + x$ (the Klein curve).

In this case, $D(x) = -4x^9 - 27x^2$, whose only square factor is x^2, so again the proper fractions integral over x have the form $\frac{a + by + cy^2}{x}$, where a, b, and c are rational numbers. Application of Newton's polygon in the case $a = 0$ leads easily to three unambiguous truncated solutions of $y^3 + x^3 y + x = 0$, namely, $y = \gamma \sqrt[3]{x}$, where γ is a cube root of -1. Substitution of $y = -s + \cdots$ for y and of s^3 for x in $a + by + cy^2$ gives a series divisible by $x = s^3$ only if $a = b = c = 0$, so $1, y, y^2$ is an integral basis over x. The orders of the first two are 0 and 2, respectively. The third, call it $w = y^2$, is a root of the characteristic polynomial of

$$\begin{bmatrix} 0 & 1 & 0 \\ 0 & 0 & 1 \\ -x & -x^3 & 0 \end{bmatrix}^2 = \begin{bmatrix} 0 & 0 & 1 \\ -x & -x^3 & 0 \\ 0 & -x & -x^3 \end{bmatrix};$$

therefore, $w^3 + 2x^3 w^2 + x^6 w - x^2 = 0$, from which it is clear that the order of w at $x = \infty$ is 3. (Division by x^9 gives an equation showing that w/x^3 is integral over $1/x$, but division by x^6 gives one that shows that w/x^2 is not integral over $1/x$.) That $1, y, y^2$ is a normal basis follows from the observation that $1, \frac{y}{x^2}, \frac{y^2}{x^3}$ is an integral basis over $u = \frac{1}{x}$, because division of $y^3 + x^3 y + x = 0$ by x^6 gives $v^3 + uv + u^5 = 0$, where $v = \frac{y}{x^2}$, and because, as is easily shown, $1, v = \frac{y}{x^2}, \frac{v^2}{u} = \frac{y^2}{x^3}$ is an integral basis over u. Since $\lambda_1 = 0$, $\lambda_2 = 2$, and $\lambda_3 = 3$, it follows that $\mathbf{Q}$ is the field of constants, and the genus is $(0+2+3)-(3-1) = 3$.

Example 3: $\chi(x, y) = (x^2 + y^2)^2 - 2(x + y)^2$ (see Essay 4.3).

As was noted in Essay 4.3, the algebraic analysis of this example should begin with the observation that the root field of $\chi(x, y)$ contains a square root of 2 in the form of the element $\frac{x^2 + y^2}{x + y}$, which enables one to treat the root field as the root field of the polynomial $x^2 + y^2 - \sqrt{2}(x+y)$, whose degree in y is 2 instead of 4.

(The irrational constants in the root field, if there are any, can be found by constructing one solution y of $\chi(x,y) = 0$ for one rational value α of x; the field of constants $\mathbf{A}$ needed to express such a solution must contain all constants in the root field, because the solution makes it possible to express any element of the root field as a power series—possibly with some negative exponents—with coefficients in $\mathbf{A}$, and in particular to express any constant in the root field as an element of $\mathbf{A}$. For example, when $\alpha = 0$ the roots of $\chi(0,y) = y^4 - 2y^2$ yield two unambiguous truncated solutions $y = \pm\sqrt{2} \cdot x^0$ and the truncated solution $y = 0 \cdot x^0$, whose ambiguity is 2. If the ambiguous solution $y = 0 \cdot x^0$ is used as an input to Newton's polygon, the output is the truncated solution $y = -1 \cdot x$, with ambiguity 2. If this truncated solution is the input, there are two unambiguous outputs $y = -x \pm \sqrt{2} \cdot x^2$, for each of which $\sqrt{2}$ must be adjoined. Thus, $\mathbf{A} = \mathbf{Q}(\sqrt{2})$ for any one of the four infinite series solutions for $\alpha = 0$, and no cleverness is needed to discover the irrational constant $\sqrt{2}$ in the root field. For any $\chi(x,y)$, the construction of a single unambiguous solution $x = \alpha + s^m$, $y = \beta_0 + \beta_1 s + \cdots$ of $\chi(x,y) = 0$ gives a number field $\mathbf{A}$ that contains, for the same reason, all constants in the root field of $\chi(x,y)$; factorization of $\chi(x,y)$ over such an $\mathbf{A}$ will then show the extent to which the adjunction of constants can reduce the degree in y of $\chi(x,y)$, or, more precisely, will determine the degree of the root field as an extension of $\mathbf{A}(x)$.)

The elements 1, $\sqrt{2}$, y, $\sqrt{2}y$ are easily shown to be a normal basis in which the λ's are 0, 0, 1, 1, respectively, so the genus is $\frac{1}{c}\sum\lambda_i - (n_0 - 1) = \frac{1}{2}(0 + 0 + 1 + 1) - (2-1) = 0$. When $\mathbf{Q}$ is replaced by $\mathbf{Q}(\sqrt{2})$, 1 and y are a normal basis in which the λ's are 0 and 1 respectively, and the genus is $(0+1) - (2-1) = 0$.

Of course, the genus is 0 geometrically, because the curve is a circle, which is birationally equivalent to a line.

Example 4: $\chi(x,y) = y^2 + x^4 - 1$ (the elliptic curve mentioned in Essay 4.2).

Here the trace of 1 is 2, and the trace of y is 0, so the trace of $y^2 = 1 - x^4$ is $2(1 - x^4)$ and $D(x) = 4(1 - x^4)$. Since this polynomial has distinct roots, 1, y is an integral basis. The order of 1 at $x = \infty$ is of course 0, and the order of y is 2 (because division of $y^2 + x^4 - 1$ by x^4 gives a polynomial in $\frac{y}{x^2}$ and $\frac{1}{x}$). Since 1 and $\frac{y}{x^2}$ are an integral basis relative to $\frac{1}{x}$, as is easily shown, 1 and y are a *normal* basis and the field of constants is $\mathbf{Q}$, which implies that the genus is $(0 + 2) - (2 - 1) = 1$.

Example 5: $\chi(x,y) = y^2 + x^6 - 1$ (a frequently cited hyperelliptic curve).

By considerations similar to those in the last example, $D(x) = 4(1 - x^6)$ has distinct roots, so 1 and y form an integral basis. The orders at $x = \infty$ are 0 and 3 respectively, and this basis is a normal basis. Therefore the genus is $(0 + 3) - (2 - 1) = 2$.

Example 6: $\chi(x,y) = y^2 - f(x)$, where $f(x)$ is a polynomial of degree $2n-1$ or $2n$ with distinct roots (a general hyperelliptic curve).

As in the previous examples, 1 and y are a normal basis for which the orders at $x = \infty$ are 0 and n, so the genus is $(0+n) - (2-1) = n-1$, as is implied by the passage from Abel's memoir quoted in Essay 4.1.

Essay 4.6 Holomorphic Differentials

Given an algebraic curve C, the method of the preceding essay determines its genus g regarded as in Essay 4.3 as the codimension of the subvarieties of C^N swept out by algebraic variations of N points on the curve. The objective of the present essay is to express this idea in terms of differential equations

$$(1) \qquad \sum_{i=1}^{N} h_j(x_i, y_i) dx_i = 0 \qquad (j = 1, 2, \ldots, g)$$

describing these subvarieties of C^N. Here the differentials $h_j(x, y)dx$ for $j = 1$, $2, \ldots, g$ are to be a basis, over the field of constants, of the space of **holomorphic differentials** on the curve, a concept that is to be defined. The equations (1) state that algebraic variations satisfy g infinitesimal conditions, where g is the dimension of the space of holomorphic differentials; therefore, not only do the algebraic variations partition C^N into subvarieties of codimension g, but this partition is expressed by g explicit differential equations.

In these equations, (x_i, y_i) for $i = 1, 2, \ldots, N$ are given solutions of $\chi(x_i, y_i) = 0$, where $\chi(x, y) = 0$ is the equation of the curve C. The heuristic idea of (1) is the following: If (1) correctly describes the possible algebraic variations of N points, it certainly describes the possible algebraic variations of *fewer* than N points: Just add conditions $dx_i = 0$ for a certain number of the points. Therefore, there is no loss of generality in assuming that N is the number $n_0\nu$ of zeros of an element of $\Theta(x^\nu)$ for some large ν. (Here n_0 again denotes the degree of the root field as an extension of the field obtained by adjoining all its constants to $\mathbf{Q}(x)$, or, in the notation used before, $n_0 = n/c$.) As has been shown, the most general element of $\Theta(x^\nu)$ is given by an explicit formula θ that contains $N - g + 1$ unknown constants, call them $a_1, a_2, \ldots, a_{N-g+1}$ (and in fact contains them linearly); the conditions $\chi(x, y) = 0$ and $\theta(x, y, a_1, a_2, \ldots, a_{N-g+1}) = 0$ define, implicitly, N solutions (x_i, y_i) of $\chi(x_i, y_i) = 0$ as functions of $a_1, a_2, \ldots, a_{N-g+1}$, where $N = n_0\nu$. Since multiplication of θ by a constant does not change its common zeros with χ, one of the parameters in θ, say a_{N-g+1}, can be set equal to 1. Then the N moving points depend on $N - g$ parameters, and they sweep out a subvariety of codimension g. In principle, the equations (1) result from *implicit differentiation* of the defining equations $\chi = 0$, $\theta = 0$ of the N moving points (x_i, y_i) in the following way.

For fixed values of the x's, y's and a's, the $2N$ relations $d\chi(x_i, y_i) = 0$, $d\theta(x_i, y_i, a) = 0$ give $2N$ homogeneous, linear equations in the $3N - g$ differentials dx_i, dy_i, da_j, whose coefficients are rational functions of the $3N - g$ variables. The relation $d\chi(x_i, y_i) = 0$ involves just one pair of values (x_i, y_i) and $a_1, a_2, \ldots, a_{N-g}$, so it can be used (provided x_i is a local parameter at (x_i, y_i)) to express each dy_i in terms of the corresponding dx_i and da_1, $da_2, \ldots, da_{N-g}$ and in this way to reduce the differential equations to just N equations in $2N - g$ differentials dx_i and da_i. These equations can be solved, in the generic case, to express $da_1, da_2, \ldots, da_{N-g}$ in terms of the dx_i and to eliminate them, leaving g relations among the dx_i. *These g relations are the required differential equations* (1) because they describe the relations satisfied by the dx_i when the parameters a_i are allowed to vary. In other words, these are the infinitesimal relations satisfied by algebraic variations of the N points (x_i, y_i).

In practice, the actual elimination of the da_i to find the relations among the dx_i seems impractical, even in the simplest examples. Instead, the derivation of the equations (1) will depend on observing that the holomorphic differentials, the ones that express the crucial relations (1), are the differentials that *have no poles*. Heuristically, such differentials lead to relations (1) in the following way.

If $\theta(x, y)$ has $n_0\nu$ zeros on the curve $\chi(x, y) = 0$ and no zeros where $x = \infty$, and if $h(x, y)\, dx$ has no poles—even when $x = \infty$—then the differential $\frac{h(x,y)\, dx}{\theta(x,y)}$ has poles only at the $n_0\nu$ zeros of $\theta(x, y)$. Thus, one can make use of the fact that *the sum of the residues of a differential is zero* to find that

$$\sum_{\text{zeros of } \theta} \left(\text{residue of } \tfrac{h(x,y)\, dx}{\theta(x,y)} \text{ at that zero of } \theta \right) = 0.$$

As a function on the curve $\chi(x, y) = 0$, $\theta(x, y)$ can be regarded, locally, as a function of x near each of its zeros (provided these zeros avoid places on the curve where x is not a local parameter), so $\frac{d\theta}{dx}$ is meaningful at each zero of $\theta(x, y)$ on the curve. When this derivative is not zero, its reciprocal is the residue of $\frac{dx}{\theta(x,y)}$ at the zero of θ because $\theta(x, y) = a_1 x + a_2 x^2 + \cdots$ implies that this residue is, by definition, $\frac{1}{a_1}$ when $a_1 \neq 0$. Thus

$$0 = \sum_{\text{zeros of } \theta} \left(\text{residue of } \tfrac{h(x,y)\, dx}{\theta(x,y)} \text{ at that zero of } \theta \right) = \sum_{\text{zeros of } \theta} h(x_i, y_i) \frac{dx_i}{d\theta},$$

where dx_i for each i is the infinitesimal change in x_i that results from an infinitesimal change $d\theta$ in θ. In other words, if the $n_0\nu$ points where θ is zero are moved to the nearby points where θ is $d\theta$, then the $n_0\nu$ changes dx_i in the x-coordinates of the intersection points satisfy $\sum h(x_i, y_i)\, dx_i = 0$, as was to be shown, provided the zeros of $\theta(x, y)$ are at points where both x and θ are local parameters on the curve. That this necessary condition for the dx_i to result from an algebraic variation of the intersection points is also a sufficient condition follows from—or at any rate is made plausible by—the fact that *the number of linearly independent holomorphic differentials is g,* so that the system of differential equations (1) describes a subvariety containing the algebraic variations that has the same dimension $N - g$ (at generic points) as the subvariety of algebraic variations and that therefore must coincide with it.

With this geometric motivation, the remainder of this essay will (a) define the notion of "holomorphic differential" in a precise algebraic way that accords with the notion of "no poles," (b) prove that the dimension of the holomorphic differentials as a vector space over the field of constants is g, (c) prove that the sum of the residues of a differential is zero, and (d) flesh out the implicit differentiation sketched above to reach the conclusion $\sum h_j(x_i, y_i)\, dx_i = 0$.

As before, let $\chi(x, y)$ be an irreducible polynomial in two indeterminates with integer coefficients that contains both indeterminates and is monic in the indeterminate y. Let K denote the root field of $\chi(x, y)$. A **differential** in K is an expression of the form $f(x, y)\, dx$, where $f(x, y)$ is an element of K and dx is merely a symbol. More precisely, $f(x, y)\, dx$ is a differential *expressed with respect to the parameter x*; it is easy to guess how a differential expressed with respect to the parameter x might be expressed with respect to another parameter of K, but in this essay all differentials will be expressed with respect to the preferred parameter x.

As before, the root field K of $\chi(x, y)$ will be regarded as an extension not of $\mathbf{Q}(x)$, the field of rational functions in x, but of $K_0(x)$, the field of rational functions in x with coefficients in the field of constants K_0 of K. (The symbol K_0 thus replaces the symbol $\Theta(x^0)$ as the notation for the field of constants of the root field.) As before, let n_0 be the degree of K as an extension of $K_0(x)$. By the definition of the genus, the dimension of $\Theta(x^\nu)$ as a vector space over K_0 is $n_0\nu - g + 1$ for all sufficiently large ν.

In this essay, instead of differentials $f(x, y)\, dx$ themselves, their *traces* will be considered; the **trace** of $f(x, y)\, dx$ is by definition the differential $\mathrm{tr}_x(f(x, y))dx$, where dx is a symbol and $\mathrm{tr}_x(f(x, y))$ is the element of $K_0(x)$ that is the trace of $f(x, y)$ with respect to the field extension $K \supset K_0(x)$; in other words, $\mathrm{tr}_x(f(x, y))$ is the trace of the $n_0 \times n_0$ matrix that represents multiplication by $f(x, y)$ with respect to the basis $1, y, y^2, \ldots, y^{n_0-1}$ of K over $K_0(x)$ (or, for that matter, with respect to *any* basis of K over $K_0(x)$). The heuristic idea behind this definition is that $\mathrm{tr}_x(f(x, y)\, dx)$ is the sum over all n_0 values at x of the differential $f(x, y)\, dx$, which, being symmetric in the n_0 values of y for any given x, is a rational function of x alone.

Holomorphic differentials were described above as differentials without poles. Certainly, the trace of a holomorphic differential must therefore be dx times an element of $K_0(x)$ without poles; in other words, if $h(x, y)\, dx$ is holomorphic, then $\mathrm{tr}_x(h(x, y))$ must be a polynomial in x. However, this necessary condition should not be expected to be sufficient, because $f(x, y)\, dx$ might have two poles at the same value of x that cancel when the sum is taken over all y. In the case of two canceling poles, however, one would expect to be able to choose an element $\theta(x, y)$ of K that was zero at just one of the poles and that had no poles for finite values of x, so that $\theta(x, y)f(x, y)\, dx$ would be a differential that had no poles where $f(x, y)\, dx$ did not and for which the poles that canceled in $\mathrm{tr}_x(f(x, y)\, dx)$ no longer canceled in $\mathrm{tr}_x(\theta(x, y)f(x, y)\, dx)$. Therefore, such a differential would not satisfy the *stronger* necessary condition for a differential $f(x, y)\, dx$ to be holomorphic: For every $\theta(x, y)$ that is integral over x, $\mathrm{tr}_x(\theta(x, y)f(x, y))$ is a polynomial in x. But this necessary condition, too, should not be expected to be sufficient, because it would not detect poles of $f(x,y)\,dx$ at places where $x = \infty$.

For these reasons, a differential $f(x,y)\,dx$ will be said to be **holomorphic for finite** x if $\operatorname{tr}_x(\theta(x,y)f(x,y))$ is a polynomial in x whenever $\theta(x,y)$ is integral over x, and will be said to be **holomorphic** if it is holomorphic for finite x and if $f\,d(\frac{1}{u}) = -\frac{f}{u^2}\,du$ is holomorphic for finite u. (Here $d(\frac{1}{u}) = -\frac{du}{u^2}$ is a definition. It will be justified by Corollary 1 below. Since tr_u is the same as tr_x when $u = \frac{1}{x}$—both are found using a basis of the field over $\mathbf{Q}(x) = \mathbf{Q}(u)$—to say that $-\frac{f}{u^2}\,du$ is holomorphic for finite u means that $\operatorname{tr}_u(\theta \cdot \frac{f}{u^2}) = \operatorname{tr}_x(x^2\theta f)$ is a polynomial in $u = \frac{1}{x}$ whenever θ is integral over u.)

Theorem. *Construct the holomorphic differentials for a given $\chi(x,y)$ and prove that their dimension as a vector space over the field of constants K_0 is the genus of the root field of $\chi(x,y)$.*

Proof. Let the construction that was used to determine the genus in Essay 4.5 be used to construct a normal basis $y_1, y_2, \ldots, y_{n_0}$ of the root field as an extension of the field $K_0(x)$ of rational functions of x with coefficients in the field of constants K_0 and to construct nonnegative integers $\mu_1, \mu_2, \ldots, \mu_{n_0}$ for which an element of the root field is in $\Theta(x^\nu)$ if and only if its unique expression in the form $\phi_1(x)y_1 + \phi_2(x)y_2 + \cdots + \phi_{n_0}(x)y_{n_0}$, where the coefficients are in $K_0(x)$, has coefficients that are *polynomials* in x with coefficients in K_0 and the degrees of these polynomials satisfy $\deg \phi_i + \mu_i \leq \nu$.

Let S_1 denote the symmetric $n_0 \times n_0$ matrix of polynomials in x with coefficients in K_0 whose entry in the ith row of the jth column is $\operatorname{tr}_x(y_iy_j)$, where the trace is taken relative to the extension $K \supset K_0(x)$. (In other words, this entry is the trace of the matrix that represents multiplication by y_iy_j relative to the basis $y_1, y_2, \ldots, y_{n_0}$ of K over $K_0(x)$.) The symmetric bilinear form "the trace of the product" is represented by S_1 in the sense that if $h = h_1y_1 + h_2y_2 + \cdots + h_{n_0}y_{n_0}$ and $\theta = \theta_1y_1 + \theta_2y_2 + \cdots + \theta_{n_0}y_{n_0}$ are the representations of two elements h and θ of K relative to this basis, then $\operatorname{tr}_x(h\theta) = [h]S_1[\theta]$ where $[h]$ represents the row matrix whose entries are h_1, $h_2, \ldots, h_{n_0}$, and $[\theta]$ represents the column matrix whose entries are $\theta_1, \theta_2, \ldots, \theta_{n_0}$.

With this notation, to say that $h\,dx$ is holomorphic is to say that $[h]S_1[\theta]$ is a polynomial of degree at most $\nu - 2$ whenever the ith entry θ_i of the column matrix $[\theta]$ is a polynomial whose degree is at most $\nu - \mu_i$, because $\operatorname{tr}_x(h\theta)$ must be a polynomial in x, while $\operatorname{tr}_x(-x^2h \cdot \frac{\theta}{x^\nu}) = -\frac{\operatorname{tr}_x(h\theta)}{x^{\nu-2}}$ must be a polynomial in $\frac{1}{x}$. (Note that the h_i need not be polynomials.) In other words, the row matrix $[h]S_1$ has the property that its product with a column matrix $[\theta]$ is a polynomial of degree at most $\nu - 2$ when the ith entry of θ is a polynomial of degree at most $\nu - \mu_i$. If one takes all entries but one of $[\theta]$ to be zero and that one to be a polynomial of degree $\nu - \mu_i$ for some large ν, one sees that the ith entry of $[h]S_1$ must be a rational function whose product with any polynomial of degree $\nu - \mu_i$ is a polynomial of degree at most $\nu - 2$. Thus, the ith entry of $[h]S_1$ must be a polynomial of degree at most $\mu_i - 2$ when $\mu_i \geq 2$ and must be zero if μ_i is 0 or 1. In other words, $[h]$ must have the form $[c]S_1^{-1}$, where c is a row matrix whose ith entry is a polynomial in x of degree at most $\mu_i - 2$ with coefficients in K_0. (In particular, the ith entry is zero when μ_i is 0 or 1.) This formula $[h] = [c]S_1^{-1}$ completely describes the holomorphic differentials $h\,dx$. The number of constants in the coefficients of the entries of $[c]$ is the sum of the numbers $\mu_i - 1$ over all values of i for which $\mu_i > 0$. Since exactly one μ_i is zero (because $K_0 = \Theta(x^0)$ consists of all elements $\phi_1 y_1 + \phi_2 y_2 + \cdots + \phi_{n_0} y_{n_0}$ in which $\phi_i = 0$ when $\mu_i > 0$ and ϕ_i is constant when $\mu_i = 0$), it follows that the number of arbitrary constants in this formula for h is $(\sum \mu_i) - (n_0 - 1)$, which is the genus, as was to be shown.

The proof that the sum of the residues of any differential $f(x,y)\,dx$ is zero reduces, by virtue of the *definition* of the sum of the residues as the sum of the residues of the rational differential $\operatorname{tr}_x(f(x,y))dx$, to the same statement for *rational* differentials $\frac{p(x)}{q(x)}dx$, where $p(x)$ and $q(x)$ are polynomials with coefficients in some algebraic number field K_0 and $q(x) \neq 0$. To define the sum of the residues of such a differential, it will be convenient to assume that the denominator $q(x)$ splits into linear factors over K_0, although, as will be seen, the *sum* of the residues can be expressed rationally in terms of the coefficients of $p(x)$ and $q(x)$ even when this condition is not fulfilled. By the method of partial fractions, one can see that if $q(x) = \prod(x - a_i)^{e_i}$, where the a_i are distinct constants, then

$$\frac{p(x)}{q(x)} = P(x) + \sum_i \sum_{\sigma=1}^{e_i} \frac{\rho_{i\sigma}}{(x - a_i)^\sigma}$$

for suitable constants $\rho_{i\sigma}$. (One can assume without loss of generality that $\deg p < \deg q$, so that $P(x) = 0$. Multiplication of both sides of the required equation

$$\frac{p(x)}{q(x)} = \sum_i \sum_{\sigma=1}^{e_i} \frac{\rho_{i\sigma}}{(x - a_i)^\sigma}$$

by $q(x) = \prod(x - a_i)^{e_i}$ gives an equation of the form $p(x) = \sum \rho_{i\sigma} A_{i\sigma}(x)$ in which the polynomials $A_{i\sigma}(x)$ have degree less than $k = \deg q$ and depend only on $q(x)$. This gives an inhomogeneous $k \times k$ system of linear equations satisfied by the k required coefficients $\rho_{i\sigma}$. When $p(x) = 0$, these equations have only the trivial solution,* so for any $p(x)$ of degree less than k they have

---

* Multiplication of $\sum_{i=1}^{\mu} \frac{\phi_i(x)}{(x-a_i)^{\nu_i}} = 0$, where $a_1, a_2, \ldots, a_\mu$ are distinct algebraic numbers and $\deg \phi_i < \nu_i$ for each i, by $\prod_{i=1}^{\mu}(x - a_i)^{\nu_i}$ gives an equation $\sum_{i=1}^{\mu} \psi_i(x) = 0$ in which the $\psi_i(x)$ are polynomials. All but one of these polynomials is divisible by $(x - a_1)^{\nu_1}$, so the remaining one must also be divisible by $(x - a_1)^{\nu_1}$, from which it follows that $\phi_1(x)$ must be zero. In the same way, $\phi_i(x) = 0$ for each i.

a unique solution.) The **residue at** $x = a$ of $\frac{p(x)}{q(x)}dx$ is defined to be ρ_{i1}, the coefficient of $\frac{1}{x-a_i}$ in the partial fractions expansion of $\frac{p(x)}{q(x)}$, when a is one of the roots a_i of $q(x)$; otherwise, the residue at $x = a$ of $\frac{p(x)}{q(x)}dx$ is zero.

Note that the residue at $x = a$ of $\left(\frac{p(x)}{q(x)} + \frac{p_1(x)}{q_1(x)}\right) dx$ is the sum of the residues at $x = a$ of $\frac{p(x)}{q(x)}dx$ and $\frac{p_1(x)}{q_1(x)}dx$. (The partial fractions decomposition of a sum is the sum of the partial fractions decomposition when terms with the same denominators are combined.)

The conventional statement that *the sum of the residues of a rational differential is zero* assumes that the "residue at $x = \infty$" is included in the sum. In this way, the conventional statement can be seen, as the corollary below shows, as a method of *evaluating* the sum of the residues of $\frac{p(x)}{q(x)}dx$ over all *finite* values of a. This evaluation is in fact quite easy:

Proposition.   *The sum of the residues at a of a rational differential $\frac{p(x)}{q(x)}dx$ over all finite values of a is*

$$\lim_{x \to \infty} \frac{x \cdot r(x)}{q(x)} = \left.\frac{u^{e-1}r(\frac{1}{u})}{u^e q(\frac{1}{u})}\right|_{u=0} \qquad (e = \deg q),$$

where $r(x)$ is the remainder when $p(x)$ is divided by $q(x)$, where the limit on the left is merely a mnemonic standing for the expression on the right, and the expression on the right denotes the quotient of constants in which the denominator is the leading coefficient of q and the numerator is the coefficient of x^{e-1} in $r(x)$.

Proof. Since the residues of $\left(P(x) + \frac{r(x)}{q(x)}\right) dx$ are the same as those of $\frac{r(x)}{q(x)}dx$, one can assume without loss of generality that the quotient $\frac{p(x)}{q(x)}$ in the given differential is a proper fraction; i.e., one can assume $r(x) = p(x)$. The residue of $\frac{\rho}{(x-a)^e}$ is ρ if $e = 1$ and 0 if $e > 1$, so for fractions of the particular form $\frac{r(x)}{q(x)} = \frac{\rho}{(x-a)^e}$ the residue is given by the formula $\lim_{x\to\infty} \frac{x \cdot r(x)}{q(x)}$. The theorem therefore follows from the observation that if $\frac{r(x)}{q(x)}$ and $\frac{r_1(x)}{q_1(x)}$ are proper fractions, then

$$\frac{x \cdot r(x)}{q(x)} + \frac{x \cdot r_1(x)}{q_1(x)} = \frac{x \cdot r(x)q_1(x) + x \cdot r_1(x)q(x)}{q(x)q_1(x)},$$

so the same is true of their limits as $x \to \infty$, interpreted as in the statement of the theorem. (Note also that $\lim_{x\to\infty} \frac{x \cdot r(x)}{q(x)}$ is unchanged if a common factor is canceled from numerator and denominator.)

Corollary 1.   *The sum of the residues of $\frac{p(x)}{q(x)}dx$ over all finite values of x is minus the residue at $x = \infty$, which residue is by definition the residue at $u = 0$ of*

$$\frac{p(\frac{1}{u})}{q(\frac{1}{u})}d(\frac{1}{u}) = -\frac{p(\frac{1}{u})}{u^2 q(\frac{1}{u})}du.$$

(The expression on the left is a mere mnemonic that takes advantage of the formula $d(\frac{1}{u}) = -\frac{du}{u^2}$ of elementary calculus.)

Deduction. What is to be shown is that the value of

$$\frac{u^{e-1}r(\frac{1}{u})}{u^e q(\frac{1}{u})}$$

at $u = 0$ is the residue at $u = 0$ of

$$\frac{p(\frac{1}{u})}{u^2 q(\frac{1}{u})}du = \frac{u^{e-1}p(\frac{1}{u})}{u \cdot u^e q(\frac{1}{u})}du.$$

Since this differential has the form

$$\frac{P(u)}{uQ(u)}du, \quad \text{where} \quad \frac{P(u)}{Q(u)}$$

is a proper fraction in which $Q(0) \neq 0$, this conclusion follows immediately from the definition.

Corollary 2. *When the residue of $\frac{p(x)}{q(x)}dx$ at $x = \infty$ is defined as in Corollary 1, the sum of the residues of a rational differential is zero.*

These algebraic facts make possible a plausible implicit differentiation of $\chi(x, y) = 0$ and $\theta(x, y, a_1, a_2, \ldots, a_{N-g}) = 0$ that leads to

$$(2) \qquad \sum_{i=1}^{N} h_j(x_i, y_i)dx_i = 0 \qquad (j = 1, 2, \ldots, g)$$

when the dy's and da's are eliminated.

As before, there is no loss of generality in assuming that $N = n_0\nu$ for some large ν and that the (x_i, y_i) are the intersection points $\chi(x_i, y_i) = 0$, $\theta(x_i, y_i) = 0$ for some fixed $\theta = a_1\theta_1 + a_2\theta_2 + \cdots + a_{N-g+1}\theta_{N-g+1}$, where the θ_i are a basis of $\Theta(x^\nu)$ over K_0 and $a_1, a_2, \ldots, a_{N-g+1}$ are fixed constants. In addition, it will be assumed that the chosen θ is in "general position" in the sense that x is a local parameter at each of the N intersection points (x_i, y_i) and θ has poles of order ν at each of the n points where $x = \infty$.

Each of the N intersection points (x_i, y_i) implies a pair of differential equations

$$(3) \qquad \qquad \qquad \chi_x dx_i + \chi_y dy_i = 0,$$
$$\theta_x dx_i + \theta_y dy_i + \theta_1 da_1 + \theta_2 da_2 + \cdots + \theta_{N-g+1}da_{N-g+1} = 0,$$

where subscripts x and y denote partial derivatives, and these partial derivatives are to be evaluated at the point $(x_i, y_i, a_1, a_2, \ldots, a_{N-g+1})$ at which a_1,

$a_2, \ldots, a_{N-g+1}$ have the given values that determine the N points (x_i, y_i), and x_i and y_i are the coordinates of one of these points.

Elimination of dy_i from the pair of equations (3) gives the single equation $dx_i + Q(\theta_1 da_1 + \theta_2 da_2 + \cdots + \theta_{N-g+1} da_{N-g+1}) = 0$ in which Q denotes the quotient $\frac{\chi_y}{\chi_y \theta_x - \theta_y \chi_x}$. This quotient is in fact the reciprocal of the derivative of θ with respect to x (eliminate dy from the equations $\chi_x dx + \chi_y dy = 0$ and $\theta_x dx + \theta_y dy = d\theta$, a computation that assumes x is a parameter on the curve at the point in question). Otherwise stated, it is the residue of the differential $\frac{dx}{\theta}$ at this zero of the denominator θ, because it is the value of the quotient $\frac{x - x_i}{\theta(x,y)}$ at the point (x_i, y_i) where numerator and denominator, taken separately, are both zero. (It is natural to think of this number as a limit, but of course it can be described algebraically as the value of the rational function of x and y when it is put in canonical form—a numerator in which y has degree less than $n = \deg_y \chi$ and a denominator that is a polynomial in x alone that is relatively prime to the numerator.)

Therefore, if each equation $dx_i + Q(\theta_1 da_1 + \theta_2 da_2 + \cdots + \theta_{N-g+1} da_{N-g+1}) = 0$ is multiplied by the value $h(x_i, y_i)$ of h at the corresponding point (x_i, y_i) and all N of these equations are added, the result is $\sum h(x_i, y_i) dx_i + C_1 da_1 + C_2 da_2 + \cdots + C_{N-g+1} da_{N-g+1} = 0$, where the coefficient C_j of da_j is the sum over all N zeros of θ on the curve $\chi = 0$ of $h\theta_j$ times the residue of $\frac{dx}{\theta}$ at that point. It is to be shown that each such coefficient C_j is zero.

Since neither θ_j nor $h\,dx$ has poles for finite x, the differential $\theta_j h\, dx / \theta$ has residues for finite x *only* at the zeros (x_i, y_i) of θ, and these residues are the values at (x_i, y_i) of $\theta_j h$ times the residue of dx/θ at (x_i, y_i). In short, C_j is the sum of all residues of the differential $\theta_j h\, dx / \theta$ at points where x is finite. Therefore, it is minus the sum of the residues at $x = \infty$ of the differential $\theta_j h\, dx / \theta$. Since θ_j has order at most ν at $x = \infty$ (it is in $\Theta(x^\nu)$) and θ has order ν at $x = \infty$ (by assumption), θ_j / θ is finite at $x = \infty$, so $\theta_j h\, dx / \theta$ has no pole at $x = \infty$, which implies $C_j = 0$ and $\sum h(x_i, y_i) dx_i = 0$, as was to be shown.

Example 1: $\chi(x, y) = y^3 + x^3 y + x$ (the Klein curve).

As was seen in Essay 4.5, 1, y, y^2 are a normal basis over $\mathbf{Q}(x)$ for which the λ's are 0, 2, 3. Therefore $(h_1 + h_2 y + h_3 y^2) dx$ is a holomorphic differential if and only if $[h_1 \ \ h_2 \ \ h_3] S = [0 \ \ a \ \ bx + c]$, where a, b, and c are rational numbers and the matrix S, which has $\operatorname{tr}(y^{i+j-2})$ in the ith row of the jth column, is easily found to be

$$
\begin{bmatrix}
3 & 0 & -2x^3 \\
0 & -2x^3 & -3x \\
-2x^3 & -3x & 2x^6
\end{bmatrix}.
$$

When $c = 1$ and $a = b = 0$, this gives a 3×3 homogeneous linear system whose solution is $[h_1 \ \ h_2 \ \ h_3] = \frac{1}{4x^9 + 27x^2}[4x^6 \ \ -9x \ \ 6x^3]$. Thus, $h = \frac{4x^6 - 9xy + 6x^3 y^2}{4x^9 + 27x^2}$, which can be written more simply as $h = \frac{1}{3y^2 + x^3}$. It is easy to see that the solution in which $b = 1$ and $c = a = 0$ is x times this one, and the solution in which $a = 1$ and $b = c = 0$ is y times this one, which leads to the formula

$$\frac{c + bx + ay}{3y^2 + x^3} \, dx$$

for the most general holomorphic differential on this curve. The formula has three parameters a, b, c because the genus is 3. (For an easier derivation of this formula, see the examples of Essay 4.8.)

Example 2: $\chi(x, y) = y^2 - f(x)$, where $f(x)$ is a polynomial of degree $2n - 1$ or $2n$ with distinct roots (a general hyperelliptic curve).

As was seen in Essay 4.5, 1 and y are a normal basis for which the orders at $x = \infty$ are 0 and n. (The matrix $S(x)$ is $\begin{bmatrix} 2 & 0 \\ 0 & 2f(x) \end{bmatrix}$, whose determinant $D(x) = 4f(x)$ has distinct roots, so 1, y is an integral basis over x. Division of $y^2 - f(x) = 0$ by x^{2n} gives $(\frac{y}{x^n})^2 - \frac{f(x)}{x^{2n}} = 0$, which, when $v = \frac{y}{x^n}$ and $u = \frac{1}{x}$, is a curve of the same form $v^2 - F(u) = 0$ of which 1, v is an integral basis over u. It follows that 1, y is a normal basis relative to x in which the order of y at $x = \infty$ is n.) Therefore, $(h_1 + h_2 y) \, dx$ is a holomorphic differential if and only if

$$\begin{bmatrix} h_1 & h_2 \end{bmatrix} \begin{bmatrix} 2 & 0 \\ 0 & 2f(x) \end{bmatrix} = \begin{bmatrix} 0 & q(x) \end{bmatrix},$$

where $q(x)$ is a polynomial of degree at most $n - 2$. Thus, $h \, dx = \frac{q(x) y \, dx}{2f(x)} = \frac{q(x) dx}{2y}$ is the most general holomorphic differential, where $q(x)$ is a polynomial of degree at most $n - 2$. The genus is $n - 1$.

Essay 4.7 The Riemann–Roch Theorem

Dedekind and Weber say in their classic treatise that the Riemann–Roch theorem, in its usual formulation, determines the number of arbitrary constants in a function with given poles [14, §28]. Indeed, that is exactly the way Roch himself formulated the theorem [58], as his title "On the Number of Arbitrary Constants in Algebraic Functions" indicates. The answer, a formula for the dimension of the vector space of rational functions with (at most) given poles, is a corollary of the theorem of this essay, which describes the *principal parts* of rational functions on an algebraic curve.

Let $f(x,y)$ be a rational function on a curve $\chi(x,y) = 0$, say $f(x,y) = p(x,y)/q(x)$, where p and q are polynomials with integer coefficients and f is regarded as an element of the root field of $\chi(x,y)$. The **principal parts of f at finite values of** x are, by definition, the terms with negative exponents in the expansions of f in powers of $x-\alpha$ for algebraic numbers α. Such expansions are obtained by applying Newton's polygon to expand y in $n = \deg_y \chi$ ways in (possibly fractional) powers of $x - \alpha$, substituting these expansions in $p(x,y)$, and multiplying the result by the expansion of $1/q(x)$ in powers of $x-\alpha$; they can contain negative powers of $x - \alpha$ only if the expansion of $1/q(x)$ does, which is to say, only if α is a root of $q(x)$. The principal parts of $f(x,y)$ thus amount simply to a list of the roots α of $q(x)$ and, for each of them, a list of the terms, if any, with negative exponents in the n series found by substituting expansions of y in powers of $x - \alpha$ in $f(\alpha + (x - \alpha), y)$.

One can define the **principal parts of f at $x = \infty$** as the principal parts at $u = 0$ when $u = \frac{1}{x}$, but for the sake of simplicity this essay will deal only with rational functions that are finite at $x = \infty$, so that there are *no* principal parts at $x = \infty$. Specifically, the only functions considered will be those of the form $f(x,y) = p(x,y)/q(x)$, where $p(x,y)$ is in $\Theta(x^\nu)$ for $\nu = \deg q$. Expansion of numerator and denominator of $f(x,y) = \frac{p(x,y)/x^\nu}{q(x)/x^\nu}$ in powers of $1/x$ then gives a quotient of power series in $1/x$ in which neither numerator nor denominator contains terms with negative exponents (the numerator is integral over $1/x$) and the denominator is not zero when $1/x = 0$, so the expansions of $f(x,y)$ in powers of $\frac{1}{x}$ contain no terms with negative exponents.

For values α of x at which $\chi(x,y) = 0$ *ramifies*—which is to say that at least one of the expansions of y in powers of $x-\alpha$ involves fractional powers—the principal parts of a function satisfy an obvious consistency requirement, namely, since one solution $y = \beta_0 + \beta_1 s + \beta_2 s^2 + \cdots$ in which $s = \sqrt[m]{x - \alpha}$ for $m > 1$ implies $m - 1$ other solutions obtained by multiplying s by some mth root of 1 other than 1, it must be true that a term of any one of the corresponding expansions of $f(x,y)$ determines the term with the same exponent in any of the other $m - 1$ expansions: One needs merely to change s to ωs for a suitable root of unity ω.

For these reasons, a **set of proposed principal parts** of a rational function on the curve $\chi(x,y) = 0$ will be defined to consist of (1) an algebraic number field **A**, (2) a finite set of elements $\alpha_1, \alpha_2, \ldots, \alpha_\mu$ of **A**, and (3) for each of the α_i and for each of the n ways of expanding y in powers of $x - \alpha_i$ an expression of the form $\gamma_1 (x - \alpha_i)^{-1} + \gamma_2 (x - \alpha_i)^{-2} + \cdots + \gamma_l (x - \alpha_i)^{-l}$, where l is a positive integer and the γ's are in $\mathbf{A}$, except that in the case of expansions of y that are in powers of $\sqrt[m]{x - \alpha_i}$ with $m > 1$ the expressions must take the form $\gamma_1 (\sqrt[m]{x - \alpha_i})^{-1} + \gamma_2 (\sqrt[m]{x - \alpha_i})^{-2} + \cdots + \gamma_l (\sqrt[m]{x - \alpha_i})^{-l}$ and must satisfy the consistency requirement just described. (A natural way to handle this consistency requirement is to prescribe the expression $\gamma_1 (\sqrt[m]{x - \alpha_i})^{-1} + \gamma_2 (\sqrt[m]{x - \alpha_i})^{-2} + \cdots + \gamma_l (\sqrt[m]{x - \alpha_i})^{-l}$ for just *one* expansion of y in powers of $\sqrt[m]{x - \alpha_i}$ in each set of m and to derive the others from it.)

The problem is to determine whether, for a given set of proposed principal parts, there is a rational function on the curve that is finite where $x = \infty$ and that has the stated principal parts. The answer given by the theorem below is that the holomorphic differentials give simple necessary and sufficient conditions for there to be such a function.

If $h \, dx$ is a holomorphic differential and if $f(x, y) = \frac{p(x,y)}{q(x)}$ is finite when $x = \infty$, then $\mathrm{tr}_x(fh)$ on the one hand is a rational function of x whose denominator $q(x)$ has degree ν and whose numerator $\mathrm{tr}_x(ph)$ has degree at most $\nu - 2$, so the sum of the residues of $\mathrm{tr}_x(fh)$ over all finite values of the variable is zero, and on the other hand is a rational function whose residue at any finite value α of x is a linear function of the principal parts of f. In this way, a certain linear function of the principal parts of f is necessarily zero. Explicitly, the following lemma can be used to express the residue of $\mathrm{tr}_x(fh)$ at $x = \alpha$ in terms of the principal parts of f:

Lemma. *Let $g(x, y)$ be a rational function on the curve $\chi(x, y) = 0$. The expansion of $\mathrm{tr}_x(g)$ in powers of $x - \alpha$ is the sum of the n expansions in (possibly fractional) powers of $x - \alpha$ obtained by substituting the n solutions of $\chi(x, y) = 0$ at $x = \alpha$ in $g(x, y)$.*

Proof. Let the main theorem of Part 1 be used to construct a minimal splitting polynomial, call it $F(x, z)$, of $\chi(x, y)$ regarded as a polynomial in y with coefficients in $\mathbf{Z}[x]$ and let $\hat{K}$ be the root field of $F(x, z)$, which is to say that $\hat{K}$ is the splitting field of $\chi(x, y)$. Finally, let $\bar{z}$ be an expansion of a solution z of $F(x, z) = 0$ in (possibly fractional) powers of $x - \alpha$, say $\bar{z} = \delta_0 + \delta_1 s + \delta_2 s^2 + \cdots$, where $x = \alpha + s^m$. (Let m be determined by the condition that the indices i of terms in $\bar{z}$ in which $\delta_i \neq 0$ have no common divisor greater than 1.) The substitution $x = \alpha + s^m$, $z = \bar{z}$ embeds $\hat{K}$ in the field of quotients of the ring of power series in s with coefficients in $\mathbf{A}$, where $\mathbf{A}$ is an algebraic number field containing α that is constructed by the Newton polygon algorithm. Let $\mathbf{A}\langle s \rangle$ denote this field of quotients; it is, in effect, the ring of formal power series in s with coefficients in $\mathbf{A}$ enlarged to include power series with a finite number of terms with negative exponents, because the reciprocal of a power series $\gamma_i s^i + \gamma_{i+1} s^{i+1} + \cdots$ in which the term of lowest degree has degree i can be expressed as a power series $\frac{1}{\gamma_i} s^{-i} + \cdots$ in which there are i or fewer terms with negative exponents.

In short, the splitting field $\hat{K}$ of $\chi(x, y)$ can be regarded[*] as a subfield of $\mathbf{A}\langle s \rangle$ for a sufficiently large algebraic number field $\mathbf{A}$ under an embedding that carries x to $\alpha + s^m$. Let $g_1, g_2, \ldots, g_\mu$ be the distinct images of $g(x, y)$ under the Galois group of $\hat{K}$. The irreducible polynomial $\psi(X)$ with coefficients in[†] $\mathbf{Q}(x)$ of which $g(x, y)$ is a root is then $\prod_{i=1}^{\mu}(X - g_i)$, and $\operatorname{tr}_x(g)$ is $-j$ times the coefficient of $X^{\mu-1}$ in $\psi(X)$, where j is the degree of the root field K of $\chi(x, y)$ as an extension of its subfield generated by g, because the matrix that represents multiplication by g relative to a basis of K over $\mathbf{Q}(x)$ can be arranged as a $j \times j$ matrix of $\mu \times \mu$ blocks in which the blocks off the diagonal are all zero and the diagonal blocks are all the matrix whose first $\mu-1$ rows are the last $\mu - 1$ rows of I_μ and whose last row is the negatives of the coefficients of ψ (except the leading coefficient 1) listed in reverse order. Thus, since the coefficient of $X^{\mu-1}$ is $-(g_1 + g_2 + \cdots + g_\mu)$, the expansion of $\operatorname{tr}_x(g)$ in powers of s is given by $\operatorname{tr}_x(g) = j(g_1 + g_2 + \cdots + g_\mu)$ when the g_i are represented as elements of $\mathbf{A}\langle s \rangle$. What is to be shown, then, is that substitution of the n expansions of y in powers of s, along with substitution of $\alpha + s^m$ for x, into $g(x, y)$ gives each of the μ expansions g_i for $i = 1, 2, \ldots, \mu$ exactly j times.

The Galois group of $\hat{K}$ expresses each root z of $F(x, z)$ as a polynomial in one such root with coefficients in $\mathbf{Q}(x)$. Substitution of $\overline{z}$ in these polynomials, together with substitution of $\alpha + s^m$ for x, gives $\deg_z F(x, z)$ distinct embeddings of $\hat{K}$ in $\mathbf{A}\langle s \rangle$. The possible expansions of y in powers of s and the possible expansions g_i of $g(x, y)$ all occur as images of y or $g(x, y)$, respectively, under these embeddings. The action of the Galois group on the embeddings implies the desired conclusion that each g_i occurs for the same number of different expansions of y.

(Note in particular that all fractional powers of $x - \alpha$ cancel when the sum $g_1 + g_2 + \cdots + g_\mu$ is computed. This is a clear consequence of the "consistency requirement" described above, because the sum $1 + \omega^k + \omega^{2k} + \cdots + \omega^{(m-1)k}$ is zero whenever ω is an mth root of unity and $\omega^k \neq 1$.)

Thus, the residue of $\operatorname{tr}_x(fh)\, dx$ at any α is the sum of the coefficients of $(x - \alpha)^{-1}$ over all n expansions of fh in powers of $s = \sqrt[m]{x - \alpha}$. Not only does this sum depend linearly on the principal parts of f, but the entries in the matrix that describes it are coefficients in the expansion of h in powers of s.

---

[*] Such an embedding of an algebraic field of transcendence degree 1 in $\mathbf{A}\langle s \rangle$ is analogous to an embedding of an algebraic number field in the field of complex numbers (see Essay 5.1). As in the latter case, the field $\hat{K}$ loses much of its constructive meaning when it is regarded merely as a subfield of $\mathbf{A}\langle s \rangle$, because elements of $\mathbf{A}\langle s \rangle$ are infinite series. An element of $\hat{K}$ is a root of a polynomial whose coefficients are rational functions of x, so the infinite series that represents it as an element of $\mathbf{A}\langle s \rangle$ can be specified by giving enough terms to determine an unambiguous truncated solution of the equation in question, after which all later terms are determined by Newton's polygon.

[†] As always, $\mathbf{Q}(x)$ denotes the field of quotients of $\mathbf{Z}[x]$, which is to say the field of rational functions in x.

Explicitly, if $h = h_0 + h_1 s + h_2 s^2 + \cdots$ and if $f = \gamma_1 s^{-1} + \gamma_2 s^{-2} + \cdots + \gamma_l s^{-l}$, then the coefficient of $(x - \alpha)^{-1}$ in fh is $h_0 \gamma_m + h_1 \gamma_{m+1} + h_2 \gamma_{m+2} + \cdots + h_{l-m} \gamma_l$, a linear function of γ_1, γ_2, ..., γ_l. When this formula is summed over all principal parts of f it gives, for each holomorphic differential $h\,dx$, an explicit linear function of the principal parts of f that must be zero.

Theorem. *If a set of proposed principal parts satisfies the condition just described for each holomorphic differential $h_i dx$ in a basis $h_1 dx$, $h_2 dx$, ..., $h_g dx$ of the holomorphic differentials on $\chi(x, y) = 0$, then it in fact gives the principal parts of some rational function on the curve. In short, these g necessary conditions for a set of proposed principal parts to be the principal parts of a function are sufficient.*

Proof. Let a set of proposed principal parts be called **subordinate to** a polynomial $q(x)$ if each α for which it specifies a polynomial $\gamma_1 s^{-1} + \gamma_2 s^{-2} + \cdots + \gamma_l s^{-l}$ is a root of $q(x)$ and if, moreover, the multiplicity of α as a root of $q(x)$ is at least l/m, so that multiplication of $\gamma_1 s^{-1} + \gamma_2 s^{-2} + \cdots + \gamma_l s^{-l}$ by $q(x) = q(\alpha + s^m)$ makes all exponents nonnegative. Every set of proposed principal parts is subordinate to some $q(x)$, so it will suffice to prove that the theorem holds for all sets of proposed principal parts subordinate to a given $q(x)$. Moreover, those subordinate to $q(x)$ are also subordinate to $q(x)r(x)$ for any polynomial $r(x)$, so one can assume without loss of generality that $q(x)$ is a polynomial of high degree.

If $f(x, y)$ is finite at $x = \infty$ and its principal parts are subordinate to $q(x)$, then $q(x)f(x, y)$ has order at most $\deg q$ at $x = \infty$ and has no poles for finite x, which is to say that $f(x, y) = p(x, y)/q(x)$, where $p(x, y)$ is in $\Theta(x^\nu)$ for $\nu = \deg q$. But for large ν the set of such functions $f(x, y)$ is a vector space of dimension $n\nu - g + 1$ over the field of constants. Functions that differ by a constant have the same principal parts and conversely, so the vector space of principal parts that actually occur is seen in this way to have dimension $n\nu - g$ for large ν.

On the other hand, the dimension of the space of proposed principal parts subordinate to $q(x)$ can be found in the following way. If α is a root of $q(x)$ of multiplicity μ, and if none of the n expansions of y in powers of $x - \alpha$ involve fractional powers, the proposed principal parts subordinate to $q(x)$ contain $n\mu$ coefficients corresponding to this α, μ coefficients in each expansion (those of $(x - \alpha)^{-1}$, $(x - \alpha)^{-2}$, ..., $(x - \alpha)^{-\mu}$). The same formula μn holds even when some expansions involve fractional powers, because the proposed principal part corresponding to an expansion in powers of $\sqrt[m]{x - \alpha}$ contains m times as many coefficients in the required range, but by the "consistency requirement" the coefficients of just *one* determines those of a set of m of them. Therefore, a set of proposed principal parts subordinate to $q(x)$ contains $n\nu$ unknown coefficients γ, where $n = \deg_y \chi$ and $\nu = \deg q$.

In short, *the principal parts that actually occur are a subspace of codimension g in the $n\nu$-dimensional space of proposed principal parts subordinate to*

$q(x)$ when $\nu = \deg q$ is sufficiently large. Since the conditions imposed by the holomorphic differentials—the sum of the residues of $\operatorname{tr}_x(fh)$ is zero for all holomorphic differentials $h\,dx$—are expressed by g homogeneous linear conditions on the coefficients of the principal parts, the actually occurring ones account for *all* of those that satisfy the necessary conditions *provided* the necessary conditions are independent (because then they determine a subspace of codimension g). In short, it will suffice to prove that every polynomial divides one for which the g necessary conditions are independent as conditions on sets of proposed principal parts subordinate to that polynomial.

The g homogeneous linear conditions imposed by the holomorphic differentials on proposed principal parts subordinate to $q(x)$ are expressed by a $g \times (n\nu)$ matrix of elements of $\mathbf{A}$, call it C_q. What is to be shown is that every polynomial divides a polynomial $q(x)$ for which the rank of C_q is g. This will be done by showing that if the rank of C_q is less than g and if β is any element of $\mathbf{A}$ that is not a root of $q(x)$, then replacing $q(x)$ with $(x - \beta)q(x)$ increases the rank of C_q, except for very extraordinary coincidences in the choice of β which can easily be avoided.

In fact, changing $q(x)$ to $(x - \beta)q(x)$ increases the degree of $q(x)$ by 1 and therefore adds n columns to C_q. Each of the new columns contains the g values of the coefficient h_i of one of the basis $h_1 dx$, $h_2 dx$, $\ldots$, $h_g dx$ of holomorphic differentials at one of the n points on the curve at which $x = \beta$. More precisely, let β be required to be an element of $\mathbf{A}$ (β can be taken to be a positive integer) for which $\chi(\beta, X)$ has distinct roots; then each column of the new C_q corresponds to one of the roots γ of $\chi(\beta, X)$ and it contains the values of $h_1, h_2, \ldots, h_g$ when (β, γ) is substituted for (x, y). It is to be shown that if the original C_q has rank less than g, then the extended C_q has rank greater than that of the original, except under extraordinary circumstances.

The ranks of the original and the extended C_q are unchanged by a change of basis of the holomorphic differentials. If the rank of the original C_q is less than g, then there are constants $c_1, c_2, \ldots, c_g$, not all zero, such that multiplication of the original C_q on the left by the row matrix with entries $c_1, c_2, \ldots, c_g$ gives a row of zeros. If a new basis of the holomorphic differentials is used in which the first holomorphic differential is $\sum_{i=1}^{g} c_i h_i dx$, the original C_q becomes a matrix whose first row is zero and the extended C_q becomes a matrix in which the first row contains $n\nu$ zeros and n new entries that are the values of the new h_1 at the n points $(x, y) = (\beta, \gamma)$. Thus, the extended C_q has greater rank unless all n of these values are zero.

Clearly, it would be an extraordinary coincidence if a value of β chosen at random were to result in even one value of h_1 that was zero, much less n of them. Since the number of zeros of the rational function h_1 (which is nonzero and does not have a pole at (β, γ) by the choice of β) is finite, one can easily find a new β for which the first row of the extended C_q contains a nonzero entry, and therefore find a β for which the rank of C_q is increased.

This theorem determines exactly which proposed principal parts are actual principal parts and therefore solves the Riemann–Roch problem of determining the dimension of the space of rational functions with prescribed poles with, at most, prescribed multiplicities. In the notation and terminology of the proof, one can say that *the vector space of functions whose principal parts are subordinate to $q(x)$ has dimension $n\nu - \rho + 1$*, where ρ is the rank of C_q, because the proposed principal parts are a space of dimension $n\nu$, those that actually occur satisfy ρ independent conditions, and the linear function that carries functions to their principal parts has a one-dimensional kernel.

This formula $n\nu - \rho + 1$ gives the answer only in the special cases in which, roughly speaking, a pole is allowed at one point only if a pole is allowed at all other points where x has the same value (multiplicities counted). A more general case of the formula can be stated by introducing a little more terminology. Let one set of proposed principal parts be said to be **subordinate** to another if they make use of the same algebraic number field $\mathbf{A}$, if each α of the first also occurs in the second, and if for each proposed expansion of y in powers of $x - \alpha$, the terms in the corresponding expression $\gamma_1 (\sqrt[m]{x - \alpha_i})^{-1} + \gamma_2 (\sqrt[m]{x - \alpha_i})^{-2} + \cdots + \gamma_l (\sqrt[m]{x - \alpha_i})^{-l}$ (where in most cases m is 1) of the first all have exponents at least as great (bearing in mind that -1 is greater than -2) as the smallest exponent of a nonzero term in the corresponding expression of the second. In short, the first set of proposed principal parts calls for no poles of greater multiplicity than are called for by the second. Let the **number of coefficients** in a set of proposed principal parts be the number of elements of $\mathbf{A}$ that need to be specified to describe a set of proposed principal parts that is subordinate to it (bearing in mind that if it calls for terms with fractional exponents, then the coefficients of *one* of the m expressions $\gamma_1 (\sqrt[m]{x - \alpha_i})^{-1} + \gamma_2 (\sqrt[m]{x - \alpha_i})^{-2} + \cdots + \gamma_l (\sqrt[m]{x - \alpha_i})^{-l}$ determine those in the other $m - 1$).

Corollary 1 (Riemann–Roch Theorem). *The functions whose principal parts are subordinate to a given set of proposed principal parts form a vector space of dimension $N - \rho + 1$, where N is the number of coefficients in the given set of proposed principal parts and ρ is the rank of the $g \times N$ matrix that describes the necessary and sufficient conditions of the theorem.*

Deduction. The sets of principal parts subordinate to the given set form a space of dimension N; the necessary and sufficient conditions describe a subspace of codimension ρ, which is the space of possible principal parts of actual functions; and the space of functions with these principal parts has dimension one greater, because two functions with the same principal parts differ by a constant.

The theorem also implies that the conditions (1) of Essay 4.6 satisfied by algebraic variations of N points on a curve are *sufficient,* with a few added assumptions, for a proposed variation to be algebraic:

Corollary 2. *Let (x_i, y_i) for $i = 1, 2, \ldots, N$ be pairs of algebraic numbers that satisfy $\chi(x_i, y_i) = 0$, and suppose that $\chi(x_i, X)$ has n distinct roots for each x_i, $i = 1, 2, \ldots, N$, so that Newton's polygon gives a unique power series solution $y = y_i + \beta(x - x_i) + \cdots$ of $\chi(x_i, y)$ for each i. For any list of nonzero* algebraic numbers $\delta_1, \delta_2, \ldots, \delta_N$ that satisfy $\sum_{i=1}^{N} h(x_i, y_i)\delta_i = 0$ for all holomorphic differentials, there is a rational function f on $\chi(x, y) = 0$ whose zeros are precisely at the points (x_i, y_i) and whose expansion $f = \gamma_i(x - x_i) + \cdots$ in powers of $x - x_i$ at each such point shows that $\frac{dx}{df}|_{(x_i, y_i)} = \delta_i$ in the sense that $\delta_i = \frac{1}{\gamma_i}$.*

More picturesquely, prescribed infinitesimal changes dx_i in the x-coordinates of the points are generated by changing f from 0 to df, which changes x_i to $x_i + \delta_i df$, provided the prescribed changes satisfy the necessary conditions $\sum h\, dx_i = 0$ for all holomorphic differentials $h\, dx$.

Deduction. The required function f is found by using the theorem to construct a function θ finite at $x = \infty$ whose principal parts are $\frac{\delta_i}{x - x_i}$ for $i = 1, 2, \ldots,$ N and setting $f = \frac{1}{\theta}$. Then f is zero only at the (x_i, y_i), and at these points $\frac{1}{f} = \frac{\delta_i}{x - x_i} + \cdots$, from which $f = \frac{1}{\delta_i}(x - x_i) + \cdots$ follows.

Corollary 3. *When (x_i, y_i) for $i = 1, 2, \ldots, N$ are points on $\chi(x, y) = 0$ as in Corollary 2, the rational functions on $\chi(x, y) = 0$ that have simple poles, at most, at these N points and no other poles form a vector space of dimension $N - g + 1 + \mu$, where g is the genus of the curve and μ is the dimension of the vector space of holomorphic differentials that are zero at all N points.*

Deduction. By Corollary 2, the dimension is $N - \rho + 1$, where ρ is the rank of the $g \times N$ matrix whose columns correspond to the N given points and whose g entries in each column are the values at the corresponding point of the coefficients h_i of a basis $h_1 dx, h_2 dx, \ldots, h_g dx$ of the holomorphic differentials. What is to be proved is that in this case $\rho = g - \mu$, which follows immediately from the observation that μ is the dimension of the kernel of the linear function from $\mathbf{A}^g$ to $\mathbf{A}^N$ given by multiplication of row matrices of length g on the right by the matrix.

In particular, the space of functions described in Corollary 3 has dimension at least $N - g + 1$.

---

* It is natural to exclude $\delta_i = 0$, because the points (x_i, y_i) for which $\delta_i = 0$ can be omitted from the list.

Essay 4.8 The Genus Is a Birational Invariant

So far in these essays, the genus of the field of rational functions on an algebraic curve $\chi(x, y) = 0$ has been described in ways that used the special element x of the field, first when the description was in terms of the dimension of the vector space $\Theta(x^\nu)$, then when it was in terms of the dimension of the space of holomorphic differentials $h\, dx$ over the field of constants. However, the geometric motivation of the concept in terms of algebraic variations leads one to believe that the genus depends only on the *curve*, not on the parameter x used to describe the curve. If z is any element of the root field of $\chi(x, y)$ that is not a constant, one can construct* an element w of this root field such that every element of the field can be expressed rationally in terms of z and w and such that z and w satisfy a relation of the form $\chi_1(z, w) = 0$, in which $\chi_1(z, w)$ is an irreducible polynomial with integer coefficients that is monic in w. In short, the root field of $\chi(x, y)$ can also be described as the root field of $\chi_1(z, w)$. What is to be expected, and what will be proved in this essay, is that the genus is the same whether x or z is the special parameter used to define it. In other words, the genus depends only on the field of rational functions on the curve, which is what it means to say that the genus is a *birational invariant*.

The needed connection is obvious from the point of view of differential calculus: The rule $h\, dx \leftrightarrow \left(h\frac{dx}{dz} \right) dz$ establishes a one-to-one correspondence between differentials expressed with respect to x and differentials expressed with respect to z. The heuristic meaning of "holomorphic" is "no poles," so this correspondence between differentials with respect to x and those with respect to z should be expected to put the *holomorphic* differentials in the two cases in one-to-one, linear correspondence, implying that the dimension of these spaces of holomorphic differentials—the genus in the two cases—should be the same.

It is easy to give algebraic meaning to $\frac{dx}{dz}$. Let x and z be given elements of the root field K of $\chi(x, y)$. There is† an irreducible polynomial $\phi(X, Z)$ in two indeterminates with integer coefficients—it is uniquely determined, up to

---

* See Essay 2.2. The field $\mathbf{Q}(z)$ of rational functions in z is isomorphic to a subfield of the root field, provided z is not a constant. Adjunction of x and y to $\mathbf{Q}(z)$ gives an explicit extension of $\mathbf{Q}(z)$ that is isomorphic to the root field of $\chi(x, y)$. By the theorem of the primitive element, such a double adjunction can be obtained by a simple adjunction, and one can describe a simple adjunction as the root field of an irreducible monic polynomial with coefficients in $\mathbf{Z}[z]$ in the usual way.

† Since the root field is an extension of $\mathbf{Q}(x)$ of finite degree n, the powers 1, z, z^2, ..., z^n are linearly dependent over $\mathbf{Q}(x)$, which is to say that one can find a nonzero polynomial of degree at most n with coefficients in $\mathbf{Q}(x)$ of which z is a root. The needed relation $\phi(x, z) = 0$ is found by clearing denominators and passing to an irreducible factor if necessary. Because the root field of $\chi(x, y)$ contains $\mathbf{Q}(x)$, the relation $\phi(x, z) = 0$ must involve z. To say that z is a *parameter*—that it is not a constant—means that ϕ also involves x; if this is not the case, the

its sign, by x and z—with the property that $\phi(x, z) = 0$ in K. The derivative of x with respect to z is found algebraically by implicit differentiation: differentiation of $\phi(x, z) = 0$ gives $\phi_x(x, z)dx + \phi_z(x, z)dz = 0$ and therefore gives

$$\frac{dx}{dz} = -\frac{\phi_z(x, z)}{\phi_x(x, z)},$$

where the subscripts indicate partial derivatives. The theorem to be proved states that when $\frac{dx}{dz}$ is defined algebraically in this way, $h\,dx$ *is holomorphic if and only if* $h\frac{dx}{dz}dz$ *is*, when, naturally, one determines whether $h_1 dz$ is holomorphic for $h_1 = h\frac{dx}{dz}$ by dealing with it as a differential in the root field of $\chi_1(z, w)$ rather than the root field of $\chi(x, y)$.

The notion of "principal parts" of an element of K that was defined in the preceding essay will play an important role in the proof, but now the dependence of these principal parts on x needs to be emphasized: The **principal parts of f relative to a parameter** x are the terms with negative exponents[*] in all possible expansions of f in (possibly fractional) powers of $x - \alpha$ for algebraic numbers α. (Possible principal parts at $x = \infty$ will be ignored.) As was shown in the last essay, for a given f one can find a polynomial $q(x)$ with the property that the only possible principal parts of f relative to x occur when α is a root of $q(x)$. Therefore, the determination of the principal parts relative to x is a finite—and usually quite simple—calculation. One can then use the following lemma to determine whether $f\,dx$ is holomorphic for finite x:

Lemma 1.   *A differential $f\,dx$ is holomorphic for finite x if and only if all terms of all principal parts of f relative to x have exponents greater than* -1.

This criterion says roughly that $f\,dx$ has no poles, because at a point of the curve where $x = \alpha$ there is a local parameter s for which $x - \alpha = s^m$ for some m; to say that $f\,dx$ has no pole at this place where $s = 0$ is to say that $f\,d(s^m) = mfs^{m-1}ds$ has no pole, which is to say that multiplication of f by s^{m-1} clears its denominator, or, what is the same, that its expansion in powers of $x - \alpha = s^m$ contains no terms whose exponents are less than or equal to -1.

Proof. Suppose first that all terms of all principal parts of f relative to x do have exponents greater than -1. If θ is integral over x, then its image under any embedding of K in $\mathbf{A}\langle s \rangle$ that takes x to $\alpha + s^m$ for some m is a series

---

denominator of $\frac{dx}{dz}$ is zero and the derivative is not defined (x is not a function of z).

[*] This definition is imprecise in that it ignores the question of *multiplicities*. In Essay 4.7, $n = \deg_y \chi$ distinct embeddings of the root field of χ in $\mathbf{A}\langle s \rangle$ were constructed for each given α. A given f may have fewer than n distinct images, so the same principal parts may occur for more than one embedding. Obviously, Lemma 1 is not affected by the way in which multiplicities are treated.

that contains no powers of s with negative exponents. Therefore, it contains no powers of $x - \alpha = s^m$ with negative exponents, so the image of $f\theta$ under any such embedding contains no terms in which the exponent on $x - \alpha$ is -1 or less. As was seen in Essay 4.7, the expression of $\operatorname{tr}_x(f\theta)$ as a power series in $x - \alpha$ is the sum of the n images of $f\theta$ in $\mathbf{A}\langle s \rangle$. Therefore, it is a series in which no term has an exponent less than or equal to -1. Thus, since it is a series expansion of a rational function of x, which implies that it contains no terms with fractional exponents, it is a power series in $x - \alpha$. Since this is true for every α, it follows that $\operatorname{tr}_x(f\theta)$ must in fact be a polynomial in x whenever θ is integral over x. In short, $f\,dx$ is holomorphic for finite x which completes the proof of Lemma 1.

Conversely, suppose that some embedding of the root field K of $\chi(x,y)$ in $\mathbf{A}\langle s \rangle$ that carries x to $\alpha + s^m$—where α is an algebraic number and m is a positive integer—carries f to a series in which the exponent on s is less than or equal to $-m$, so that the exponent on $x - \alpha$ is less than or equal to -1. Let $z_1, z_2, \ldots, z_n$ be an integral basis of the root field over x, and let $\theta = \sum c_i z_i$, where $c_1, c_2, \ldots, c_n$ are constants to be determined. Consider the terms in the n expansions of θ in (possibly fractional) powers of $x - \alpha$ in which the exponents are less than 1. When the n expansions do not involve fractional powers, the expansion of each z_i has just one term—the constant term—in which the exponent is less than 1 in each of the n embeddings, and the same is true of θ; in this case, the n constant terms in the series representations of θ are the entries of the column matrix Mc where c is the column matrix with entries $c_1, c_2, \ldots, c_n$ and M is the $n \times n$ matrix whose jth column contains the n constant terms in the images of z_j under the n embeddings.

When one or more of the n expansions do involve fractional exponents—when the curve is ramified at $x = \alpha$—a similar statement is true. An embedding involving powers of $s = \sqrt[m]{x - \alpha}$ in which $m > 1$ implies $m - 1$ others in which s is replaced by ωs for the mth roots of unity ω other than 1. There are precisely m terms (some of which may be zero) in any one of these series in which the exponent on $x - \alpha$ is less than 1, namely, the terms in s^0, s^1, $\ldots$, s^{m-1}. Since the sum of the various values of m is n, it follows that all coefficients of all n expansions of θ in which the exponents on $x - \alpha$ are less than 1 are determined by just n such coefficients, namely, the coefficients of a *selection* of the expansions when just one expansion is selected from each set of m related expansions. The same is true of each z_j, and all coefficients of θ are determined by those given by a formula Mc, as before, in which the jth column of M gives the coefficients of the selected expansions of z_j.

(For example, in the case of the integral basis 1, y, y^2/x of the root field of $y^3 - xy + x^3$ over x, it was shown in Essay 4.4 that there are three expansions of y when $\alpha = 0$, namely, $y = \pm\sqrt{x} + \cdots$ and $y = 0 + \cdots$, where the omitted terms are divisible by x. Therefore, every θ integral over x has three expansions, but the one corresponding to $y = -\sqrt{x} + \cdots$ can be derived from the one corresponding to $y = \sqrt{x} + \cdots$ by changing $\sqrt{x}$ to $-\sqrt{x}$. When just two expansions, those corresponding to $y = \sqrt{x} + \cdots$ and $y = 0 + \cdots$, are selected, the selected expansions of $\theta = c_1 + c_2 y + c_3 y^2/x$ are given by the formula

$$Mc = \begin{bmatrix} 1 & 0 & 0 \\ 0 & 1 & 0 \\ 1 & 0 & 1 \end{bmatrix} \begin{bmatrix} c_1 \\ c_2 \\ c_3 \end{bmatrix},$$

where the first two rows contain the coefficients of 1 and $\sqrt{x}$ in the first selected expansion and the last row contains the coefficient of 1 in the second.)

That the matrix M is invertible can be proved as follows: If $Mc = 0$, then the n expansions of θ contain no terms in which the exponent on $x - \alpha$ is less than 1, so $\theta/(x - \alpha)$ is integral over x. If any c_i were nonzero, one of the coefficients $c_i/(x-\alpha)$ in the representation of $\theta/(x-\alpha)$ relative to the integral basis z_1, z_2, ..., z_n would not be a polynomial, contrary to the definition of an integral basis. Therefore, $Mc = 0$ implies $c = 0$, so the square matrix M is invertible. In other words, the coefficients in the terms of the expansion of θ in which the exponent on $x - \alpha$ is less than 1 can be given arbitrarily chosen values (subject to the relations among m such expansions when $m > 1$ that were just noted) by taking the column matrix c to be the column matrix of chosen values multiplied on the left by M^{-1}.

Because the principal parts of f are assumed to contain a nonzero term in which the exponent on $x - \alpha$ is less than or equal to -1, the least such exponent has the form $e = -i - \frac{j}{m}$, where $i \geq 1$ and $0 \leq j < m$. Let $\theta = \sum c_i z_i$ be chosen so that the coefficient of $(x - \alpha)^{j/m}$ in the expansion of θ in the embedding in $\mathbf{A}\langle s \rangle$ that gives rise to the nonzero term with exponent e in the expansion of f is nonzero, but all other expansion coefficients of terms with exponent less than 1 are zero. Then the expansion of $f\theta$ in m embeddings begins $\gamma(x - \alpha)^{-i} + \cdots$, where $\gamma \neq 0$, while in the remaining embeddings the expansion of $f\theta$ contains no terms in which the exponent on $x - \alpha$ is less than or equal to $-i$. Therefore, the sum of the n expansions of $f\theta$ is $m\gamma(x - \alpha)^{-i} + \cdots$, where $i \geq 1$. In particular, this sum is not a polynomial in x.

The bilinear form "the trace of the product" from $K \times K$ to $\mathbf{Q}(x)$ is described, relative to the integral basis z_1, z_2, ..., z_n, by an $n \times n$ symmetric matrix S of polynomials in x with integer coefficients, namely, the matrix whose entry in the ith row of the jth column is $\mathrm{tr}_x(z_i z_j)$. To say that $f\,dx$ is holomorphic for finite x means simply that all entries of $[f]S$ are polynomials in x when $[f]$ denotes the row matrix whose entries are the coefficients that represent f in the integral basis z_1, z_2, ..., z_n. If this were the case, the sum of the expansions of $f\theta$, which is $[f]S[c]$ where $[c]$ contains the coefficients of $\theta = \sum c_i z_i$ as above, would also be a polynomial in x. Since it is not, $f\,dx$ must not be holomorphic for finite x, which completes the proof of Lemma 1.

Theorem. *Let z be a parameter in the root field of $\chi(x,y)$ and let $\frac{dx}{dz}$ be the element of the root field defined using implicit differentiation as above. A differential $f\,dx$ is holomorphic if and only if $f\frac{dx}{dz}\,dz$ is holomorphic.*

Proof. The reciprocal of $\frac{dx}{dz}$ is $\frac{dz}{dx}$, so it will suffice to prove that "$f\,dx$ is holomorphic" implies "$f\frac{dx}{dz}\,dz$ is holomorphic."

Let δ be a given algebraic number and let an embedding of K in $\mathbf{A}\langle\sigma\rangle$ be given that carries z to $\delta + \sigma^\mu$ for some $\mu > 0$. It is to be shown that if $h\,dx$ is holomorphic, then the image of $h \cdot \frac{dx}{dz}$ under this embedding contains no terms in which the exponent on $z - \delta$ is less than or equal to -1, or, what is the same, that all exponents in the expansion of $(z - \delta) \cdot h \cdot \frac{dx}{dz}$ are positive.

Assume first that the image of x under the given embedding has no terms in which the exponent on σ is negative; say it is $\alpha + \alpha'\sigma + \alpha''\sigma^2 + \cdots$. In this case, let m be the exponent of the first nonzero term in the expansion of $x - \alpha$. (There is such a term because x is not a constant.) When an mth root ϵ_1 of the reciprocal of $\alpha^{(m)}$ is adjoined to $\mathbf{A}$, if necessary, the following lemma constructs a substitution $\sigma = \epsilon_1 s + \epsilon_2 s^2 + \epsilon_3 s^3 + \cdots$ that carries $x = \alpha + \alpha^{(m)}\sigma^m + \cdots$ to $\alpha + s^m$.

Lemma 2. *Given a nonzero power series $A_m x^m + A_{m+1}x^{m+1} + A_{m+2}x^{m+2} + \cdots$ in which the coefficients are algebraic numbers and the first nonzero term contains x to the power $m > 0$, and given an mth root C_1 of $1/A_m$, construct an infinite series $x = C_1 s + C_2 s^2 + C_3 s^3 + \cdots$ with algebraic number coefficients whose substitution in the series results in s^m.*

Proof. Substitution of $x = C_1 s + C_2 s^2 + C_3 s^3 + \cdots$ in $A_m x^m + A_{m+1}x^{m+1} + A_{m+2}x^{m+2} + \cdots$ gives $B_m s^m + B_{m+1}s^{m+1} + B_{m+2}s^{m+2} + \cdots$ where $B_m = A_m C_1^m = 1$, $B_{m+1} = mA_m C_1^{m-1}C_2 + A_{m+1}C_1^{m+1}, \ldots$. The formula for B_{m+i} when $i > 0$ contains the terms $mA_m C_1^{m-1}C_{i+1}$ and $A_{m+i}C_1^{m+i}$; the remaining terms in the formula constitute a polynomial in $C_1, C_2, \ldots, C_i$ and $A_m, A_{m+1}, \ldots, A_{m+i-1}$ with integer coefficients. Thus, the requirement $B_{m+i} = 0$ for $i > 0$ is the statement that C_{i+1} is a polynomial in $C_1, C_2, \ldots, C_i$ and $A_m, A_{m+1}, \ldots, A_{m+i}$ divided by $mA_m C_1^{m-1} = m/C_1$. Since $A_m = 1/C_1^m$, it follows that each successive C_{i+1} can be expressed rationally in terms of $C_1, A_{m+1}, A_{m+2}, \ldots, A_{m+i}$. The series $C_1 s + C_2 s^2 + C_3 s^3 + \cdots$ constructed in this way has the required property.

Because the given embedding $K \to \mathbf{A}\langle\sigma\rangle$ followed by the substitution $\sigma = \epsilon_1 s + \epsilon_2 s^2 + \cdots$ carries x to $\alpha + s^m$, and because $h\,dx$ is holomorphic, the resulting embedding $K \to \mathbf{A}\langle s\rangle$ carries h to a series in s in which no term has an exponent less than or equal to $-m$ on s. Otherwise stated, all exponents in the expansion of $(x - \alpha) \cdot h$ in powers of s are positive. Since this expansion is found by substituting the expansion of σ in powers of s into the expansion of $(x - \alpha) \cdot h$ in powers of σ, it follows *that all exponents in the expansion of $(x - \alpha) \cdot h$ in powers of σ are positive.*

Let this expansion be multiplied by the expansion of $\frac{dx}{dz} \cdot \frac{z-\delta}{x-\alpha}$ in powers of σ. On the one hand, the result is $(z-\delta) \cdot h \cdot \frac{dx}{dz}$. On the other hand, if $\phi(x, z) = 0$ is the equation satisfied by x and z, then $\phi\left(\alpha + \alpha^{(m)}\sigma^m + \cdots, \delta + \sigma^\mu\right)$ is identically zero, so differentiation with respect to σ gives $\phi_x(x, z)(m\alpha^{(m)}\sigma^{m-1} +$

$\cdots) + \phi_z(x,z)\mu\sigma^{\mu-1} = 0$, where x and z stand for their expansions as power series in σ and the omitted terms are divisible by σ^m. Multiplication by σ then gives $\phi_x(x,z)\left(ma^{(m)}(x-\alpha) + \cdots\right) + \phi_z(x,z)\mu(z-\delta) = 0$, where the omitted terms are divisible by σ^{m+1}. Division by $\phi_x(x,z)$ (which is not zero, because z is not a constant) times $x - \alpha$ gives $ma^{(m)} + \cdots - \mu\frac{dx}{dz} \cdot \frac{z-\delta}{x-\alpha} = 0$, where the omitted terms are all divisible by σ. This equation shows that the expansion in powers of σ of $\frac{dx}{dz}\frac{z-\delta}{x-\alpha}$ is the constant $\frac{m}{\mu}a^{(m)}$ plus terms in σ. Therefore, $(z-\delta)\cdot h\cdot\frac{dx}{dz} = ((x-\alpha)\cdot h)\left(\frac{m}{\mu}a^{(m)} + \cdots\right)$ is a product of two series in σ, one with positive exponents and one with no negative exponents, which shows that *all terms in the expansion of* $(z-\delta)\cdot h\cdot\frac{dx}{dz}$ *in powers of* σ *have positive exponents,* a conclusion that holds for any embedding $K \to \mathbf{A}\langle\sigma\rangle$ that carries z to $\delta + \sigma^\mu$ and carries x to a series with no negative exponents.

If an embedding that carries z to $\delta + \sigma^\mu$ carries x to a series with some negative exponents, it carries $u = \frac{1}{x}$ to a series in which all exponents are positive. Since $\frac{h}{u^2}\cdot du$ is holomorphic for finite u by virtue of the assumption that $h\,dx$ is holomorphic, it follows that all exponents in the expansion of $(z-\delta)\cdot\frac{h}{u^2}\cdot\frac{du}{dz}$ in powers of σ are positive. By the chain rule, $\frac{dx}{dz} = \frac{dx}{du}\frac{du}{dz} = \frac{-1}{u^2}\cdot\frac{du}{dz}$ when $x = \frac{1}{u}$, so it follows that all exponents in the expansion of $(z-\delta)\cdot h\cdot\frac{dx}{dz} = (z-\delta)\cdot h\cdot\frac{-1}{u^2}\cdot\frac{du}{dz}$ in powers of σ are positive in this case too.

Thus, Lemma 1 implies that $h\cdot\frac{dx}{dz}\cdot dz$ is holomorphic for finite z.

By the same token, $h\cdot\frac{dx}{dv}\cdot dv$ is holomorphic for finite v for any parameter v and in particular when $v = \frac{1}{z}$. Therefore $h\cdot\frac{dx}{dz}\cdot\frac{-1}{v^2}\cdot dv$ is holomorphic for finite $v = \frac{1}{z}$, which completes the proof that $h\cdot\frac{dx}{dz}\cdot dz$ is holomorphic.

Corollary.   *The genus is a birational invariant.*

The determination of the genus can be accomplished by finding holomorphic differentials, for which the following proposition is useful.

An algebraic curve $\chi(x,y) = 0$ is **nonsingular for finite** x if no pair (α,β) of algebraic numbers satisfies all three conditions $\chi(\alpha,\beta) = 0$, $\chi_x(\alpha,\beta) = 0$, and $\chi_y(\alpha,\beta) = 0$.

Proposition.   *If* $\chi(x,y) = 0$ *is nonsingular for finite* x, *then* $h\,dx$ *is holomorphic for finite* x *if and only if* $h\cdot\chi_y$ *is integral over* x.

In other words, when $\chi(x,y) = 0$ is nonsingular for finite x, the differentials holomorphic for finite x are those of the form $\frac{\phi(x,y)\,dx}{\chi_y(x,y)}$, where $\phi(x,y)$ is integral over x.

Proof. First assume that $h\,dx$ is holomorphic. By the proposition of Essay 4.5, it will suffice to prove that the image of $h\cdot\chi_y$ in each embedding of K in $\mathbf{A}\langle s\rangle$ that carries x to $\alpha + s^m$, and carries rational numbers to themselves, is without negative exponents.

When $\chi_y(\alpha,\beta) \neq 0$, β is a simple root of $\chi(\alpha,y)$, which implies, as was shown in Essay 4.4, that $x = \alpha + s$, $y = \beta$ is an unambiguous truncated solution of $\chi(x,y) = 0$. Such a truncated solution implies an infinite series solution $y = \beta + \beta'(x - \alpha) + \beta''(x - \alpha)^2 + \cdots$. The corresponding embedding $K \to \mathbf{A}\langle s \rangle$ does not involve fractional powers of $x - \alpha$. The assumption that $h\,dx$ is holomorphic implies that the image of h in $\mathbf{A}\langle s \rangle$ contains no exponents less than or equal to -1, so all exponents are greater than or equal to zero. The same is true of the image of χ_y—it is a polynomial in x and y and is therefore integral over x—so the image of $h \cdot \chi_y$ under the embedding has no negative exponents, as was to be shown.

Otherwise, $\chi_x(\alpha, \beta) \neq 0$, because the curve is nonsingular for finite x. In this case, the polynomial $\Phi_0(t)$ in $\chi(\alpha + s, \beta + t) = \Phi_0(s) + \Phi_1(s)t + \cdots + t^n$ is divisible by s but not s^2, so the Newton polygon algorithm leads to a "polygon" with one segment from $(0, 1)$ to a point where $j_i = 0$; call it $(\tau, 0)$. The ambiguity of the truncated solution $x = \alpha + s$, $y = \beta$ is then τ, and the output of Newton's polygon is τ unambiguous truncated solutions $x = \alpha + s_1^\tau$, $y = \beta + \sqrt[\tau]{\zeta_0} \cdot s_1$ (one solution for each of the τ possible values of $\sqrt[\tau]{\zeta_0}$). By Lemma 2, the infinite series expansion $y - \beta = \sqrt[\tau]{\zeta_0} \cdot s_1 + \beta'' s_1 + \cdots$ implies an infinite series expansion $s_1 = \epsilon_1(y - \beta) + \epsilon_2(y - \beta)^2 + \cdots$, whose substitution in $\sqrt[\tau]{\zeta_0} \cdot s_1 + \beta'' s_1 + \cdots$ gives $y - \beta$ and whose substitution in the embedding $K \to \mathbf{A}\langle s_1 \rangle$ therefore gives an embedding $K \to \mathbf{A}\langle y - \beta \rangle$ that carries y to $\beta + (y - \beta)$. Because $h \cdot \frac{dx}{dy} \cdot dy$ is holomorphic, it follows that the image of $h \cdot \frac{dx}{dy} = -h \cdot \frac{\chi_y}{\chi_x}$ under this embedding has no exponents less than or equal to -1. It has no fractional exponents, so all exponents in the expansion of $h \cdot \frac{\chi_y}{\chi_x}$ in powers of $y - \beta$ are at least zero. Therefore, the same is true of its expansion in powers of s_1. Since χ_x is a polynomial in $x = \alpha + s_1^\tau$ and $y = \beta + \sqrt[\tau]{\zeta_0} \cdot s_1 + \cdots$, the expansion of $h \cdot \chi_y = h \cdot \frac{\chi_y}{\chi_x} \cdot \chi_x$ in powers of s_1 has no terms with negative exponents, as was to be shown. Thus, the proof that "$h\,dx$ is holomorphic" implies "$h \cdot \chi_y$ is integral over x" is complete.

To prove, conversely, that all differentials of the form $\frac{\phi}{\chi_y}\,dx$ in which ϕ is integral over x are holomorphic for finite x it will suffice to prove that $\frac{1}{\chi_y}\,dx$ is holomorphic for finite x. Certainly for any embedding $x = \alpha + s^m$, $y = \beta + \beta's + \beta''s^2 + \cdots$ for which $\chi_y(\alpha, \beta) \neq 0$ the expansion of $\frac{1}{\chi_y}$ in powers of s contains no negative exponents. All other embeddings have the form $x = \alpha + s_1^\tau$, $y = \beta + \beta's_1 + \beta''s_1^2 + \cdots$, where $\beta' \neq 0$—as was just seen—by virtue of the assumption that $\chi(x, y)$ has no singularities for finite x. Differentiation of $\chi(\alpha + s_1^\tau, \beta + \beta's_1 + \cdots) = 0$ gives $\chi_x(\alpha + s_1^\tau, \beta + \beta's_1 + \cdots) \cdot \tau s_1^{\tau-1} + \chi_y(\alpha + s_1^\tau, \beta + \beta's_1 + \cdots) \cdot (\beta' + \cdots) = 0$. Thus, $\frac{1}{\chi_y} = -\frac{\beta' + \cdots}{\chi_x \cdot \tau \cdot s_1^{\tau-1}}$. Since $\chi_x(\alpha, \beta) \neq 0$, it follows that $-\tau + 1$ is the least exponent in the expansion of $\frac{1}{\chi_y}$ in powers of s_1. Since $s_1^{-\tau+1} = (y - \beta)^{-1+\frac{1}{\tau}}$, the desired conclusion that the principal parts of $\frac{1}{\chi_y}$ relative to this embedding $K \to \mathbf{A}\langle s \rangle$ contain no exponents that are -1 or less follows.

Example 1: $\chi(x, y) = y^2 + x^4 - 1$ (the elliptic curve mentioned in Essay 4.2)

Since $\chi_y = 0$ implies $y = 0$ and $\chi_x = 0$ implies $x = 0$, this curve is nonsingular for finite x, because $(\alpha, \beta) = (0, 0)$ does not satisfy $\beta^2 + \alpha^4 - 1 = 0$.

Therefore, the differentials holomorphic for finite x are those of the form $\frac{\phi\,dx}{2y}$, where ϕ is integral over x. The substitution $x = \frac{1}{u}$, $y = \frac{v}{u^2}$ puts this curve in the form $v^2 + 1 - u^4 = 0$ and puts $dx/2y$ in the form $-(\frac{1}{u^2})du/2(\frac{v}{u^2}) = -du/2v$, so $\phi\,dx/2y$ is holomorphic if and only if ϕ is integral over x *and* over $u = 1/x$, which is to say, if and only if ϕ is constant.

Example 2: $\chi(x,y) = y^3 + x^3y + x$ (the Klein curve)

This curve is nonsingular for finite x, because $3\beta^2 + \alpha^3 = 0$ and $3\alpha^2\beta + 1 = 0$ imply that $\beta = -\frac{1}{3\alpha^2}$ and $\alpha^3 = -3\beta^2 = -\frac{1}{3\alpha^4}$, so that $\alpha^7 = -\frac{1}{3}$; therefore $\beta^3 + \alpha^3\beta + \alpha \neq 0$, because $\alpha^3\beta + \alpha = \alpha^3 \cdot \frac{-\frac{1}{3\alpha^2}}{1} + \alpha = \frac{2\alpha}{3}$ is not the negative of $\beta^3 = -\frac{1}{27\alpha^6} = -\frac{\alpha}{27\cdot(-1/3)} = \frac{\alpha}{9}$.

Since $\chi_y(x,y) = 3y^2 + x^3$, every holomorphic differential can be written $\frac{\phi\,dx}{3y^2+x^3}$ for some ϕ integral over x. Because $1, y, y^2$ is an integral basis over x (Essay 4.5), ϕ must be a polynomial in x and y. Determining the holomorphic differentials on the Klein curve therefore amounts to determining the polynomials $\phi(x,y)$ in x and y for which $\frac{\phi\,dx}{3y^2+x^3}$ is holomorphic. Such a differential is holomorphic for finite x, and the problem is to determine the conditions under which it is without poles at $x = \infty$.

The substitution $x = \frac{1}{u}$, $y = \frac{v}{u^3}$ transforms the curve into $v^3 + u^3v + u^8 = 0$, a curve with a singularity at $(u,v) = (0,0)$. The first step of the Newton polygon algorithm in the case in which the value of u is 0 calls for setting $u = s$ and $v = 0 + t$, which leads to $s^8 + s^3t + t^3$. The polygon is based on the points $(0,8)$, $(1,3)$, and $(3,0)$, so it consists of two segments $5i + j = 8$ and $3i + 2j = 9$. The first segment furnishes an unambiguous truncated solution $u = s$, $v = -s^5$, which implies an infinite series solution, and the second furnishes the remaining two solutions in the form of the unambiguous truncated solutions $u = \sigma^2$, $v = \pm i\sigma^3$.

The expression of $\frac{dx}{3y^2+x^3}$ relative to the first of these is

$$\frac{dx}{3y^2+x^3} = \left(-\frac{du}{u^2}\right) \cdot \frac{1}{3\cdot\frac{v^2}{u^6} + \frac{1}{u^3}} = \frac{-s^4ds}{(3s^{10}+\cdots)^2 + s^3} = (-s+\cdots)ds.$$

Therefore, there is no pole for this embedding if this differential is multiplied by $x = \frac{1}{s} + \cdots$ or by any power of $y = \frac{v}{u^3} = -s^2 + \cdots$. The expression of $\frac{dx}{3y^2+x^3}$ relative to the second is

$$\frac{dx}{3y^2+x^3} = \frac{-2d\sigma}{\sigma^3} \cdot \frac{1}{3\cdot\frac{(-\sigma^6+\cdots)}{\sigma^{12}} + \frac{1}{\sigma^6}} = \frac{2d\sigma}{\sigma^3} \cdot \frac{\sigma^6}{2+\cdots} = (\sigma^3 + \cdots)\,d\sigma.$$

Therefore, it remains finite for this embedding if it is multiplied by $x = \frac{1}{\sigma^2}$ or by $y = \pm\frac{i}{\sigma^3} + \cdots$ but not if it is multiplied by any polynomial of higher degree in x or y.

In conclusion, the holomorphic differentials in this case are

$$\frac{(a + bx + cy)\,dx}{3y^2 + x^3},$$

as was already found in Essay 4.6.

5

Miscellany

Essay 5.1 On the So-Called Fundamental Theorem of Algebra

> *Nor does this account of infinity rob the mathematicians of their study; for all that it denies is the actual existence of anything so great that you can never get to the end of it. And as a matter of fact, mathematicians never ask for or introduce an infinite magnitude; they only claim that the finite line shall be of any length they please; and it is possible to divide any magnitude whatsoever in the same proportion as the greatest magnitude.—* Aristotle [3, p. 261].

Regrettably, the name "fundamental theorem of algebra" has become firmly attached to the statement that *the field of complex numbers is algebraically closed*—that is, the statement that a polynomial in one indeterminate x whose coefficients are complex numbers can be written in the form $a_0 \prod_{j=1}^{n}(x - x_j)$, where x_1, x_2, ..., x_n are complex numbers and a_0 is the leading coefficient of $f(x)$. From a constructive point of view, this "theorem" is *false* for the following reason.

A "complex number" by definition has the form $a + bi$, where a and b are real numbers and $i = \sqrt{-1}$. A real number is a convergent sequence of rational numbers. Thus, computations with real numbers are computations with convergent sequences, so they are computations with *approximations,* and *it is not constructively true that a real number is either zero or nonzero.* For example, the sequence of partial sums of the series $\sum_{i=1}^{\infty} \frac{n_i}{10^i}$, where $n_i = 1$ if i is an even number greater than 2 that is not the sum of two primes and $n_i = 0$ for all other numbers i, is a well-defined real number, call it r, because its decimal expansion to a huge number of places is easily computable, and its computation to *any* number of places is a finite calculation. For as far as the computation has been carried, r appears to be zero. However, the statement that $r = 0$ amounts simply to the Goldbach conjecture that every even number greater than 2 is a sum of two primes, a statement that up to the present time no one has been able to prove or disprove.* To write $rx^2 + x$ in the form $a_0 \prod(x - x_j)$ one must determine whether the Goldbach conjecture is true: If it is, then $rx^2 + x = x$, but otherwise, $rx^2 + x = rx(x + \frac{1}{r})$. At the present time, then, no known construction finds the representation of $rx^2 + x$ in the required form.

As will be shown in this essay, the theorem *is* true constructively when the coefficients of the given polynomial are known *exactly*.

Theorem. *Given a polynomial $f(x)$ whose coefficients are complex rational numbers—numbers of the form $u + vi$ in which u and v are rational numbers and $i = \sqrt{-1}$—express it in the form $a_0 \prod_{j=1}^{n}(x - x_j)$, where x_1, x_2, $\ldots$, x_n are complex numbers and a_0 is its leading coefficient.*

If $\overline{f}(x)$ is the complex conjugate of $f(x)$, then $f(x)\overline{f}(x)$ is a polynomial with rational coefficients, so the fundamental theorem of Essay 1.2 provides an irreducible, monic polynomial $G(y)$ with integer coefficients that splits $f(x)\overline{f}(x)(x^2 + 1)$. When the term i in the coefficients of $f(x)$ is identified with one of the two square roots of -1 in the root field of $G(y)$, $f(x)$ becomes a polynomial with coefficients in the root field that splits into linear factors, because the polynomial $f(x)\overline{f}(x)$ of which it is a factor splits into linear factors. Therefore, $f(x) \equiv a_0 \prod(x - \rho_j(y)) \bmod G(y)$ where the $\rho_j(y)$ are elements of the root field. Because the root field is isomorphic to the subfield $\mathbf{Q}[y^*]$ of the complex numbers when y^* is a complex root of $G(y)$, these observations imply that the theorem to be proved is a corollary of the following simpler theorem:

Main Theorem. *Given a monic, irreducible polynomial with integer coefficients, construct a complex root.*

(More generally, let the field $\mathbf{Q}[i]$ of complex rational numbers be replaced by the root field of a monic, irreducible polynomial with integer coefficients. The Main Theorem implies that such a root field is isomorphic to a subfield of the complex numbers. In other words, every algebraic number field is isomorphic to a subfield of the complex numbers. However, because algebraic computations can be done *exactly* in a root field, such subfields of the complex numbers are very special; in them, as in $\mathbf{Q}[i]$, algebraic computation can be done exactly. If $f(x)$ is a polynomial with coefficients in such a subfield of the complex numbers, then $f(x)$ can be written as a factor of a polynomial (its norm) with rational coefficients, and the method just explained shows that $f(x)$ can be written in the form $a_0 \prod(x - x_j)$, where the x_j are complex numbers, but where the x_j are in fact contained in some algebraic number field viewed as a subfield of the complex numbers.)

---

* If one accepts Kronecker's criterion quoted in the epigraph to Essay 1.5, one must say that the definition of "positive real number" lacks a firm foundation because no method can be given for determining whether the real number r is positive.

Lemma. *Given a monic, irreducible polynomial $f(x)$ with integer coeffi-cients, and given a positive rational number ϵ, find a rational complex number x_0 for which $|f(x_0)| < \epsilon$.*

Proof. Let n be the degree of the given polynomial $f(x)$. Because $\frac{f'(x)}{f(x)} - \frac{n}{x}$ can be written as a quotient of polynomials in which the numerator has degree $n-1$ and the denominator has degree $n+1$, $\frac{f'(x)}{f(x)} = \frac{n}{x} + O(|x|^{-2})$ as $|x| \to \infty$; that is, there is a constant K for which $\left|\frac{f'(x)}{f(x)} - \frac{n}{x}\right| < \frac{K}{|x|^2}$ for all sufficiently large $|x|$. It follows that if R is a positive integer and S is the square in the complex plane whose vertices are $\pm R \pm Ri$, then

$$\int_{\partial S} \frac{f'(x)}{f(x)} \, dx = \int_{\partial S} \frac{n}{x} \, dx + O\left(\frac{1}{R}\right)$$

as $R \to \infty$. Since $\int_{\partial S} \frac{n}{x} \, dx = 2\pi i \cdot n$ for all R, it follows that there is a positive integer R, call it R_0, for which $\int_{\partial S_0} \frac{f'(x)}{f(x)} \, dx \neq 0$, where S_0 is the square in the complex plane whose vertices are $\pm R_0 \pm R_0 i$.

Because $\frac{f(x)-f(y)}{x-y}$ is a polynomial in x and y with integer coefficients, there is an integer M such that $\left|\frac{f(x)-f(y)}{x-y}\right|$ is at most M on the square S_0. For such an integer, $|f(x) - f(y)| \leq M|x - y|$ for all x and y in S_0. Let T be an integer larger than $\frac{4M}{\epsilon}$ and let S_0 be partitioned into subsquares that are $\frac{1}{T}$ on a side. If x and y are complex numbers in the same subsquare, then $|f(x) - f(y)| \leq \sqrt{2}\frac{1}{T}M < \frac{1}{2}\epsilon$. Thus, if $|f(x)| \geq \epsilon$ at the midpoint of a subsquare, then $f(x)$ is nonzero throughout the subsquare, which means that $\log f(x)$ is defined as a function of x throughout the subsquare; since the derivative of $\log f(x)$ is $\frac{f'(x)}{f(x)}$, it follows that the integral of $\frac{f'(x)}{f(x)} \, dx$ around the boundary of each such subsquare is zero. If $|f(x)| \geq \epsilon$ were true for the midpoints of *all* the subsquares, then $\int_{\partial S_0} \frac{f'(x)}{f(x)} \, dx$, which is the sum of the integrals around the boundaries of all the subsquares, would be zero, contrary to the choice of S_0. Thus, the finite set of midpoints of the subsquares contains at least one complex rational number x_0 for which $|f(x_0)| < \epsilon$, as required.

(Note that the coordinates of the midpoints are rational, so $|f(x)|^2$ at the midpoints can be computed exactly and compared to the rational num-ber ϵ^2. Obviously, the amount of computation required to find an x_0 by this construction could be huge even for a rather simple $f(x)$ and a moderate ϵ. The method described here is not a *practical* way of finding an x_0—which in most cases would be easily accomplished by simple bisection methods to get a rough estimate, followed by Newton's method—but a way that can be succinctly described and is a finite calculation.)

Proof of the Main Theorem. Let $f(x)$ be the given monic, irreducible polyno-mial with integer coefficients. Since $|f(x)| \to \infty$ as $|x| \to \infty$, a positive integer N can be chosen for which $|x| \geq N$ implies $|f(x)| \geq 1$.

By the Euclidean algorithm, there are polynomials $\alpha(x)$ and $\beta(x)$ with rational coefficients for which $\alpha(x)f(x) + \beta(x)f'(x)$ divides both $f(x)$ and $f'(x)$. Because $f(x)$ is irreducible and the degree of $f'(x)$ is less than that of $f(x)$, $\alpha(x)f(x) + \beta(x)f'(x)$ must be a nonzero rational number, so it can be assumed without loss of generality to be 1. Let A and B be positive integers for which $|2\alpha(x)| < A$ and $|2\beta(x)| < B$ throughout the disk $|x| \le N$ in the complex plane.

Finally, let C be a positive integer that is an upper bound for the modulus of the polynomial $\frac{f'(x)-f'(y)}{x-y}$ when x and y are complex numbers whose moduli are less than $N + 1$.

Use the lemma to find a rational complex number x_0 for which $|f(x_0)|$ is less than both $\frac{1}{A}$ and $\frac{1}{4B^2C}$ and define a sequence $x_1, x_2, \ldots, x_n, \ldots$ by the formula

$$x_{n+1} = x_n - \frac{f(x_n)}{f'(x_0)}.$$

This sequence converges, as is proved by the estimate

$$(1) \qquad |x_{n+1} - x_n| < \frac{1}{2}|x_n - x_{n-1}| \qquad \text{(for } n = 1, 2, \ldots)$$

which will be proved inductively using the estimate

$$(2) \quad |f'(x) - f'(x_0)| < \frac{1}{2B} \qquad \text{(for } x \text{ on the line segment from } x_{n-1} \text{ to } x_n).$$

Because $2 = 2\alpha(x_0)f(x_0) + 2\beta(x_0)f'(x_0)$ and because $|f(x_0)| < \frac{1}{A} \le 1$, the estimate $|x_0| < N$ holds, so $2 = |2\alpha(x_0)f(x_0) + 2\beta(x_0)f'(x_0)| \le A \cdot \frac{1}{A} + B \cdot |f'(x_0)|$, which implies $\frac{1}{|f'(x_0)|} < B$. Thus, because $x_1 - x_0 = -\frac{f(x_0)}{f'(x_0)}$, the estimate $|x_1 - x_0| < B \cdot \frac{1}{4B^2C} = \frac{1}{4BC}$ holds. In particular, because $|x_0| < N$, the line segment from x_0 to x_1 lies inside the disk for which the estimate $|f'(x) - f'(x_0)| < C \cdot |x - x_0|$ applies, and $|f'(x) - f'(x_0)|$ on this segment is at most $C \cdot \frac{1}{4BC} = \frac{1}{4B}$, which implies (2) in the case $n = 1$. When (2) is used to estimate the modulus of

$$x_2 - x_1 = x_1 - x_0 - \frac{f(x_1) - f(x_0)}{f'(x_0)} = \frac{1}{f'(x_0)} \int_{x_0}^{x_1} (f'(x_0) - f'(x))dx$$

one obtains $|x_2 - x_1| < B|x_1 - x_0| \cdot \frac{1}{2B} = \frac{1}{2}|x_1 - x_0|$, which proves (1) in the case $n = 1$.

Now if (1) holds for all numbers less than n, then $|x_n - x_0| \le |x_n - x_{n-1}| + |x_{n-1} - x_{n-2}| + \cdots + |x_1 - x_0| < \left(\frac{1}{2^{n-1}} + \frac{1}{2^{n-2}} + \cdots + 1\right)|x_1 - x_0| < 2 \cdot \frac{1}{4BC} = \frac{1}{2BC}$. It follows that all values of x on the segment from x_{n-1} to x_n satisfy $|x| < N + 1$, so $|f'(x_0) - f'(x)| < C \cdot \frac{1}{2BC}$, which proves the estimate (2) for n. When this estimate is applied to the formula

$$x_{n+1} - x_n = \frac{1}{f'(x_0)} \int_{x_{n-1}}^{x_n} (f'(x_0) - f'(x))dx$$

it gives (1) for this n (the modulus of the constant in front is less than B and the modulus of the integrand is less than $\frac{1}{2B}$).

By (1), the sequence of complex rational numbers x_0, x_1, x_2, ... satisfies the Cauchy criterion, which is to say that it defines a complex number. More-over, $|x_{n+1} - x_n| = \frac{|f(x_n)|}{|f'(x_0)|}$, so $|f(x_n)| = |f'(x_0)||x_{n+1} - x_n| < \frac{1}{2^n}|f'(x_0)||x_1 - x_0|$ approaches 0 as $n \to \infty$, which is to say that the complex number defined by the sequence is a root of f.

Alternative Proof (See Gauss, [30]). Let $f(x)$ be a monic, irreducible poly-nomial with integer coefficients. If $\deg f$ is odd, a simple bisection argument proves that f must have not just a *complex* root but even a *real* root. (Since $f(N)$ is positive and $f(-N)$ is negative for all sufficiently large N, one can construct an interval on which $f(x)$ changes sign. Repeated bisections—note that each bisection point is rational, so the value of f there can be determined exactly—produce nested intervals that become arbitrarily short on which f changes sign. The sequence of upper ends of these intervals converges, and its limit is a root of f.) The idea of the proof is to use the quadratic formula to reduce the problem of finding a complex root of a given polynomial whose degree is even to the problem of finding a complex root of a polynomial whose degree is divisible fewer times by 2.

Specifically, given a monic, irreducible f with integer coefficients whose degree is divisible ν times by 2, an auxiliary monic, irreducible polynomial with integer coefficients will be constructed with two properties: (1) Its de-gree is divisible fewer than ν times by 2. (2) A complex root of the original polynomial can be constructed if a complex root of the auxiliary polynomial is given.

Let the given monic, irreducible polynomial with integer coefficients be written in the form $f(x) = \prod(x - \rho_j)$ and define $F(u,v) = \prod_{j,k}(u - (\rho_j + \rho_k)v + \rho_j\rho_k)$, where the product is over all pairs (j, k) of indices for which $1 \leq j < k \leq n$. Here the ρ_i are elements of some Galois algebraic number field, so the coefficients of F, being symmetric in the ρ_i, are integers. The aux-iliary polynomial will be $F(u, V)$, where V is an integer for which $F(u, V)$ (a polynomial in u with integer coefficients) is relatively prime to its derivative in the sense that there are polynomials $\alpha(u)$ and $\beta(u)$ with rational coefficients for which $\alpha(u)F(u, V) + \beta(u)\frac{\partial F}{\partial u}(u, V) = 1$. The expression of $f(x)$ in the form $\prod(x - \rho_j)$ of course requires the construction of a splitting field for $f(x)$, but this construction is needed only to provide the rationale for the following con-struction of $F(u, v)$ using computations with symmetric polynomials (Essay 2.4). According to its definition, $F(u, v)$ is the polynomial in u in which the coefficient of $u^{\binom{n}{2}-i}$ is $(-1)^i$ times the ith elementary symmetric polynomial in the $\binom{n}{2}$ expressions $(\rho_j + \rho_k)v - \rho_j\rho_k$, where ρ_j and $\rho_k \neq \rho_j$ are in the splitting field. These are polynomials in v whose coefficients are symmetric polynomials in the ρ_j, so $F(u, v)$ is a polynomial in u and v whose coefficients are symmetric polynomials in the ρ_j. Since any symmetric polynomial is a polynomial with integer coefficients in the *elementary* symmetric polynomials, and since the ith elementary symmetric polynomial in ρ_1, ρ_2, ..., ρ_n is $(-1)^i$ times the coefficient of x^{n-i} in $f(x)$, $F(u,v)$ is expressed in this way as a polynomial in two indeterminates with integer coefficients without using the splitting field in the calculation, except as a rationale.

That an integer V can be chosen in such a way that $F(u,V)$ is relatively prime to its derivative in the sense described can be proved as follows: Again, a splitting field for $f(x)$ is needed to justify the proof but is not needed for the calculations that constitute the proof. For any integer V, the Euclidean algorithm for polynomials with coefficients in $\mathbf{Q}$ can be applied to $F(u,V)$ and $\frac{\partial F}{\partial u}(u,V)$ to find a nonzero polynomial of the form $\Delta(u) = \alpha(u)F(u,V) + \beta(u)\frac{\partial F}{\partial u}(u,V)$ that divides both $F(u,V)$ and $\frac{\partial F}{\partial u}(u,V)$. An integer V is to be found for which this "greatest common divisor" $\Delta(u)$ has degree 0. Of course the computation of $\Delta(u)$ involves only rational numbers, but it can be regarded as a calculation with polynomials whose coefficients are in a splitting field for $f(x)$. When it is regarded in this way, $\Delta(u)$ can fail to be of degree 0 only if it is divisible by at least one of the linear factors $u - (\rho_j + \rho_k)V + \rho_j\rho_k$ of $F(u,V)$ with coefficients in that field. Thus, it can fail to be of degree 0 only if one of the roots $(\rho_j + \rho_k)V - \rho_j\rho_k$ of $F(u,V)$ in this field is also a root of $\frac{\partial F}{\partial u}(u,V)$. But the formula for the derivative of a product implies that $\frac{\partial F}{\partial u}((\rho_j + \rho_k)V - \rho_j\rho_k, V)$ is equal to plus or minus the product of the $\frac{1}{2}\binom{n}{2}\left(\binom{n}{2} - 1\right)$ differences of the roots of $F(u,V)$, or, what is the same, plus or minus the value when $v = V$ of the polynomial that is the product of the $\frac{1}{2}\binom{n}{2}\left(\binom{n}{2} - 1\right)$ linear polynomials in v with coefficients in the splitting field of $f(x)$ that are differences of distinct roots of $F(u,v)$ as a polynomial in u. This product is a polynomial in v of degree $\frac{1}{2}\binom{n}{2}\left(\binom{n}{2} - 1\right)$, which is nonzero, because the factors are all nonzero (if $\rho_j + \rho_k = \rho_m + \rho_n$ and $\rho_j\rho_k = \rho_m\rho_n$, then $(x - \rho_j)(x - \rho_k) = (x - \rho_m)(x - \rho_n)$, which implies when both $j < k$ and $m < n$ that $j = m$ and $k = n$, because f is irreducible and therefore has distinct roots), so it is a nonzero polynomial in v with rational coefficients. The number of roots of such a polynomial is at most equal to its degree, so there is certainly an integer V that is not a root.

If the degree n of $f(x)$ is even—so that $\nu > 0$—the degree $\binom{n}{2}$ of $F(u,V)$ is divisible exactly $\nu - 1$ times by 2. Therefore, at least one of its irreducible factors is divisible fewer than ν times by 2. The Main Theorem will therefore be proved if $F(u,V)$ is shown to have property (2) stated above that a complex root of $F(u,V)$ (a complex root of an irreducible factor of $F(u,V)$ is of course a root of $F(u,V)$) enables one to construct a complex root of $f(x)$.

Since $F(u,v)$ has the special form $F(u,v) = \prod_{i=1}^{n}(\alpha_i + \beta_i u + \gamma_i v)$, the polynomial in three variables $F\left(u + \frac{\partial F}{\partial v}w, v - \frac{\partial F}{\partial u}w\right)$ can be written in the form $F(u,v)\phi(u,v,w)$. In fact, if one defines $Q_i(u,v)$ to be $\frac{F(u,v)}{\alpha_i+\beta_i u+\gamma_i v}$, then $F(u,v) = (\alpha_i + \beta_i u + \gamma_i v)Q_i(u,v)$, so $\frac{\partial F}{\partial u} = \beta_i Q_i + (\alpha_i + \beta_i u + \gamma_i v)\frac{\partial Q_i}{\partial u}$ and $\frac{\partial F}{\partial v} = \gamma_i Q_i + (\alpha_i + \beta_i u + \gamma_i v)\frac{\partial Q_i}{\partial v}$, and a typical factor of $F\left(u + \frac{\partial F}{\partial v}w, v - \frac{\partial F}{\partial u}w\right)$ is $\alpha_i + \beta_i\left(u + \frac{\partial F}{\partial v}w\right) + \gamma_i\left(v - \frac{\partial F}{\partial u}w\right) = \alpha_i + \beta_i u + \gamma_i v + w\left(\beta_i \frac{\partial F}{\partial v} - \gamma_i \frac{\partial F}{\partial u}\right)$. Since

$\beta_i \frac{\partial F}{\partial v} - \gamma_i \frac{\partial F}{\partial u} = (\alpha_i + \beta_i u + \gamma_i v) \left(\beta_i \frac{\partial Q_i}{\partial v} - \gamma_i \frac{\partial Q_i}{\partial u} \right)$, this factor is divisible by $\alpha_i + \beta_i u + \gamma_i v$, and the entire product is divisible by $F(u, v)$. (Explicitly, this calculation proves that the quotient $\phi(u, v, w)$ is $\prod \left(1 + w(\beta_i \frac{\partial Q_i}{\partial v} - \gamma_i \frac{\partial Q_i}{\partial u}) \right)$.)

In the case of the auxiliary polynomial $F(u, V)$ constructed for a given $f(x)$ as above, it follows that if U is a complex root of $F(u, V)$ and if U' and V' are the complex values of $\frac{\partial F}{\partial u}$ and $\frac{\partial F}{\partial v}$, respectively, when $(u, v) = (U, V)$, then $F(U + V'w, V - U'w)$, a polynomial in w with complex coefficients, is identically zero. Since $U' \neq 0$ by the construction of V, one can set $w = \frac{V-x}{U'}$ in this identity to find that $F(U + V' \frac{V-x}{U'}, x)$, a polynomial in x with complex coefficients, is identically zero. The quadratic formula can be applied to the equation $U + V' \frac{V-x}{U'} = x^2$ to find a complex root, call it X. This complex number is a root of $f(x)$, because $0 = F(X^2, X) = \prod_{j,k} (X^2 - (\rho_j + \rho_k)X + \rho_j \rho_k) = \prod_{j,k} (X - \rho_j)(X - \rho_k) = f(X)^{n-1}$.

Thus, if every monic, irreducible polynomial with integer coefficients whose degree is divisible fewer than ν times by 2 has a complex root, the same is true of every monic, irreducible polynomial whose degree is divisible exactly ν times by 2, and the proof is complete.

Both proofs show clearly that the algebraic issues involved in the theorem are handled by the theorem of Essay 1.2 and have nothing to do with complex numbers. The complex numbers enter only because every algebraic number field—which is to say, in the terminology of Essay 2.2, every algebraic field of transcendence degree 0—is isomorphic to a subfield of the complex numbers. However, from a constructive point of view, to regard the field elements as complex numbers instead of algebraic quantities is a mistake, because it replaces quantities for which *exact* computation is possible with quantities for which computation involves *limits*. In short, the so-called fundamental theorem of algebra results when the "Main Theorem" above is used to debase the meaning of the theorem of Essay 1.2.

Essay 5.2 Proof by Contradiction and the Sylow Theorems

The widely held belief that "proofs by contradiction" are "not constructive" is mistaken. One can constructively prove that a certain construction is impossible by proving that *if it were possible, then another construction that is clearly impossible would also be possible.* Perhaps the oldest such proof is the proof that $\sqrt{2}$ is irrational:

Let a unit of length be fixed, and suppose an isosceles right triangle ABC could be given whose sides AB and BC and hypotenuse AC had whole number lengths. One could then construct a smaller isosceles right triangle with the same property in the following way. Let D be the point on the hypotenuse AC for which the length AD is equal to the length AB of the sides of ABC. Let E be the point where the perpendicular to the hypotenuse AC through the point D intersects the side BC. Then CDE would be an isosceles right triangle smaller than ABC whose sides and hypotenuse had whole number lengths, as can be seen in the following way:

The side CD of CDE is the difference of the lengths AC and $AD = AB$, both of which would be whole number lengths, so it would be a whole number length. Moreover, the new triangle CDE is an isosceles right triangle because angle DCE is angle ACB and angle CDE is right, so the new triangle is similar to the original one. Finally, when AE is joined, the right triangles ADE

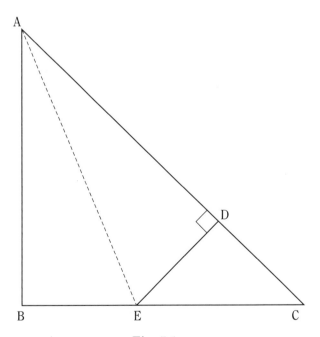

Fig. 5.1.

and ABE are congruent because the hypotenuse and one side are equal, so the lengths DE and EB are equal, and the hypotenuse CE, being the difference of the lengths CB and $EB = ED = CD$ would be a difference of whole number lengths and would therefore be a whole number length.

Thus, if there were such an isosceles right triangle, one would be able to construct an infinite decreasing sequence of positive whole numbers, for example, the whole number lengths of the hypotenuses of the sequence of isosceles right triangles constructed in this way. Since there can be no such sequence, there can be no such isosceles right triangle.

By the Pythagorean theorem, such an isosceles right triangle is the same as a pair of whole numbers s and d for which $d^2 = 2s^2$, so another way to state the same conclusion is to say that no rational number d/s can have 2 as its square.

The main Sylow theorem, Theorem 1 below, can be proved by the same method, which Fermat called the method of **infinite descent**.

Theorem 1. *If d divides the order $|G|$ of a finite group G, and if d is a power of a prime number, then G has a subgroup of order d.*

Lemma. *If the order of an abelian group is divisible by a prime number p, then the group contains an element of order p.*

Note that the lemma is a very special case of Theorem 1.

Proof. Suppose a finite abelian group G is given whose order is divisible by a prime p but that contains no element of order p. It will be shown that one would then be able to construct a *smaller* group of the same type.

Choose an element a of G that is not the identity ($|G|$ is divisible by $p > 1$, so G contains elements other than 1) and let $q > 1$ be the order of a. If p divided q, then $(a^{q/p})^p = 1$ would imply, since G contains no element of order p, that $a^{q/p} = 1$, contrary to the definition of q. Therefore, p does not divide q. Because p is prime, p and q are relatively prime, the Euclidean algorithm implies that some multiple of q is one more than a multiple of p, say $uq = vp + 1$. Let C be the subgroup of G generated by a, and let G' be the quotient group G/C. (Since G is abelian, C is a normal subgroup.) Let b be an element of G that represents an element of G' whose pth power is 1. Then $b^p = a^i$ for some i, and $(a^{iv}b)^p = a^{ivp+i} = a^{iuq} = 1^{iu} = 1$. Since G has no elements of order p, it follows that $a^{iv}b = 1$. Therefore, b is a power of a, which means that it represents the identity element of G'. Therefore, G' contains no element of order p. The order of $|G'|$ is $|G|/q$, which is divisible by p because $|G|$ is divisible by p and q is not. Thus, G', like G, is abelian and has order divisible by p but contains no element of order p. Since $|G'| < |G|$, the impossibility of such a group G follows from the observation that the existence of such a G would imply the existence of an infinite sequence of positive integers $|G| > |G'| > |G''| > \cdots$, which is impossible.

Proof of Theorem 1. As in the proof of the lemma, a counterexample will be shown to imply a smaller counterexample. Thus, suppose G is a counterexample, which is to say that G is a finite group for which there is a power of a prime p^k that divides $|G|$ but that is not the order of any subgroup of G. It is to be shown that from such a group one can construct a smaller such group.

Let Z be the center of G. If p divides $|Z|$, a smaller counterexample can be constructed as follows: In this case, the lemma implies (because Z is abelian) that Z contains an element of order p, call it a. Let C be the subgroup generated by a. Since G has no subgroup of order p^k and C has order p, k must be at least 2. Since C is a normal subgroup, $G' = G/C$ is a group. The order of G' is $|G|/p$, so p^{k-1} divides $|G'|$. If H' is a subgroup of G' of order m, then the elements of G that represent elements of H' form a subgroup of G of order mp. Therefore, since G has no subgroup of order p^k, G' has no subgroup of order p^{k-1}, which shows that G' is a smaller counterexample than G.

If p does not divide $|Z|$, a smaller counterexample can be constructed even more easily. Since G is the union of Z and all the conjugacy classes of G that contain more than one element (Z is the union of the conjugacy classes that contain one element), since $|G|$ is divisible by p, and since $|Z|$ is not divisible by p, the number of elements in at least one conjugacy class with more than one element must not be divisible by p. If g is in such a conjugacy class, then, because the number of elements in the conjugacy class of g is $|G|/|Z(g)|$, where $Z(g)$ is the centralizer of g, $|Z(g)| < |G|$ and p divides $|Z(g)|$ as many times as it divides $|G|$. Thus, p^k divides $|Z(g)|$ and $Z(g)$ has no subgroup of order p^k (because G has no such subgroup), so $Z(g)$ is a smaller counterexample.

Thus, Theorem 1 follows by the principle of infinite descent.

For the sake of completeness, the other Sylow theorems will be deduced from Theorem 1.

Theorem 1 shows that if p is a prime divisor of $|G|$, then G has a subgroup of order p^e, where p^e is the largest power of p that divides $|G|$. Such a subgroup is called a **Sylow p-subgroup** of G.

Theorem 2. *If H is a Sylow p-subgroup of a finite group G, and if K is a subgroup of G whose order is a power of p, then some conjugate of K is contained in H.*

In particular, any two Sylow p-subgroups of G are conjugate.

Theorem 3. *Let s be the number of Sylow p-subgroups of G. Then s divides $|G|$ and $s \equiv 1 \bmod p$.*

(Theorems 2 and 3 are true for primes that do not divide the order of G when one adopts the natural convention that in this case the only Sylow p-subgroup is $\{1\}$. Similarly, Theorem 1 is trivially true for the zeroth power 1 of any prime.)

Proof of Theorem 2. The subgroup K acts on the left cosets of H in G by left multiplication, and this action partitions the left cosets into orbits. (To say that g and g' represent cosets in the same orbit means that kg is in the coset represented by g' for some k in K, or to put it another way, $(kg)^{-1}g'$ is in H for some k in K.) The number of elements in any orbit divides the order of K, which is a power of p, so the number of elements in any orbit is p^i for some $i \geq 0$. The sum over all orbits of these numbers p^i is the number of left cosets of H in G, which, because H is a Sylow p-subgroup, is not divisible by p. Therefore, $p^i = 1$ for at least one orbit. In other words, for at least one g in G, the coset represented by kg is the same as the coset represented by g for all k in K. In short, $g^{-1}kg$ is in H for all k in K. Since $g^{-1}Kg$ is a subgroup conjugate to K, the theorem follows.

Proof of Theorem 3. Let G act on the s Sylow p-subgroups by conjugation. By Theorem 2, this action is transitive: There is just one orbit of size s. Therefore, s divides $|G|$.

Let H be a Sylow p-subgroup of G, and let H act on the s Sylow p-subgroups by conjugation. This action partitions these s subgroups into orbits. Suppose the number of elements in the orbit of H' is 1. Then $hH'h^{-1} = H'$ for all h in H. In other words, the normalizer of H' in G, call it N, contains H. Since N also contains H' and both H and H' are Sylow p-subgroups of N (p can divide $|N|$ no more times than it divides $|G|$), Theorem 2 implies that H and H' are conjugate in N, which is to say that $nH'n^{-1} = H$ for some n in N. But $nH'n^{-1} = H'$ by the definition of N. Therefore, $H' = H$, which shows that only one orbit—the orbit of H—consists of a single element. The number of elements in any orbit divides the order of H and is therefore a power of p. Therefore, all orbits other than the one with 1 element have p^i elements, where $i \geq 1$, from which $s \equiv 1 \bmod p$ follows.

Essay 5.3 Overview of 'Linear Algebra'

So ist es nicht erstaunlich, daß ein großer Teil der modernen algebrais-
chen Lehrbücher sich der abstrakten Richtung angeschlossen hat, welche
im Bereich der Forschung so große Erfolge zu verzeichnen hatte. Jedoch
mehr als einmal hatte ich Gelegenheit zu beobachten, daß dies im Bere-
ich der Lehre nicht durchweg der Fall ist. (It is therefore not surprising
that a great many modern textbooks have followed the abstract direction
that has registered such great successes in the realm of research. However, I have had more than one opportunity to observe that in the realm
of teaching this is not invariably the case.)—N. Chebotarev [8, Author's
preface to the translation]

Some years ago, Sheldon Axler published a book with the audacious title
Linear Algebra Done Right [4]. I was probably more struck by the audacity of
his title than most readers, because only a few years earlier I had published
my own book called *Linear Algebra,* in which the subject had in fact been
done right, but I had never thought to say so in the title.

Of course, Axler's idea of doing it right turned out to have nothing to do
with mine.

Without doubt the most attractive quality of mathematics is its appar-
ent lack of subjectivity. "It must be easy for you mathematicians to grade
papers," my friends often tell me, "because in mathematics there's only one
right answer." In mathematics it can even happen that the student is right
and the teacher wrong, and the teacher can be forced to admit it (usually,
we hope, cheerfully). The other side of this pleasant coin is that mathematics
attracts people who have a great need for certainty and encourages them to
develop into rigidly dogmatic thinkers.

The charge is made against advocates of constructive mathematics—it
was made against Kronecker, against Brouwer, against Bishop—that they are
dogmatists who implacably advocate unreasonably extreme views. But what
distinguishes them from their accusers is neither the extremity of their views
nor the tenacity with which they hold them, but the mere fact that their views
differ from those of their accusers. The feeling on both sides too often is, "I am
not convinced by your arguments because your arguments are unconvincing;
you are not convinced by my arguments because you are dogmatic."

Of course mathematicians feel that mathematics is pure reason and therefore immune to such controversy. But there are plenty of controversies in
mathematics. How else can Axler's and my difference regarding linear algebra
be described? His choice of title is—I assume—intended as a joke, just as I am
joking when I say that my linear algebra book had already done it right. But,
in both cases, not really. And I expect that if you ask the first mathematician
who comes along which of us is right, the reply will be that *both* are wrong,
and the right way to do linear algebra is . . . .

So, having established that it is a mere matter of opinion, let me explain,
if not why I am right, at least why my opinion differs from Axler's. His main

goal is to *avoid determinants,* for the reason that the formula for determinants is difficult to motivate, contrary to the modern style of mathematics and, as Axler shows, avoidable. I agree with him that the formula for determinants is daunting, but I believe that determinants, like a boulder in the path, need to be dealt with, not avoided. They are central to linear algebra—specifically to the solution of systems of linear equations—and the sooner students can be brought to use them and be comfortable with them, the better. My main goal in *Linear Algebra,* by contrast, was to deal with the subject in an *algorithmic* way that I have found through teaching makes sense to students and gives them the tools they need to solve problems in linear algebra. (Also, the book defines determinants without the formula in a natural way that is explained below.)

The early chapters are largely devoted to the following theorem. Let two $m \times n$ matrices of integers be called **equivalent** if one can be transformed into the other by a sequence of steps in which a row is added to or subtracted from an adjacent row or a column is added to or subtracted from an adjacent column.

Theorem. *Given two $m \times n$ matrices of integers, determine whether they are equivalent.*

Let an $m \times n$ matrix of integers be called **strongly diagonal** if it is diagonal (that is, the entry in the ith row of the jth column is zero whenever $i \neq j$), if each diagonal entry is a multiple of its predecessor on the diagonal, and if the diagonal entries are nonnegative, except that the entry in the lower right corner may be negative when the matrix is square. The theorem is proved by giving an algorithm that transforms a given $m \times n$ matrix of integers into a strongly diagonal one and proving that two strongly diagonal matrices are equivalent only if they are equal. In short, the theorem is proved by showing that strongly diagonal form is a canonical form for matrices with respect to this equivalence relation. (Strongly diagonal form is very close to what is often called **Smith normal form** in honor of H. J. S. Smith.)

The algorithm is simple. (In the book it is given in two stages: the rules for reducing to diagonal form in Chapter 2 and the additional rules for reducing to strongly diagonal form in Chapter 5.) The hard part of the proof is the proof that if two *square* diagonal matrices are equivalent, then the products of their diagonal entries are equal. That the *absolute values* of these products are equal is comparatively easy to prove, so the proof comes down to showing that the *signs* are the same. The investigation of the sign of the product of the diagonal entries in equivalent diagonal square matrices motivates the definition of the **determinant** of a square matrix as the product of the diagonal entries of an equivalent diagonal matrix; the main thing to be proved then becomes the theorem needed to make this definition valid, namely, the statement that if two square diagonal matrices are equivalent, then the products of their diagonal entries are the same. In fact, the difficult point of the proof can be put even more starkly: Let J be the strongly diagonal matrix that is the $n \times n$ identity matrix I_n with the last diagonal entry changed to -1. Prove that J is *not* equivalent to I_n.

At first glance, this theorem seems to have little to do with linear algebra as it is generally thought of (vector spaces, linear maps, bases, etc.), but it provides an algorithmic solution to the core problem of linear algebra: Given an $m \times n$ matrix and a column matrix Y of length m, find all column matrices X of length n for which $AX = Y$. In linear algebra courses, the matrices are usually assumed to have real number entries, but the limit process inherent in the notion of a real number has nothing to do with linear algebra per se, and a more reasonable assumption is that the entries are *rational* numbers. The denominators can be cleared in order to translate the problem into one in which A and Y have *integer* entries, and the problem becomes that of finding all solutions X of $AX = Y$ with rational entries (with, naturally, a preference for solutions whose entries are integers).

If matrices A and B are equivalent, the solution of $AX = Y$ is equivalent to the solution of $BX' = Y'$, because the column operations used to transform A into B can be regarded as invertible transformations of X into X', while the row operations are invertible transformations of Y into Y'. Therefore, it suffices to solve $AX = Y$ for *diagonal* matrices, which can be done *by inspection*. For example, $DX = Y$ for a diagonal matrix D (not necessarily square) has a solution X for every Y if and only if D has a nonzero entry in each row, which implies, in particular, that the number m of rows of D is no greater than the number n of columns. Similarly, DX determines X if and only if D has a nonzero entry in each column, which implies, in particular, that $m \geq n$. Thus, the equation $AX = Y$ can be *inverted* to express Y as a function of X only when $m = n$, whether or not A is diagonal. Moreover, a square matrix is invertible if and only if the product of the diagonal entries of an equivalent diagonal matrix (which, at this point of the development, has not yet been shown to be independent of the choice of the equivalent diagonal matrix) is nonzero.

If students found it helpful to think of mathematics in terms of sets and functions, this could all be told to them in the usual way: An $m \times n$ matrix describes a particular kind of function—a linear function—from $\mathbf{Q}^n$ to $\mathbf{Q}^m$. It can be onto only if $n \geq m$ and can be one-to-one only if $n \leq m$. If a linear function is both one-to-one and onto, then its inverse function is a linear function, which is to say that the square matrix of coefficients of the given function has an inverse matrix.

Time after time in teaching the course I have decided that I *surely* could explain these facts of linear algebra in terms of linear functions in a way that the students would find helpful, and time after time the effort has failed. The statement that a function is one-to-one seems indistinguishable from the definition of a function for most students. Confusion about the difference between the statement that f is a function from $\mathbf{Q}^n$ to $\mathbf{Q}^m$ and the statement that f is *onto* $\mathbf{Q}^m$ is compounded by the fact that different mathematicians mean different things when they talk about the "range" of a function. Class discussions bog down in terminological and conceptual issues that have nothing to do with linear algebra. These experiences have convinced me that the set-function conceptualization is *not* helpful for students of linear algebra. Perhaps it will work the other way around—a knowledge of linear algebra may help teach notions of sets and functions—but in my experience it does not work the way it is currently supposed to.

The above theorem and related topics form the substance of the first six chapters. Chapter 7 is on Moore–Penrose generalized inverses. For every $m \times n$ matrix A of rational numbers, there is a unique $n \times m$ matrix of rational numbers B for which AB and BA are both symmetric and the equations $ABA = A$ and $BAB = B$ both hold. This matrix B is the Moore–Penrose generalized inverse of A, or the "mate" of A, as I call it for short. Clearly, if B is the mate of A, then A is the mate of B. The main property of mates is that BY is the best solution, in the least squares sense, of the equation $AX = Y$ for any column matrix Y of length m. (More precisely, when $\|M\|^2$ denotes the sum of the squares of the entries of a matrix M, $\|Y - AX\|^2$ attains its minimum value when $X = BY$ and, among all column matrices X of length n for which this minimum is attained, $X = BY$ is the one for which $\|X\|^2$ is smallest.)

Chapter 8 generalizes the theorem stated above from the case of matrices of integers to the case of matrices of polynomials in one indeterminate x with rational coefficients. In this case, rather than restricting to addition or subtraction of an adjacent row or column, one allows subtraction of any *multiple* of a row or column from an adjacent row or column, where the multipliers are polynomials in x with rational coefficients. (In the case of integer matrices, the two definitions are the same because subtraction of an arbitrary integer multiple can be achieved by repeated additions or subtractions.) The condition that the nonzero diagonal entries of a strongly diagonal matrix must be positive is replaced by the condition that they must be *monic* unless, again, they occur in the lower right corner of a square matrix. Once again, every matrix is equivalent to a strongly diagonal matrix, and strongly diagonal matrices are equivalent only if they are equal. In this way, the problem of determining whether two given matrices are equivalent is solved.

This solution leads to the proof of the following important theorem of intermediate linear algebra: Two $n \times n$ matrices of rational numbers A and B are **similar** if there is an invertible $n \times n$ matrix of rational numbers P for which $A = P^{-1}BP$.

Theorem. *Given two $n \times n$ matrices of rational numbers, determine whether they are similar.*

Proof. It is not difficult to prove* that A is similar to B if and only if $xI - A$ is equivalent to $xI - B$ when both are regarded as matrices whose entries are polynomials in x with rational coefficients. (Here I denotes the $n \times n$ identity matrix.) Therefore, the algorithm of Chapter 8 for solving this latter problem solves the problem of the theorem.

The **characteristic polynomial** of a square matrix A is the determinant of $xI - A$. As follows from what has just been said, similar matrices have the same characteristic polynomial. The converse of this statement is false, as is shown by the fact that the matrix $A = \begin{bmatrix} 1 & 1 \\ 0 & 1 \end{bmatrix}$ is not similar to the 2×2 identity matrix. In fact, the strongly diagonal matrix equivalent to $xI - A$ in this case is $\begin{bmatrix} 1 & 0 \\ 0 & (x-1)^2 \end{bmatrix}$, not $\begin{bmatrix} x-1 & 0 \\ 0 & x-1 \end{bmatrix}$. The **minimum polynomial** of a square matrix A is the last diagonal entry of the strongly diagonal matrix equivalent to $xI - A$. The minimum polynomial of a matrix is easily shown to be the greatest common divisor of the polynomials of which it is a root (when the constant term of the polynomial is interpreted as a multiple of A^0, and A^0 is interpreted as I). For example, in the case of the matrix A just considered, A is a root of $f(x)$ if and only if $f(x)$ is a multiple of $x^2 - 2x + 1$. By the above, similar matrices have the same minimum polynomial, but this necessary condition for two matrices to be similar is still not sufficient. Such considerations lead to the study of the **elementary divisors** of a matrix A, which are certain powers of irreducible polynomials (the elementary divisors of I are $x - 1$ and $x - 1$, while $\begin{bmatrix} 1 & 1 \\ 0 & 1 \end{bmatrix}$ has just one elementary divisor $(x - 1)^2$) whose product is the characteristic polynomial; they are easily described in terms of the strongly diagonal matrix equivalent to $xI - A$, and they do determine the similarity class of A. The elementary divisors are closely related to the **rational canonical form** of a matrix. The **Jordan canonical form** of a matrix, a subject that is much taught, and in my opinion overemphasized, in intermediate linear algebra courses, is the rational canonical form if one works over the complex numbers rather than the rational numbers, or, better, if one works over an algebraic extension of **Q** that splits the characteristic polynomial of the matrix.

A matrix is **diagonalizable** if it is similar to a diagonal matrix, or, what is the same, if its elementary divisors all have degree 1. The methods of Chapter 9 not only make it possible to determine whether a given matrix is diagonalizable, they make it possible, when it is diagonalizable, to construct a similar diagonal matrix. However, this solution of the problem "determine whether a given matrix is diagonalizable" does not solve the problem of "diagonalizing" symmetric matrices in the sense of the spectral theorem because it tells only whether a symmetric matrix of rational numbers is similar to a diagonal matrix of *rational* numbers.

---

* This proof is marred by a misstatement in the first—and so far only—printing of the book. On p. 92, E should be assumed to have polynomial entries and D to have rational number entries.

Chapter 10 is devoted to the (finite-dimensional) spectral theorem, which states that symmetric matrices are similar to diagonal matrices of *real* numbers. In the strict sense, this is not a theorem of linear algebra because it involves *limits* in an essential way: The equivalent diagonal matrix normally contains irrational numbers. This topic warrants an essay of its own.

Essay 5.4 The Spectral Theorem

I remember trying unsuccessfully to concoct a constructive proof of the spectral theorem for symmetric matrices as long ago as 1964. It was only while writing my linear algebra book in the early 1990s that I realized that the *eigenvectors* would follow easily once the *eigenvalues* had been constructed, and that the eigenvalues could be described constructively, in most cases, as places where the characteristic polynomial changed sign. This does not cover the case of multiple eigenvalues, but while I was developing the ideas in Chapter 9 of *Linear Algebra* it became clear to me that the important polynomial is not the *characteristic polynomial* of the symmetric matrix but its *minimum polynomial*, which in the case of a symmetric matrix has no multiple roots. In this way, the geometrically fascinating principal axes theorem for symmetric matrices reappeared as the rather modest assertion that *the minimum polynomial of a symmetric matrix changes sign a number of times equal to its degree*. Once this is known to be true, simple bisection determines all of the eigenvalues as real numbers, and simple linear algebra over a splitting field of the minimum polynomial suffices to determine the corresponding eigenvectors. Once the problem was reduced in this way, Kronecker's work on Sturm's theorem helped me solve it finally to my satisfaction. The solution was included in the last chapter of *Linear Algebra*, where, as far as I know, no one has ever read it. Here it is once again, with a few simplifications and improvements.

Theorem. *Given a symmetric matrix S whose entries are integers, let $f(x)$ be its minimum polynomial, and let m be the degree of $f(x)$. Construct $m + 1$ rational numbers $x_0 < x_1 < x_2 < \cdots < x_m$ with the property that $f(x_{i-1})$ and $f(x_i)$ have opposite signs for $i = 1, 2, \ldots, m$.*

Proof. If $m = 1$, then $f(x) = x + c$, and one can simply set $x_0 = -N$ and $x_1 = N$ for a sufficiently large number N. Assume, therefore, that $m > 1$.

The function* $\operatorname{tr}(g(S)h(S))$, defined for pairs of polynomials $(g(x), h(x))$ with rational coefficients, can be regarded as a symmetric bilinear function from $\mathbf{Q}[x] \bmod f(x)$ to $\mathbf{Q}$. If $h(x)$ has the property that $\operatorname{tr}(g(S)h(S)) = 0$ for all polynomials $g(x)$, then $h(x) \equiv 0 \bmod f(x)$, because $\operatorname{tr}(h(S)^2)$ is the sum of the squares of the entries of the matrix $h(S)$ (because S and therefore $h(S)$ are symmetric), so the case $g = h$ of $\operatorname{tr}(g(S)h(S)) = 0$ implies $h(S) = 0$. Therefore, the $m \times m$ matrix of rational numbers that represents this symmetric bilinear form with respect to the basis $1, x, x^2, \ldots, x^{m-1}$ of $\mathbf{Q}[x] \bmod f(x)$ is invertible, from which it follows that every linear form $\mathbf{Q}[x] \bmod f(x) \to \mathbf{Q}$ can be expressed as $g(x) \mapsto \operatorname{tr}(g(S)h(S))$ for some polynomial $h(x)$ with rational coefficients. In this way, a polynomial $h(x)$ with rational coefficients can be constructed for which $\operatorname{tr}(g(S)h(S)) = g_1$ for any

---

* Here the trace $\operatorname{tr}(M)$ of a square matrix M is of course the sum to its diagonal entries.

polynomial $g(x) = g_1 x^{m-1} + g_2 x^{m-2} + \cdots + g_m$ in which the g_i are rational numbers.

If it is stipulated that $\deg h < m$, this property determines h when $f(x)$ is given. When $h(x)$ is defined to be the polynomial determined in this way, $h(x)$ is relatively prime to $f(x)$, because a common divisor $d(x)$ of $f(x)$ and $h(x)$ of positive degree, say $f(x) = q_1(x)d(x)$ and $h(x) = q_2(x)d(x)$, where $\deg d > 0$, would imply $\deg q_1 < m$, say $\deg q_1 + t = m$, where $t > 0$, so $x^{t-1}q_1(x)$ would be a polynomial of degree $m-1$ and $\mathrm{tr}\,(S^{t-1}q_1(S)h(S))$ would be nonzero, contrary to $S^{t-1}q_1(S)h(S) = S^{t-1}q_1(S)q_2(S)d(S) = S^{t-1}q_2(S)f(S) = 0$. Therefore, $h(x)$ is invertible mod $f(x)$.

Let $s_m(x) = f(x)$, let s_{m-1} be the unique inverse of $h(x)$ mod $f(x)$ whose degree is less than m, and let later terms of the sequence $s_m(x)$, $s_{m-1}(x)$, $s_{m-2}(x)$, $\ldots$, $s_k(x)$ be defined by defining $s_i(x)$ to be the negative of the remainder when $s_{i+2}(x)$ is divided by $s_{i+1}(x)$. The sequence terminates with the last nonzero term $s_k(x)$ generated in this way. It will be shown that each s_i has the form $s_i(x) = s_i^{(i)} x^i + s_{i-1}^{(i)} x^{i-1} + \cdots + s_0^{(i)}$, where the first coefficient $s_i^{(i)}$, call it c_i, is positive. In particular, $s_1(x)$ has degree 1 with a positive leading coefficient, and the final nonzero term s_0 is a positive constant.

For any $g(x) = g_1 x^{m-1} + g_2 x^{m-2} + \cdots + g_m$, the identity

$$\mathrm{tr}\,(g(S)s_{m-1}(S)h(S)^2) = g_1$$

follows from the definitions of $h(x)$ and $s_{m-1}(x)$, because $s_{m-1}(S)h(S) = I$. Use of $s_{m-2}(x) = -s_m(x) + q_{m-1}(x)s_{m-1}(x)$, where $q_{m-1}(x)$ is the quotient in the division that defines $s_{m-2}(x)$, in $\mathrm{tr}\,(g(S)s_{m-2}(S)h(S)^2)$ gives $-0 + \mathrm{tr}\,(g(S)q_{m-1}(S)s_{m-1}(S)h(S)^2)$, which is the coefficient of x^{m-1} in $g(x)q_{m-1}(x)$, provided $g(x)q_{m-1}(x)$ has degree less than m. Since

$$s_m(x) = x^m + \cdots \quad \text{and} \quad s_{m-1}(x) = mx^{m-1} + \cdots$$

(the latter because $\mathrm{tr}\,(s_{m-1}(S)s_{m-1}(S)h(S)^2) = \mathrm{tr}\,(I) = m$), it is clear that $q_{m-1}(x)$ has degree 1 and leading coefficient $1/m$, so

$$\mathrm{tr}\,(g(S)s_{m-2}(S)h(S)^2) = g_1/m$$

whenever $g(x) = g_1 x^{m-2} + \cdots$. The general case of this formula, which states that

(1) $$\mathrm{tr}\,(g(S)s_i(S)h(S)^2) = g_1/c_{i+1}$$

(c_{i+1} is the leading coefficient of $s_{i+1}(x)$), where $g(x) = g_1 x^i + \cdots$ is a polynomial whose degree is at most i, will now be proved.

Note first that the case $i = m - 2$ already proved implies, since $s_{m-2}(x)$ has degree at most $m - 2$, that the coefficient c_{m-2} of x^{m-2} in $s_{m-2}(x)$ satisfies $c_{m-2}/m = \mathrm{tr}\,((s_{m-2}(S)h(S))^2)$, which is the sum of the squares of the entries of $s_{m-2}(S)h(S)$ and is therefore positive unless $s_{m-2}(S)h(S) = 0$; thus c_{m-2} is positive, because $h(S)$ is invertible and $s_{m-2}(S) = 0$ would imply

$\operatorname{tr}(g(S)s_{m-2}(S)h(S)^2) = 0$ for all $g(x)$, but for $g(x) = x^{m-2}$ this trace is $1/m$. Therefore, the assertion that $s_i(x)$ has degree i and positive leading coefficient is proved for $i = m, m-1, m-2$.

Similarly, if this assertion and (1) are proved for both $i+2$ and $i+1$, say $s_{i+2}(x) = c_{i+2}x^{i+2} + \cdots$ and $s_{i+1}(x) = c_{i+1}x^{i+1} + \cdots$, the same method proves them for i, provided $i \geq 0$, because if $g(x) = g_1 x^i + \cdots$, then

$$\operatorname{tr}(g(S)s_i(S)h(S)^2) = -\operatorname{tr}(g(S)s_{i+2}(S)h(S)^2) + \operatorname{tr}(g(S)q_{i+1}(S)s_{i+1}(S)h(S)^2)$$

$$= -0 + \left(g_1 \frac{c_{i+2}}{c_{i+1}}\right)\frac{1}{c_{i+2}} = \frac{g_1}{c_{i+1}},$$

because $g(x)q_{i+1}(x) = \frac{g_1 c_{i+2}}{c_{i+1}} \cdot x^{i+1} + \cdots$. In particular, $s_i(S) \neq 0$. The case $g(x) = s_i(x)$ of this identity implies that c_i/c_{i+1} is the positive rational number $\operatorname{tr}((s_i(S)h(S))^2)$, so $s_i(x)$ has degree i and a positive leading coefficient, as was to be shown.

A polynomial of degree i cannot change sign more than i times, as can be seen as follows: If $F(x)$ is a polynomial, if a and b are rational numbers for which $F(a) < F(b)$, and if c is the midpoint of the interval $[a, b]$, then the rate of increase of F on the interval is the average of the rate of increase on the two halves of the interval $[a, c]$ and $[c, b]$, which is to say that

$$\frac{F(b) - F(a)}{b - a} = \frac{1}{2}\left(\frac{F(c) - F(a)}{c - a} + \frac{F(b) - F(c)}{b - c}\right),$$

as follows from $c - a = b - c = \frac{1}{2}(b - a)$. Select the half interval on which the rate of increase is larger, or, if the two rates are the same, select the half interval on the right. Iteration of this bisection and selection rule determines a nested sequence of subintervals of $[a, b]$, each half as long as its predecessor, and therefore determines a real number (their "intersection"). Since the derivative* $F'(x)$ of $F(x)$ at this real number is the limit of a nondecreasing sequence of positive rational numbers (namely, of the values of $(F(b) - F(a))/(b - a)$, where a and b are the endpoints of the successive intervals), the real number determined by the nested intervals is one at which $F'(x)$ is positive. In this way, any interval on which $F(x)$ increases contains a real number at which $F'(x)$ is positive, and therefore contains a rational number at which $F'(x)$ is positive. Similarly, any interval on which F decreases contains a rational number at which $F'(x)$ is negative. Therefore, if a polynomial with rational coefficients of degree i changes sign at least $\sigma > 0$ times, then $i > 0$, and its derivative is a polynomial of degree $i - 1$ that changes sign at least $\sigma - 1$ times. Repetition of this argument σ times gives a polynomial of degree $i - \sigma$, so $i \geq \sigma$.

Moreover, if i nonoverlapping intervals on which a polynomial of degree i changes sign are given, bisection of each of them (but moving the midpoint

---

* The derivative of $F(x)$ is of course the coefficient of h in the polynomial $F(x + h) - F(x) = F'(x)h + \cdots$, where the omitted terms all contain h^2.

slightly if it happens to be a root of the polynomial) gives $2i$ nonoverlapping intervals; the polynomial changes sign on at least i of them and, as was just shown, on no more than i of them. Therefore, repetition of the bisection process constructs i real roots of the polynomial and shows that *the values of such a polynomial for two rational numbers a and b have opposite signs if and only if the interval $[a, b]$ contains an odd number of its i real roots,* provided neither a nor b is a root.

For each $i = 1, 2, \ldots, m$, $s_i(x)$ *has i real roots, and the number of real roots of $s_i(x)$ that are greater than a given one of them is equal to the number of real roots of $s_{i-1}(x)$ that are greater than it,* as can be proved inductively as follows:

This statement is obviously true in the case $i = 1$, because a polynomial of degree 1 has just one root, and no roots of s_0 are greater than it. Suppose now that it is true for a given i, and let $\rho_1, \rho_2, \ldots, \rho_i$ be the real roots of $s_i(x)$, in ascending order. By the inductive hypothesis, the number of real roots of $s_{i-1}(x)$ greater than ρ_j is $i - j$, so, since $s_{i-1}(x)$ is positive for all sufficiently large values of x, $s_{i-1}(\rho_j)$ has the sign $(-1)^{i-j}$. (A polynomial of degree $i - 1$ has at most $i - 1$ roots, counted with multiplicities, so the roots of $s_{i-1}(x)$ are *simple* and $s_{i-1}(x)$ changes sign at each root.) Since the formula $s_{i+1}(x) + s_{i-1}(x) = q_i(x)s_i(x)$ implies* that $s_{i+1}(x)$ and $s_{i-1}(x)$ have opposite signs at a root of $s_i(x)$, it follows that the sign of $s_{i+1}(x)$ at ρ_j is $(-1)^{i-j+1}$. Because the leading coefficient of $s_{i-1}(x)$ is positive, when ρ_{i+1} is chosen to be a large enough number and when $-\rho_0$ is chosen to be a large enough number, the same rule describes the sign of $s_{i+1}(x)$ at all of the real numbers $\rho_0, \rho_1, \ldots, \rho_{i+1}$. Since these signs alternate, it follows that $s_{i+1}(x)$ changes sign $i + 1$ times, and therefore has $i + 1$ real roots. Moreover, the jth one of these roots lies in the jth interval where $s_{i+1}(x)$ changes sign, which places it between ρ_j and ρ_{j+1} and shows that the number of roots of $s_{i+1}(x)$ greater than a given root of $s_{i+1}(x)$ is the number of ρ's greater than it, as was to be shown.

Thus, sufficiently close rational approximations to the roots of $s_{m-1}(x)$, together with a pair of values $\pm N$ for a large number N, demonstrate m changes of sign in $s_m(x) = f(x)$, as required.

Corollary (The spectral theorem). *Given a symmetric matrix S whose entries are integers, find real numbers $\rho_1, \rho_2, \ldots, \rho_m$ and symmetric matrices of real numbers $P_1, P_2, \ldots, P_m$ that satisfy*

$$S = \rho_1 P_1 + \rho_2 P_2 + \ldots + \rho_m P_m,$$

---

* A common divisor of $s_i(x)$ and $s_{i-1}(x)$ divides all of the $s_j(x)$, and therefore divides the nonzero constant s_0, and must therefore be a nonzero constant. Therefore, there are polynomials $\alpha(x)$ and $\beta(x)$ with rational coefficients for which $\alpha(x)s_i(x) + \beta(x)s_{i-1}(x) = 1$. If ρ is a real root of $s_i(x)$, then $1 = \beta(\rho)s_{i-1}(\rho)$, which implies that the real number $s_{i-1}(\rho)$ is nonzero. Similarly, a root of $s_{i-1}(x)$ is not a zero of $s_i(x)$. Thus, when ρ is a root of $s_i(x)$, the equation $s_{i+1}(\rho) + s_{i-1}(\rho) = 0$ implies that $s_{i+1}(\rho)$ and $s_{i-1}(\rho)$ have opposite signs.

$$I = P_1 + P_2 + \cdots + P_m,$$

and

$$P_i P_j = \begin{cases} P_i & \text{if } i = j, \\ 0 & \text{otherwise.} \end{cases}$$

Deduction. All that is needed is to construct the matrix that is orthogonal projection on the eigenspace corresponding to each eigenvalue ρ_i, which is to say orthogonal projection on the kernel of $S - \rho_i I$ for all roots ρ_i of the minimum polynomial of S. (To say that a matrix is an orthogonal projection means that it is symmetric and idempotent.) Each of these orthogonal projections is easy to find, because the orthogonal projection on the kernel of a symmetric matrix M is $I - Q$, where Q is orthogonal projection on the image of M, which is to say that Q is M multiplied on the right by the Moore–Penrose generalized inverse of M. Therefore, the spectral decomposition of S can be given once the Moore–Penrose generalized inverses of the matrices $I - S + \rho_i I$ are found. Note that the computation of the Moore–Penrose generalized inverse of a matrix requires *exact* computations with the entries, so it becomes possible only after a splitting field for the minimum polynomial is constructed; the interpretation of the ρ's and P's as real numbers requires an identification of the splitting field with a subfield of the field of real numbers.

Essay 5.5 Kronecker as One of E. T. Bell's "Men of Mathematics"

Kronecker laid himself out in 1891 to criticize Cantor's work to his students in Berlin, and it became clear that there was no room for them both under one roof. As Kronecker was already in possession, Cantor resigned himself to staying out in the cold.—E. T. Bell, *Men of Mathematics*, p. 570.

Discussing Lindemann's proof that π is transcendental, Kronecker asked, "Of what use is your beautiful investigation regarding π? Why study such problems, since irrational [and hence transcendental] numbers do not exist?"—ibid., p. 568.

It is a mistake to take Eric Temple Bell's book *Men of Mathematics* too seriously. Bell set out to write a popular book about the history of mathematics, and he succeeded admirably. From its publication in 1937 until today, the book has amused and inspired several generations of amateur and professional mathematicians, including mine. His outrageousness is part of his winning style. But in spicing up his stories he did create distortions, extrapolations, and outright falsehoods that have since become common "knowledge."

On the whole, the picture Bell paints of Kronecker is not negative. "His skepticism was his greatest contribution," the table of contents says of Kronecker, and at many points Kronecker is made the respectable spokesman for those who objected to the growing use of the transfinite. Kronecker would probably have preferred to have less about his philosophy and more about his mathematical achievements in the book, but his importance and his contributions are certainly not slighted. Unfortunately, the word "vicious" is used more than once to describe his criticism of others, principally Weierstrass and Cantor, and, as I have said in the Preface, I do not feel this word is justified.

I have recently come to understand what lies behind the two statements of Bell that are quoted above, and in explaining them in this essay I hope to shed some light on Kronecker and his ideas, as well as on Bell, his methods, and his lack of credibility.

In the first quotation Bell implies that it was one thing to criticize professional colleagues and quite another to do it in front of students. That was surely Cantor's view of it when he bitterly complained to W. Thomé in a letter dated 21 September 1891 that Kronecker in public lectures had told his "immature audience" that Cantor's work was "mathematical sophistry" [56, document 38].

Bell was undoubtedly referring to this letter of Cantor's when he wrote of Kronecker's 1891 criticism of Cantor, but I doubt that his evidence of the alleged criticism was as reliable as that contained in the transcript of Kronecker's 1891 lectures that was recently published [45]. Despite Cantor's claim that Kronecker had denounced a specific work of his, the published version of Kronecker's lectures does not even mention Cantor, much less the specific work Cantor cites. The word "sophistry" is indeed used [45, p. 247] in connection with a transformation of ν-dimensional space into a space of some other dimension, which Cantor might reasonably imagine to be a reference to his work, but, as the editors of the lectures point out, the lectures were given just one year after Peano published his famous curve that fills an area of the plane.

It is of course possible that Cantor knew more than we do about Kronecker's actual words; he says in the letter that he had obtained a copy of the lectures by chance, but nothing of the sort is to be found among his surviving papers. However, it is equally possible that he was overreacting to a simple statement of opinion that may not even have been directed at him and that had, in the version that has survived, no tinge of personal animosity. For his part, Cantor says Kronecker's "entire course of lectures is a muddled and superficial mix of undigested ideas, boasts, unmotivated name-calling, and rotten jokes."* If Bell's impression of Kronecker's remarks came from this characterization of them it is easy to see why he used the word "vicious," but the surviving version of the lectures in no way deserves Cantor's description.

Finally, before Kronecker's hostility can be taken, as Bell does take it, to be the cause of Cantor's spending his entire career at Halle instead of being called to Berlin, one must show that somewhere there was someone who felt Cantor was a qualified and desirable candidate for appointment at Berlin. The fact is that Kronecker died in the very year 1891 that these lectures were given, so there was very soon no question of their needing to live "under one roof." Weierstrass survived Kronecker, and when Weierstrass died he was succeeded by H. A. Schwarz, no friend of Kronecker's views, but I am unaware of any effort to bring Cantor to Berlin.

When, as a graduate student, I first read the passage in Bell's book about Kronecker's attitude toward π, I think I was as indignant as Bell wanted me to be with the claim that π might not "exist." Years later, when I encountered the same anecdote in Constance Reid's book *Hilbert,* I had come to take a great interest in Kronecker and his ideas, so this time rather than being indignant, I was puzzled and unsure that the anecdote was authentic. Neither Bell nor Reid cites a source.

Kronecker's works show no inhibition about the use of π. His papers on analytic number theory and those on elliptic functions are full of π's. In the first lecture in his course of lectures on number theory [44], he refers without apology to "the transcendental number π from geometry" and notes that it can be defined by $\frac{\pi}{4} = 1 - \frac{1}{3} + \frac{1}{5} - \frac{1}{7} + \cdots$. I am not aware that he ever expressed any reservations about any particular transcendental number. What he had reservations about was the notion that the totality of real numbers could be treated as a mathematical entity. Why would he have had any reservations about the "existence" of π, or even about the meaningfulness of Lindemann's

---

* *Die ganze Vorlesung ist ein wirres oberflächliches Gemisch von unverdauten Ideen, Prahlereien, unmotivierten Schimpfereien und faulen Witzen.*

theorem that π was transcendental? On the other hand, Bell's quotation of Kronecker could not be a mere invention.

A few years ago, I found what I took to be Bell's source in Florian Cajori's *History of Mathematics* [7], where Cajori writes, "[Kronecker] once paradoxically remarked to Lindemann: 'Of what use is your beautiful research on the number π? Why cogitate over such problems, when really there are no irrational numbers whatever?' " But Cajori gives no source either. His book was originally published in 1894, which would put it only three years away from Kronecker and make plausible the hypothesis that Cajori learned the story through word of mouth. Only much later did I read carefully the copyright page of the Chelsea reprint edition I had. There was a "second, revised and enlarged edition" in 1919, of which the Chelsea edition was a reprint. When I finally tracked down a copy of the 1894 edition in microfiche, I learned that the story of Kronecker and π had been added in 1919, twenty-eight years after Kronecker's death.

Then, in June of 2003, while writing to Professor David Rowe of Mainz about a different question, I had the happy thought to ask him whether he knew where Cajori might have heard the story. By return e-mail he was able give me what seems certain to be the correct source.[*] In 1904, Teubner published a German translation of Poincaré's *Science et Hypothèse* annotated by Lindemann [54]. One of Lindemann's notes—note (4) to page 20—cites Kronecker's advocacy of restating the theory of algebraic quantities entirely in terms of the theory of polynomials with integer coefficients (see Essay 1.1) and goes on to say:

> Später ging Kronecker noch weiter, indem er die Existenz irrationaler Zahlen leugnete; so sagte er mir in seiner lebhaften und zu Paradoxen geneigten Art einmal: "Was nützt uns Ihre schöne Untersuchung über die Zahl π? Wozu das Nachdenken über solche Probleme, wenn es doch gar keine irrationalen Zahlen gibt?" (Later, Kronecker went even further and denied the existence of irrational numbers; thus, he once said to me in his lively and paradoxical way, "Of what use to us are your beautiful researches about the number π? Why consider such problems when in fact there are no irrational numbers?")

Since Lindemann published this more than ten years after Kronecker's death, we are entitled to take his quotation marks with a grain of salt. Clearly, Lindemann felt that Kronecker was teasing him—not unkindly it would seem in view of the appearance of the word "beautiful"—but Kronecker's exact words, which are essential to an understanding of the underlying criticism, would probably have been difficult for Lindemann to recall after ten minutes, not to mention ten years.

---

[*] Their different English versions of the alleged quotation suggest that Bell may have based his telling of the story directly on Lindemann, not on Cajori's retelling.

That Kronecker would prefer to state Lindemann's result in a way that made no reference to the totality of transcendental numbers comes as no surprise. If his meaning was simply that he would prefer to state the result in a form something like, "For any polynomial $f(x)$ in one variable with integer coefficients the sequence of rational numbers $f(1 - \frac{1}{3} + \frac{1}{5} - \frac{1}{7} + \cdots + \frac{1}{4n+1})$ can be bounded away from zero," no one would be scandalized, and the statement could not be used to ridicule Kronecker's views. I certainly do not claim to know what the point of Kronecker's criticism might have been—it could have been the form in which Lindemann stated his result or the methods he used or many things in between—but it seems certain to me that he would have regarded the result as having meaning and even as having considerable interest.

For the fun of it, we can indulge Bell in his extravagant caricatures of our mathematical forebears, but we should be careful not to let them affect our understanding of the history of our subject. In particular, we should not let Kronecker's role as Bell's gadfly obscure his true-life role as a great mathematician whose works are classics.

References

[1] N. H. Abel, *Mémoire sur une propriété générale d'une classe très-étendue de fonctions transcendantes*, Mémoires présentés par divers savants à l'Académie des sciences, Paris, 1841, Oeuvres Complètes, vol. 1, 145-211, (4.1).

[2] N. H. Abel, *Sur la résolution algébrique des équations*, Oeuvres Complètes, vol. 2, pp. 217–243, (4.5).

[3] Aristotle, *The Physics*; P. H. Wicksteed and F. M. Cornford, translators, Harvard Univ. Press, Cambridge, Mass., 1957, (5.1)

[4] S. Axler, *Linear Algebra Done Right*, Springer-Verlag, New York, 1996, (5.3).

[5] E. T. Bell, *Men of Mathematics*, Simon and Schuster, New York, 1937, 1962, (5.5).

[6] N. Bourbaki, *Éléments d'Histoire des Mathématiques, 2nd edition*, Hermann, Paris, 1969 (4.3).

[7] F. Cajori, *History of Mathematics*, Macmillan, New York, 1894, 1919, Chelsea Reprint 1980, 1985, Landmarks of Science Microform (5.5)

[8] N. G. Chebotarev (Tschebotaröw), *Grundzüge der Galois'schen Theorie*, Noordhoff, Groningen-Djakarta, 1950, Translation of *Osnovie Teorii Galua*, Gosudarstvennoe Techniko-Teoreticheskoe-Isdatelstvo, Moscow–Leningrad, 1934–1937, (Synopsis, 1.7, 5.3).

[9] N. G. Chebotarev, *Newton's Polygon and its Role in the Present Development of Mathematics (Russian)*, Isaac Newton, 1643–1727 (S. I. Vavilova [transliterated Wawilow on the English version of the title page], ed.), Izdatelstvo Akademii Nauk, Moscow–Leningrad, 1943, pp. 99–126, (4.4).

[10] H. T. Colebrooke, *Algebra, with Arithmetic and Mensuration, from the Sanscrit of Brahmegupta and Bhascara*, J. Murray, London, 1817, (3.1).

[11] Gabriel Cramer, *Introduction à l'analyse des lignes courbes algébriques*, Frères Cramer, Geneva, 1750, Landmarks of Science Microform (4.4).

[12] R. Dedekind, *Über einen arithmetischen Satz von Gauss*, Mitt. Deut. Math. Ges. Prag, (1892), 1–11, Werke, vol. 2, 28–38. (2.5).

[13] R. Dedekind, *Über die Begründung der Idealtheorie*, Nachr. Kön. Ges. Wiss. Göttingen, (1895), 106–113, Werke, vol. 2, 50–58. (2.5).

[14] R. Dedekind and H. Weber, *Theorie der algebraischen Funktionen einer Veränderlichen*, Jour. für Math., **92** (1882), 181–290, Dedekinds Werke, vol. 1, 238–349. (4.5, 4.7).

[15] L. E. Dickson, *History of the Theory of Numbers*, Carnegie Institute, Washington, 1920, Chelsea reprint, 1971 (3.1).

[16] P. G. L. Dirichlet, *Vorlesungen über Zahlentheorie* (R. Dedekind, ed.), Vieweg, Braunschweig, 1863, 1871, 1879, 1894, Chelsea reprint, 1968, (3.3).

[17] H. M. Edwards, *Euler and Quadratic Reciprocity*, Mathematics Magazine, **56** (1983), 285–291, (3.5).

[18] H. M. Edwards, *Galois Theory*, Springer-Verlag, New York, 1984 (1.7, 1.9, 2.1, 2.3, 2.4).

[19] H. M. Edwards, *Divisor Theory*, Birkhäuser, Boston, 1990, (2.5).

[20] H. M. Edwards, *Linear Algebra*, Birkhäuser, Boston, 1995, (5.3, 5.4).

[21] H. M. Edwards, *Kronecker on the Foundations of Mathematics*, From Dedekind to Gödel (Jaakko Hintikka, ed.), Kluwer, 1995, pp. 45–52, (Preface, 1.1).

[22] H. M. Edwards, *Kronecker's Fundamental Theorem of General Arithmetic*, Proceedings of a conference held at MSRI, Berkeley, in April 2003, (to appear) (Synopsis).

[23] H. M. Edwards, O. Neumann and W. Purkert, *Dedekinds "Bunte Bemerkungen" zu Kroneckers "Grundzüge"*, Arch. Hist. Exact Sci., **27** (1982), 49–85, (2.5).

[24] F. Engel, *Eduard Study*, Jahres. der DMV **40** (1931), (4.2).

[25] Euclid, *The Thirteen Books of Euclid's Elements*, T. L. Heath, translator and editor, 2nd Edition, Cambridge Univ. Press, 1925, Dover reprint, 1956, (1.2, 1.4, 3.1).

[26] L. Euler, *Observationes de Comparatione Arcuum Curvarum Irrectificabilium*, Novi Comm. acad. sci. Petropolitanae, **6** (1761), 58–84, Opera, ser. 1, vol. 21, pp. 80–107, Eneström listing 252, (4.2)

[27] É. Galois, *Mémoire sur les conditions de résolubilité des équations par radicaux*, J. Math. Pures et Appl., **11**, 1846, 381–444, see [18] for other citations and for English translation. (Synopsis, 1.2, 1.9, 2.1, 2.3, 2.4).

[28] C. F. Gauss, *Disquisitiones Arithmeticae*, Braunschweig, 1801, (Synopsis, 2.5, 3.3, 3.4, 3.5, 3.6, 3.7).

[29] C. F. Gauss, *Demonstratio Nova Theorematis Omnem Functionem Rationalem Integram Unius Variabilis in Factores Reales Primi vel Secundi Gradus Resolvi Posse* (1799), Helmstadt, Werke, vol. 3, pp. 1–30, (5.1).

[30] C. F. Gauss, *Demonstratio Nova Altera Theorematis Omnem Functionem Rationalem Integram Unius Variabilis in Factores Reales Primi vel Secundi Gradus Resolvi Posse*, Comm. soc. reg. sci. Gottingensis (1815), Werke, vol. 3, 31–56, (5.1).

[31] K. Hensel and G. Landsberg, *Theorie der algebraischen Funktionen einer Variabeln*, Leipzig, 1902, Chelsea Reprint, 1965 (4.4).

[32] O. Hölder, *Über den Casus Irreducibilis*, Math. Annalen, **38** (1891), 307–312, (1.7).

[33] A. Hurwitz, *Über die Theorie der Ideale*, Nachr. kön. Ges. Wiss. Göttingen (1894), 291–298, Werke, vol. 2, 191–197 (2.5).

[34] A. Hurwitz, *Über einen Fundamentalsatz der arithmetischen Theorie der algebraischen Größen*, Nach. kön. Ges. Wiss. Göttingen (1895), 230–240, Werke, vol. 2, 198-207 (2.5).

[35] A. N. Kolmogorov and A. P. Yuskevich (ed.), *Mathematics in the 19th Century (Russian)*, Nauk, Moscow, 1978, English translation by A. Shenitzer, Birkhäuser, 1992 (1.2).

[36] A. Kneser, *Über die Gattung niedrigster Ordnung ...*, Math. Annalen, **30** (1887), 179–202, (1.7).

[37] L. Kronecker, *Über die verschiedenen Sturm'schen Reihen und ihre gegenseitigen Beziehungen*, Monatsber. Akad. Wiss. Berlin (1873), 117–154, Werke, I, 303–348, (2.4).

[38] L. Kronecker, *Über die Discriminante algebraischer Functionen einer Variablen*, Jour. für Math., **91** (1881), 301–334, Werke, II, 193–236, (4.3).

[39] L. Kronecker, *Grundzüge einer arithmetischen Theorie der algebraischen Größen*, Jour. für Math. **92** (1882), 1–122, Werke, II, 237–388, (1.1, 1.4, 1.5, 1.7, 2.2, 2.4, 4.5).

[40] L. Kronecker, *Die Zerlegung der ganzen Grössen eines natürlichen Rationalitäts-bereichs in ihre irreductibeln Factoren*, Jour. für Math. **94** (1883), 344–348, Werke, II, 409–416, (1.4).

[41] L. Kronecker, *Zur Theorie der Formen höhere Stufen*, Monatsber. Akad. Wiss. Berlin (1883), 957–960, Werke, II, 419-424, (2.5).

[42] L. Kronecker, *Ein Fundamentalsatz der allgemeinen Arithmetik*, Jour. für Math. **100** (1887), 490–510, Werke, IIIa, 209–240, (1.1, 1.2, 1.8).

[43] L. Kronecker, *Über den Zahlbegriff*, Jour. für Math. **101** (1887), 260–272, Werke, IIIa, 249–274, (1.1, 1.7).

[44] L. Kronecker, *Vorlesungen über Zahlentheorie*, Teubner, Leipzig, 1901, Reprint, Springer, New York, 1978 (3.2, 5.5).

[45] L. Kronecker, *Über den Begriff der Zahl in der Mathematik (Sur le concept de nombre en mathématiques)*, Retranscribed and annotated by J. Boniface and N. Schappacher, Revue d'histoire des mathématiques, **7** (2001), 207–275, (1.1, 5.5).

[46] E. E. Kummer, *Zur Theorie der complexen Zahlen*, Jour. für Math. **35** (1847), 319–326, Collected Papers, 1, 203-210, (3.6).

[47] E. E. Kummer, *Über die allgemeinen Reciprocitätsgesetze unter den Resten und den Nichtresten*, Math. Abh. Kön. Akad. Wiss. Berlin, 1859, Collected Papers, 1, 699–839, (3.5, 3.6).

[48] J. L. Lagrange, *Additions* to Euler's *Algebra*, republished in vol. 7 of Lagrange's *Oeuvres* and vol. 1 (1) of Euler's *Opera* (3.3).

[49] H. W. Lenstra, Jr., *Solving the Pell Equation*, AMS Notices, **49** (2002), pp. 182–192, (3.1).

[50] A. Loewy, *Algebraische Gleichungen mit reelen Wurzeln*, Math. Zeitschrift **11** (1921), 108–114, (1.7).

[51] I. Newton, *The Mathematical Papers of Isaac Newton* (D. T. Whiteside, ed.), Cambridge Univ. Press, 1969, (4.4).

[52] O. Ore, *Neils Henrik Abel, Mathematician Extraordinary*, Univ. of Minnesota, Minneapolis, 1957, Chelsea reprint, 1974 (4.1).

[53] H. Poincaré, *L'Oeuvre Mathématique de Weierstrass*, Acta Mathematica **22** (1899), 1–18, (Preface).

[54] H. Poincaré, *Wissenschaft und Hypothese* (F. Lindemann, ed.), Teubner, Leipzig, 1904, (5.5).

[55] *Princeton University Bicentennial Conferences, Series 2, Conference 2, Problems of Mathematics*, Reprinted in *A Century of Mathematics in America*, P. Duren et al., eds., AMS, Providence, 1989. (1.4).

[56] W. Purkert and H. J. Ilgauds, *Georg Cantor*, Birkhäuser, Basel, 1987, (5.5).

[57] B. Riemann, *Grundlagen für eine allgemeine Theorie der Functionen einer veränderlichen complexen Grösse*, Riemann's gesammelte mathematische Werke, 1892, Dover reprint, 1953 (4.3).

[58] G. Roch, *Ueber die Anzahl der willkürlichen Constanten in algebraischen Functionen*, Jour. f. Math. (1864), 372–376, (4.7).

[59] W. Scharlau, *Unveröffentlichte algebraische Arbeiten Richard Dedekinds aus seiner Göttinger Zeit 1855–1858*, Arch Hist. Exact Sci. **27** (1982), 335–367, (1.7).

[60] H. J. S. Smith, *Report on the Theory of Numbers*, Reports of the British Association for the Advancement of Science, 1859–1865, (3.6).

[61] R. J. Walker, *Algebraic Curves*, Princeton Univ. Press, Princeton, 1950, Springer-Verlag reprint, 1978, (4.4).

[62] A. Weil, *Number-theory and algebraic geometry*, Proceedings of the International Mathematics Congress, VI. II, 1950, pp. 90–100, Collected Papers, vol. 2, 442–452 (Preface, 2.2).

[63] A. Weil, *Number Theory*, Birkhäuser, Boston, 1984, (3.6).

Index